ELECTRONIC STRUCTURE
and
CHEMICAL BONDING

World Scientific Series in Contemporary Chemical Physics

---

*To view the complete list of the published volumes in the series, please visit:
http://www.worldscibooks.com/series/wssccp_series.shtml

World Scientific Series in Contemporary Chemical Physics – Vol. 13

ELECTRONIC STRUCTURE

———— and ————

CHEMICAL BONDING

J R Lalanne

Bordeaux I Univ., France

Co-authors:

R. Boisgard
professeur agrégé, lycée Vaucanson (Math. Sup.), Tours.

D. Chartier
professeur, lycée (2ème cycle), La Réole.

A. Ducasse
professeur, université Bordeaux I.

J. Hoarau
professeur, université Bordeaux I.

J. R. LaLanne
professeur, universit´´Bordeaux I.

M. B. Mauhourat
professeur agrégé, lycée R. Cassin (Math. Spé.) Bayonne.

C. Raballand
professeur agrégé, lycée Montaigne (Math. Spé.) Bordeaux

J. C. Rayez
professeur, université Bordeaux I.

C. Rullière
directeur de recherche, CNRS, Bordeaux.

F. Rivoal
professeur agrégé, lycée St-Louis (2ème cycle), Bordeaux

B. Veyret
directeur de recherche, CNRS, Bordeaux.

World Scientific
Singapore • New Jersey • London • Hong Kong

Published by

World Scientific Publishing Co. Pte. Ltd.
5 Toh Tuck Link, Singapore 596224
USA office: 27 Warren Street, Suite 401-402, Hackensack, NJ 07601
UK office: 57 Shelton Street, Covent Garden, London WC2H 9HE

British Library Cataloguing-in-Publication Data
A catalogue record for this book is available from the British Library.

World Scientific Series in Contemporary Chemical Physics — Vol. 13
ELECTRONIC STRUCTURE AND CHEMICAL BONDING

ISBN-13 978-981-02-2665-7
ISBN-10 981-02-2665-9

Co-authors:

R. BOISGARD, professeur agrégé, lycée Vaucanson (Math. Sup.), Tours.

D. CHARTIER, professeur, lycée (2ème cycle), La Réole.

A. DUCASSE, professeur, université Bordeaux I.

J. HOARAU, professeur, université Bordeaux I.

J.R. LALANNE, professeur, université Bordeaux I.

M.B. MAUHOURAT, professeur agrégé, lycée R. Cassin (Math. Spé.) Bayonne.

C. RABALLAND, professeur agrégé, lycée Montaigne (Math. Spé.) Bordeaux.

J.C. RAYEZ, professeur, université Bordeaux I.

C. RULLIÈRE, directeur de recherche, CNRS, Bordeaux.

F. RIVOAL, professeur agrégé, lycée St-Louis (2ème cycle), Bordeaux.

B. VEYRET, directeur de recherche, CNRS, Bordeaux.

Co-authors:

Preface

The "Service de Formation Initiale et Continue des Enseignants de Sciences Physiques" (SEFICESP) [Center for ab initio courses and further education for physical science teachers] at Bordeaux I university was set up in 1982. In the same year, a teacher-training program was organized in five high schools in the Aquitaine region, involving about 130 teachers from various teaching levels. Most of the participants had experienced difficulties in teaching the electronic structure of atoms and chemical bonding, due to a lack of proper understanding of this issue. Indeed, teachers attending courses organized by the SEFICESP in other subjects had frequently expressed similar concerns, and additional classes were often improvised on this topic to meet their needs.

In view of this strong demand, we organized a summer school in July, 1990 at Val Louron (Pyrénées); attended by over a hundred teachers. The main goals of this summer school were as follows:

1. To broaden the knowledge of teachers working in this field. We used a new approach to teaching about electrons and chemical bonding, based on quantum physics.

2. To develop new educational tools and teaching software based on this approach. The software was to be suitable for both high school and university-level courses.

3. To publish a book with a comprehensive educational and theoretical background, to provide good reference material for teachers at all levels.

This book is the result of two years' hard work. It is currently used in both undergraduate and post-graduate courses at Bordeaux I university (Physical Science & Physical Chemistry).

This book is also intended for use in other branches of higher education, including engineering schools, polytechnics, etc. It also contains some useful suggestions for high school teachers.

As previously mentioned, this book is the outcome of *a joint project and is not merely a compilation of several individual contributions.* The homogeneity of the final manuscript is one of our major achievements.

Contributions from three groups of scientists — CNRS, universities and high schools — have been included in this book. We hope that this diversity will solve some of the educational problems encountered at various teaching levels in the past.

Several appendices are included to provide a rest from the more theoretical presentations. The appendices present *cultural background* (bibliography, extracts from earlier publications, etc.) *survey results* (quite an eye-opener !) and suggested *teaching approaches* (sometimes presented as games). Several *problems* are included, with their *solutions*, for readers to check their progress.

The first part of this book is a *historical introduction*.

The second part is concerned with the theoretical background to a description of chemical bonding and electronic structure. We then introduce the *quantum basis of chemical bonding and a description of molecular symmetry*. However, this book does not intend to compete with other excellent books on quantum mechanics and group theory. We have, however, tried to maintain as high a standard of accuracy in the introductory part of this book.

The third part presents *two complementary descriptions of chemical bonding*. The first covers *the mechanical aspects*, which, to the best of our knowledge, had never been presented in a French-language book. This approach is a powerful educational tool for teaching on all levels. The second highlights the *applications and limitations of the orbital concept* in chemical bonding. This is intended to provide a proper theoretical background for an in-depth understanding of the concept of molecular orbitals, frequently used in the presentation of organic and inorganic chemical reactivity.

We would like to express our gratitude to the many colleagues who helped in the preparation of this book, especially J.C. Colson, C. Desportes, J. Faure, H. Gié, A. Pacault, J.L. Rivail, J. Serre, M. Serrero, R. Subra, S.J. Teichner, C. Vidal and P. Lalanne, who were involved in our first further education course.

We would also like to thank:

— The "Union des Physiciens", (44, Boulevard St Germain, 75270 Paris Cedex 06), who distribute all the illustration software used in this book and "Microlambda" (Z.I. Auguste, 33160 Cestas, France) for allowing us to use their computer equipment and for producing the many computer screen images used as illustrations.

— The "Archigroup" architect agency (44, Avenue de Candau, 33600 Pessac, France) for several illustrations.

— All those involved in the preparing the text, especially Mrs. S. Bertrand, M. Mondolfi, and N. Robineau, for preparing the final version of this book.

Bordeaux, August 1992.

Acknowledgements

We are most grateful to

— N. Aguado, who directed the translation of the children's drawings.

— Mrs L. Orrit, née Wolker; Mrs C. Rychlewski (Aquitaine Traduction) and Mr. W. Forst who helped translate the book.

One of us (J.R.L.) would like especially to thank Mr. B. Veyret, without whom this English edition would not have been possible.

Bordeaux, August 1995.

Contents

INDEX

Symbols

Greek alphabet used in the book
(left : capital letter ; right : small letter)

Alpha		a	Theta	Q	q	Sigma	S	s
		b		L	l			t
Beta	G	g	Lambda		m	Tau	f	j
Gamma	D	d	Mu		n	Phi		c
		e			x		Y	y
Delta		z	Nu		p	Chi		w
Epsilon		h	Xi		r	Psi		
Zeta			Pi			Omega		
Eta			Rho					

Fundamental constants
(and their values in SI system)

Electron charge	q	$-$ 1.602177	x	10^{-19}C
Avogadro constant	N	6.02214	x	10^{23} mol^{-1}
Planck constant	h	6,62608	x	10^{-34} Js
Dirac constant	$\hbar$	1.05457	x	10^{-34} Js
Rydberg constant	R	1.09737	x	10^7 m^{-1}
Atomic mass unit	u.a	1.66054	x	10^{-27} kg
Electron rest mass	m_e	9.10939	x	10^{-31} kg
Proton rest mass	m_p	1.672623	x	10^{-27} kg
Permittivity of vacuum	e_0	8.8541878	x	10^{-12} C^2 N^{-1} m^{-2}
Speed of light in vacuum	c	2.99792458	x	10^8 ms^{-1}

Conversion factors
(in SI system)

1 Angström	(Å)	=	10^{-10} m			
1 Bohr	(B)	=	5.291772	x	10^{-11} m	
1 Debye	(D)	=	3,3333	x	10^{-30} Cm	
1 electron volt	(eV)	=	1.602177	x	10^{-19} J	
1 Hartree	(H)	=	4.35975	x	10^{-18} J	

Typographical convention

italics	operator	H
bold type	vector	**r**
brackets	matrix	(H)
square brackets	column vector	$\begin{bmatrix} a \\ b \end{bmatrix}$
vertical dashes	determinant	$\begin{vmatrix} a & b \\ c & d \end{vmatrix}$

PART I

HISTORICAL SURVEY

CHAPTER I

Main events in the history
of chemical bonding

I. Historical survey

The history of chemical bonding began in the eighteenth century. Until then, the only explanations found in literature were based on the old theories of :

– *Thales of Miletus (sixth century B.C.):*

All substances originate from water (this idea was later revived by Newton).

– *Democritus (fifth century B.C.):*

Theory of atoms.

– *Aristotle(381 B.C.):*

Earth, air, fire and water are the four components of matter. This idea was revived by alchemists during the Middle Ages (see Fig. I.1).

– *Paracelsus (fifteenth century):*

Each of Aristotle elements is composed of three « principles »: salt, sulphur and mercury.

One alternative hypothesis stated that the universe was made of only seven elements: Earth, air, fire, water, salt, sulphur and mercury. Stahl's theory (eighteenth century), also known as the *« phlogistic » theory*, was the prevailing hypothesis throughout the eighteenth century.

The only proposed explanations of chemical bonding were anthropomorphic : Atoms bond because they have hooks. The main idea was that of *avidity*, directly linked to sexuality, underlying alchemy as a whole (see Table I.1).

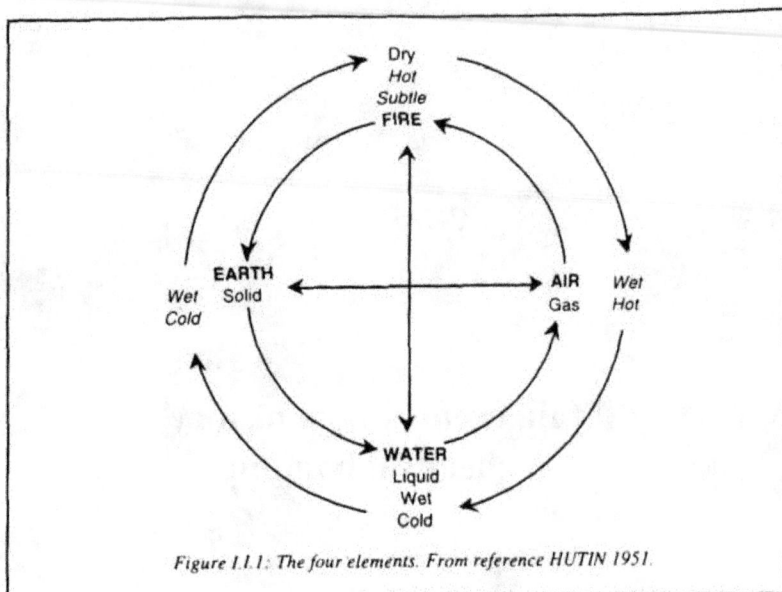

Figure 1.1.1: The four elements. From reference HUTIN 1951.

TABLE 1.1.1 — **Sexual dualism and chemical avidity**
(from reference HUTIN 1951).

Male	Female
Sperm	Menstrue
Active	Passive
Form	Matter
Soul	Body
Fire	Water
Hot-dry	Cold-wet
Sulphur	Mercury
Gold	Silver
Sun	Moon
Leaven	Unleavened dough

So, bases are presented as "round" and acids as "sharp".

Following the Lavoisier "revolution" in chemistry (1787), and the experimental introduction of atomic theory by Dalton (1800), a major advance was made in 1812 by Berzelius (professor of chemistry at the medico-surgical institute of Stockholm) who established his famous *dualistic theory*, after Davy's discovery of the laws of electrolysis (1806). The notion of *avidity* was replaced by *affinity*, still a purely qualitative concept, which only acquired its quantitative dimension after the work of De Donder, more than one century later. Berzelius' affinity was directly linked to electrostatic attractive forces.

O : large negative pole and small positive pole

K : large positive pole and small negative pole

H : no predominance of positive or negative pole

On the base of this principle, Berzelius drew up a provisional series of the elements:

— *decreasing predominance of the negative pole:* O, S, N, F, Cl, Br, I...H

— *decreasing predominance of the positive pole:* K, Na, Li, Ba, Ca, Mg,
   Al, Mn, Zn, Fe, Ni ... Cu, Ag, Hg, Pt, Au, H.

All chemical compounds were said to be divisible into two parts with opposite electrical charges.

Example: K_2SO_4 could be expressed as K_2O^+ ... SO_3^-

However, many problems remain unsolved :

— The proposed groups are not necessarily those revealed by electrolysis.

— A compound may be represented by different formulae. Example:

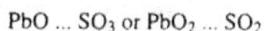

$$PbO \ ... \ SO_3 \ or \ PbO_2 \ ... \ SO_2$$

— Bonds between atoms with charges of the same sign, as in the case of SO_2 or SO_3, were still unaccounted for.

— In his theory (called the "exchange" theory), Dumas (1834) showed how it was possible to replace one atom by another, inside a molecule, without changing the chemical properties of the molecule, even though the two atoms (original and exchanged) belong to opposite Berzelius' series.

— Finally the deficiencies of the dualistic theory were clearly apparent in organic chemistry, as Berzelius had already pointed out in his work.

However, the qualitative description of pure electrostatic chemical bonds, proposed by Kossel (1916) was still based on the Berzelius theory.

The deficiencies of the electrostatic view stimulated the development of other theories. Thus, at about the same time (1852), Frankland, a professor at Edinburgh university, proposed the valence theory. Figure 1.1.2 shows some chemical bonding representations of simple molecules.

The fundamental work of Kekule (1856) on aromatic substances and the new type of valency introduced by Werner in 1892 ultimately led to the Lewis and Langmuir electronic theory of valency (1917).

However, while the Kossel theory did not give rise to much further development, Lewis' theory of valency generated important theoretical works, once Schrödinger (1925) had demonstrated how the de Broglie equation, which, in its simple form, applied only to a free-moving electron, could be plausibly adapted to the case of electrons bound in atoms (in conjunction with the Pauli exclusion principle).

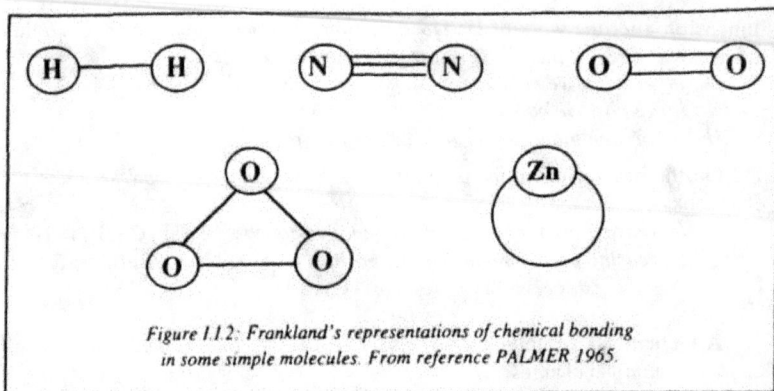

Figure 1.1.2: Frankland's representations of chemical bonding in some simple molecules. From reference PALMER 1965.

At this time, two different approaches were explored.

1. The theoretical work of Heitler and London (1927) on the hydrogen molecule, later extended to the case of more complex molecules by Slater and Pauling, led to the valence-bond theory. This theory was particularly suited to the ideas of chemists concerning the electronic structure of molecules: Molecules are assumed to be composed of atomic cores (nucleus and inner-shell electrons) linked by diatomic bonds due to peripheral electron pairs. If A and B are the two atoms and 1 and 2 the two electrons, the molecular wave function ψ which characterizes H_2 in its lowest energy level is written

$$\psi = \psi(1,2)f(1,2) \qquad \text{(I.1.1)}$$

where $\psi(1,2)$ is the symmetrical function of space

$$\psi(1,2) = \frac{1}{\sqrt{2\left(1+S_{AB}^2\right)}}\left[1s_A(1)\,1s_B(2) + 1s_B(1)\,1s_A(2)\right] \qquad \text{(I.1.2)}$$

$1s_A(1)$ is the 1s atomic orbital of atom A, associated with electron 1, and S_{AB} the overlap integral of the $1s_A$ and $1s_B$ orbitals. The antisymmetrical spin wave function, $f(1,2)$, is defined by

$$f(1,2) = \frac{1}{\sqrt{2}}\left[\alpha(1)\,\beta(2) - \alpha(2)\,\beta(1)\right] \qquad \text{(I.1.3)}$$

α and β are the eigenfunctions of the s_z operator, associated with the eigenvalues $+\hbar/2$ and $-\hbar/2$, respectively (see Eq. II.1.5).

The reader is invited to check that this expression is equivalent to the following :

$$\Psi = \frac{1}{2\sqrt{\left(1+S_{AB}^2\right)}}$$

$$\times \left\{ \begin{vmatrix} 1s_A(1)\alpha(1) & 1s_B(1)\beta(1) \\ 1s_A(2)\alpha(2) & 1s_B(2)\beta(2) \end{vmatrix} - \begin{vmatrix} 1s_A(1)\beta(1) & 1s_B(1)\alpha(1) \\ 1s_A(2)\beta(2) & 1s_B(2)\alpha(2) \end{vmatrix} \right\} \qquad \text{(I.1.4)}$$

which can also be written in the more compact form

$$\psi = \frac{1}{\sqrt{2\left(1+S_{AB}^2\right)}} \left(\left|1s_A \,\overline{1s_B}\,\right| - \left|\,\overline{1s_A}\, 1s_B\,\right|\right) \qquad (I.1.5)$$

We can extend this presentation to more complex molecules, for instance to the Li-Li molecule. The electronic structure of the Li atom is $1s^2\, 2s^1$ and the ψ wave function is the difference between two determinants, which, ignoring the normalization constant, may be expressed as

$$\psi = \left(\left|1s_A \,\overline{1s_A}\, 1s_B \,\overline{1s_B}\, 2s_A \,\overline{2s_B}\,\right| - \left|1s_A \,\overline{1s_A}\, 1s_B \,\overline{1s_B}\, 2s_A \,2s_B\,\right|\right) \qquad (I.1.6)$$

Two comments should be made:

— 1s cores do not participate in chemical bonding.
— The peripheral shell 2s is directly involved in bonding.

For more complex molecules, each pair of electrons leads to a valence bond structure, and the most general wave function is chosen as a linear combination built on these restricted structures. The variational principle (described in detail in the following chapters) is then used to minimize the energy.

The Valence-Bond theory, directly linked to the Lewis image, presents one major disadvantage that has restricted its use: It is difficult to apply in computer calculations. We should, however, point out that this situation is currently changing and that there have recently been some signs of revived interest. The V.B. method is developed in the review paper "Forty years of V.B. theory", by M. Simonetta in "Structural Chemistry and Molecular Biology", A. Rich and N. Davidson Editors, Freeman, San Francisco, 1968.

2. The rapid progress of computer assisted numerical calculation has largely favored the Molecular Orbitals (MO_S) method based on the work of Hartree and Fock (1930) and adapted to computer calculations by Roothaan in 1951. This method, supplemented by the Lennard-Jones equivalent-orbital concept in 1949, consists of a model where pairs of electrons "occupy" molecular orbitals. Associated first-electron-ionization energies can be measured with an accuracy of about 10^{-2} eV by using photoelectronic spectroscopy. Experimental and calculated values correspond to within 0.1 eV, confirming the pertinence of the Hartree-Fock treatment for these calculations.

This method, described in detail in the following chapters, makes it possible to calculate the electron density at each point in space. It is therefore possible to draw the electron density profile and visualize binding and antibinding zones, as proposed by Berlin in 1951.

In conclusion, we would like to make three remarks:

— *We have only mentioned the main events in the history of chemical* bonding. Many names have been omitted, including Hückel, Coulson and Longuet-Higgins for their work on conjugation, Pauli for the exclusion principle (1925), and Hund and Mulliken who introduced MO_S.

— We have voluntarily avoided chemical bonding in metals (which requires a solid-state physics background), Van Der Waals forces, and hydrogen bonding... .

— Finally, we shall not speak of purely "ionic" chemical bonding which is highly improbable in molecules.

II. Appendix 1: References

The history of chemical bonding may be found in the following books:

HUTIN S.— *L'alchimie.* Collection "Que sais-je ?" n° 506. Presses Universitaires de France, Paris (1951).

LOCKEMANN G.— *Histoire de la Chimie.* Dunod, Paris (1962).

MASSAIN A.— *Chimie et Chimistes.* Magnard, Paris (1952).

PALMER W.G.— *A history of the concept of valency.* Cambdrige University Press, London (1965).

and these scientific publications:

BERLIN T.— Binding Regions in Diatomic Molecules, in : *J. Chem. Phys.* **19**, 208 (1951).

FOCK V.— Naherungsmethode zur Lösung des quantenmechanischen Mehrkör Perproblems, in : *Z. Physik.* **61**, 126 (1930).

KOSSEL A.— Uber Molekulbildung als Frage des Atombaus, in : *Ann. Physik. und Chemie* **49**, 229 (1916).

LANGMUIR I.— Isomorphism, Isosterim and Covalence, in : *J.A.C.S.* **41**, 868 1543 (1919).

LENNARD JONES J.— The Molecular Orbital Theory of Chemical Valency I. The determination of molecular orbitals, in : *Proc. Roy. Soc.* **A198**, 1 (1949).

LEWIS G.N.— The Atom and the Molecule, in : *J.A.C.S.* **38**, 762 (1916).

ROOTHANN C.C.J.— New Developments in Molecular Orbital Theory, in : *Rev. Mod. Phys.* **23**, 161 (1951).

SCHRÖDINGER C.E.— Quantisierung als Eigenwertproblem, in : *Ann. Physik.* **79**, 361 (1926).

SLATER J.C.— Atomic shielding constants, in : *Phys. Rev.* **36**, 57 (1930).

III. Appendix 2: Synopsis

Some milestones in the history of chemical bonding :

19th century

1812 : BERZELIUS	*Dualistic theory*
1834 : DUMAS	*Substitution theory of valency*
1856 : KEKULE	*« On the structure of aromatic compounds »*
1892 : WERNER	*Structure of complexes*

20th century

1916 : KOSSEL	*Ionic chemical bonding*
1917 : LEWIS-LANGMUIR	*Theory of valency*
1925 : PAULI	*Exclusion principle*
1925 : SCHRODINGER	*The equation of ...*
1927 : HEITLER et LONDON	*Valence-bond theory*
1929 : SLATER	*The determinant of ...*
1930 : HARTREE et FOCK	*Self-consistent field*
1931 : PAULING	*Directional character of the covalent bond*
1949 : LENNARD JONES	*The concept of equivalent orbital*
1951 : ROOTHAN	*Adaptation of the Hartree-Fock method to computer data*
1951 : BERLIN	*"Binding" and "anti-binding" zones*

The rapid development of research has given rise to many scientific papers. The most significant are listed in Chapter IV, Appendix 11.

PART II

**MOLECULAR ELECTRONIC
STRUCTURE AND CHEMICAL BONDING
DESCRIPTIONS:
THEORETICAL BASES**

CHAPTER II

Quantum mechanics
and molecular symmetry

I. Quantum bases of chemical bonding

This section is largely based on *"Quantum Mechanics"*, by C. Cohen-Tannoudji, B. Diu, and F. Laloë, J. Wiley & sons, Inc, New York, 1976. We highly recommend it for more detailed reading on the subject.

1. Fundamental concepts

In order to study chemical bonds, we must analyze the highly complex microscopic structure of molecules. However, available experimental data on these microscopic systems are the result of measurements of physical quantities on macroscopic samples (i.e., samples which are large enough to be handled, with physical dimensions comparable to our own). It is impossible to observe or analyze these molecular structures directly, so their existence can only be postulated. This means that it becomes extremely difficult to give a mental representation of these structures. In order to build a microscopic model, we must necessarily start from our experience of the macroscopic world around us. As an initial approach, we can imagine a structure copied directly from a macroscopic structure. Then, we apply the classical laws of physics to this model. There is no *a-priori* reason why this model should be coherent with measurements obtained indirectly on the corresponding microscopic system. By analyzing the ensuing discrepancies, we can modify the model and the laws governing, in order to produce a model yielding correct results. Moreover, if this microscopic model is used to build macroscopic entities, the results obtained by applying the new laws must be coherent with those obtained by the classical laws. In other words, the new laws must contain the classical laws. This is the procedure used to describe the universe at a microscopic level, using a model based on our experience of the macroscopic world.

To understand more fully how this transition from the macroscopic to the microscopic world came about, it is very instructive to follow the historical development of quantum mechanics. Unfortunately, it was not possible to include this background in this book, due to its length and complexity. We shall present a simple description of some of the fundamental experiments that contributed to the basic ideas which eventually led to the full axiomatics of our present understanding of the microscopic world.

The classical model for the atom is a planetary model: negatively charged electrons gravitate around a positive nucleus, with mainly electrostatic interaction. We are now going to show the inconsistencies which appear when the laws of classical mechanics are applied to this model. These will enable us to derive some fundamental concepts to help us understand the physical meaning of the postulates of quantum mechanics, which we shall enumerate later.

1.1 Probabilistic and nondeterministic description

Electrons in the planetary model are particles which can be produced and isolated experimentally. The experiment described in this section will clearly shows that the laws of classical mechanics cannot be applied to these particles.

Electron diffraction

The experiment described here is not an actual experiment, like those of Davisson and Germer (1927) or G.P. Thompson (1927), but rather a thought experiment leading to exactly the same conclusions as the real experiments. In this experiment, electrons are diffracted through two slits. It is easier to compare the results of this thought experiment with those obtained by diffracting electromagnetic waves than if we had used real experiments. This comparison also gives a good indication why the concept of trajectory must be rejected for microscopic particles of matter. Real experiments on neutron diffraction also lead to similar conclusions.

S is a source producing a beam of mono-kinetic electrons [i.e., electrons with the same speed, for instance a speed on the order of a hundred electron-volts (eV)]. This beam falls on a reflecting plane, P, with two thin slits F_1 and F_2 (with hypothetical dimensions as follows : 1 nm wide and 10 nm apart). The trajectories of the electrons can be calculated according to the laws of classical mechanics, in which case a detector placed on screen e (an electron multiplier, for example) should detect impacts of particles of mass m_e exclusively between points M'_1 and M''_1 and between M'_2 and M''_2 (see Fig. II.I.1).

If the experiment is carried out with a very low particle flow, we always observe impacts of the same energy on screen e, confirming the existence of identical particles of mass m_e. However, the actual distribution of these impacts is not at all as predicted by the trajectories calculated using classical mechanics (see Fig. II.I.2).

First, impacts occur outside the regions located between M'_1 and M''_1 and between M'_2 and M''_2. The trajectories of some electrons may have been modified by reflection off the edges of the slits, which could explain the dispersion of the impacts. However, we can see that the impacts form a fringe pattern on screen e, and that these fringes disappear if one or the other of the slits is closed. If one slit is closed, the impacts have a distribution centered on the "geometrical" impact spot $M'_1 M''_1$ (or $M'_2 M''_2$ according to the slit left open). Therefore, the fringes seem to be the consequence of an interference between the particles passing through one slit and those passing through the other slit. Again, we could imagine that this interference is brought about by the collision of electrons, whose trajectory was disturbed by the sides of the slits.

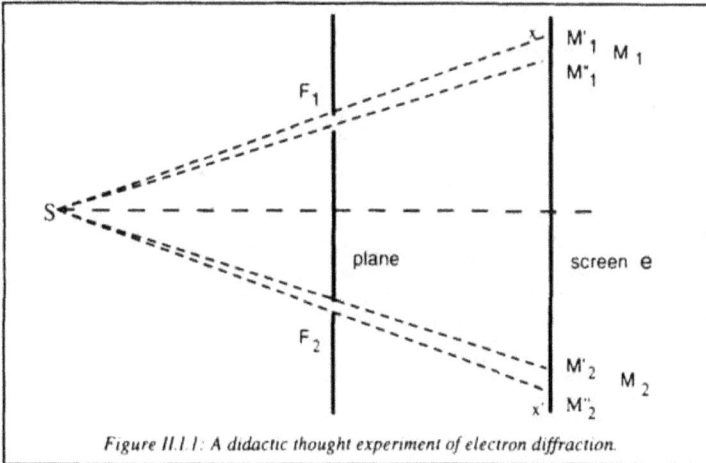

Figure II.1.1: A didactic thought experiment of electron diffraction.

However, if the electron flow is reduced to such an extent that the probability that a particle which just passed through F_1 will collide with a particle from F_2 is practically zero, the fringe pattern is still observed, provided the observation time is long enough. Clearly, we have to admit that the very concept of trajectory should be reconsidered. The microscopic particle, i.e., the electron in this experiment, does not move according to the laws known to govern the movement of macroscopic objects. Therefore, if we want to describe the movement of electrons, we must find laws other than those of classical mechanics.

Analogy with Young slits

Of course, this electron diffraction experiment immediately brings to mind the well-known Young slit experiment in the optics, involving a light source. Indeed, the experimental setup is absolutely identical, except that the source produces electromagnetic waves instead of electrons. Of course, the positions and dimensions of the slits are quite different (for example in the domain of visible light, slits typically have a width of 0.1 mm and are separated by a few tenths of a millimeter), while the detector is sensitive to radiation (photodiode or photomultiplier, for instance), but the observed results are identical. However, whereas, in the case of electrons, the fact that fringes are observed on the screen is absolutely impossible to explain by classical theories, in the case of Young slits, it is perfectly predicted by the classical laws of radiation. In the Young slit experiment, the photoelectric detector responds to the intensity $I(x,t)$ of the light-wave on screen e. Each of the slits, F_1 and F_2, diffracts the wave radiated by S, and, in turn, acts as a new light source emitting a wave towards screen e, described by electric field E_1 or E_2, respectively. The total electric field, E_t, on screen e, is expressed as

$$E_t = E_1 + E_2 \tag{II.I.1}$$

and the intensity detected by the detector is

$$I = |E_t|^2 = |E_1 + E_2|^2 = I_1 + I_2 + J_{12} \tag{II.I.2}$$

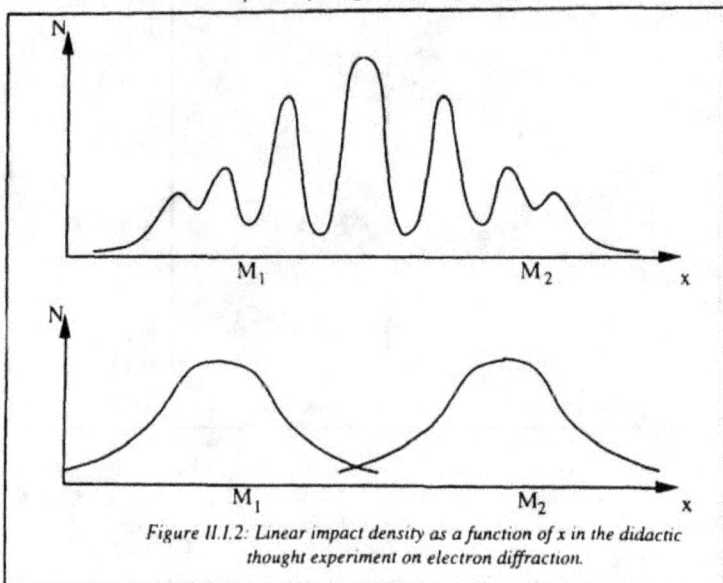

Figure II.1.2: Linear impact density as a function of x in the didactic thought experiment on electron diffraction.

J_{12} is the interference term. It is responsible for the fringes observed on the screen, and is due to the fact that the detector does not respond directly to the field itself, but is sensitive to the square of the modulus of the field. This quadratic detection makes it possible to observe phase correlations between E_1 and E_2.

Some notions of probabilistic mechanics

In order to explain the appearance of fringes in the electron-diffraction experiment, we are tempted to replace the notion of "particle" usually associated with electrons, by the notion of "wave". Electrons would then be represented as waves to which, by analogy with radiation waves, we attribute a frequency v and a wavelength λ, related to the energy E and momentum $\mathbf{p}$ of the particle as follows:

$$E = h\nu$$

$$\mathbf{p} = \hbar\mathbf{k} \quad \text{where} \quad \mathbf{k} = (2\pi/\lambda)\,\mathbf{i} \qquad\qquad \text{(II.1.3)}$$

i is the unit vector in the direction of particle velocity and $\hbar = h/2\pi$, where h is the Planck constant (this follows de Broglie's hypothesis of 1923). However, if we represent the electron by a wave, we lose its particle aspect, although this is an established experimental fact. In our experiment, we saw that the energy impacted the detector in small entities, in "grains", corresponding to particles with a mass m_e and a speed determined by the system. On the other hand, in the classical description of a wave, energy is distributed continuously so there are no "grains" of energy. This demonstrates that if one insists on describing electrons by means of a classical model, both particle description and wave descriptions must be considered simultaneously. *However, this particle-wave dichotomy simply shows that neither of the two classical descriptions is correct and that another representation must be found, naturally including both aspects, each revealed by appropriate experiments.* After all, it is not very satisfactory to have two mutually exclusive models (particles and waves) to represent the same physical entity (in our case, an electron), especially if one model is needed to explain one result of an experiment, and the other to explain another result of the same experiment.

The fundamental idea underlying the construction of a new description containing both aspects, is to keep the concept of particle, but to discard the classical laws describing it, especially the classical idea of a particle precisely localized on a well-defined trajectory. So, to describe the electron as a quantum particle, it must be associated with a function $\Psi(\mathbf{r},t)$ known as a wave function. This function describes the state of the particle. $\Psi(\mathbf{r},t)$ does not attribute a trajectory to the quantum particle. We state that $|\Psi(\mathbf{r},t)|^2$ represents the spatial probability density of finding an electron at point **r** at time t. The fringes in our experiment are then accounted for by the interference between the amplitudes of Ψ_1 and Ψ_2. corresponding to the two possible paths of the electron

$$\psi_t = \psi_1 + \psi_2 \qquad |\psi_t|^2 = |\psi_1|^2 + |\psi_2|^2 + 2R\{\psi_1\psi_2{}^*\} \quad \text{(II.1.4)}$$

where $R\{..\}$ represents the real part of the expression in brackets.

Nevertheless, the wave function ψ representing the electron must not be mistaken for a wave in the classical sense, as it describes a particle of matter with a well-defined mass. The wave function is not governed by all the classical laws concerning waves. At present, we still do not know how to build such a wave function, nor what equations govern its evolution in time. By defining a probability density $|\Psi|^2$, we no longer have the possibility that exists in classical mechanics of predicting a trajectory for the particle starting from its initial conditions. We must therefore try to predict the evolution of the probability density of the particle at a particular point r by exchanging a deterministic view for a probabilistic concept. At this point, we would like to point out that, if we were dealing with a more complex system like an atom or a molecule, instead of a single electron, we would represent the whole microscopic system under consideration (atom or molecule) by a single wave function ψ. In no case would we attribute distinct wave functions to each particle in the system, electrons and nuclei ; *the particle cannot be separated from the system.*

1.2 Simultaneous description of matter and light

We have already shown that, in the case of the Young slit experiment, the fringes observed were in accordance with the classical description of radiation. On the other hand, if this experiment were carried out with a very small light flux, individual impacts would be registered on the screen, exactly as in the case of the electron diffraction. These impacts are not at all compatible with the idea of a wave: They reveal the discontinuous distribution of the energy. In the same way, even radiation cannot be correctly described as a classical wave; it also needs an associated particle, called a *photon*. Other historical experiments had already shown an intuitive need for the photon, for instance the photoelectric effect and black-body radiation. The comparison between these two experiments — Young slits and electron diffraction — makes it easy to see the need to represent matter and radiation by the same model: A model of particles with an associated wave function characteristic of the state of the system, determining the amplitude of the probability of finding a particle at any point in space. The postulates of quantum mechanics must hold true for both the "wave" and the "particle" aspects. This is in contrast to classical physics, where radiation and matter were completely separate. The laws governing the movement of physical objects had nothing to do with those governing waves. Quantum mechanics emerges as a *unifying theory*, pulling together the two major aspects of our universe.

1.3 The problem of measurement

In order to test the need for a probabilistic theory, as described in the last section, we can consider the electron diffraction experiment, trying to measure the probability that one electron will pass through one of the slits. A strong light source is placed just behind the plane containing the slits. If an electron passes through one of the slits, analysis of the light scattered by this electron (a charged particle) will make it possible to determine through which of the two slits, F_1 or F_2, the electron traveled. (see Fig. II.I.3).

We observe that:

– One electron never goes through both slits simultaneously; there are never two perfectly simultaneous flashes from F_1 and F_2.

– The total number of electrons that passed through slits F_1 and F_2 is equal to the total number of impacts on screen e. However, we notice that, while we were counting electrons, the interference pattern on the screen disappeared. If we put out the light and stop counting the number of electrons passing through specific slits, the interference reappears. Whatever changes we make to the experimental system (changing the intensity, changing the wavelength of the light, etc.), either we do not count the electrons very accurately, in which case we obtain an interference pattern on the screen (contrast improves as the proportion of electrons accounted for is smaller); or we count practically every electron in which case interference is almost nonexistent.

Thus we can draw two essential conclusions:

– If we assume that the state of the electron when it is still at S is such that it has a certain well-defined probability of passing through one or the other of the slits, a single measurement of this electron cannot yield the probability value. The measurement only identifies whether the electron passed through slit 1 or slit 2. Nevertheless, by making several measurements on electrons in the same state, we can reconstruct the characteristics of the initial state: The distribution of the number of electrons passing through one or the other of the slits, as measured on all the electrons, is the same as the initial, monoelectronic probability.

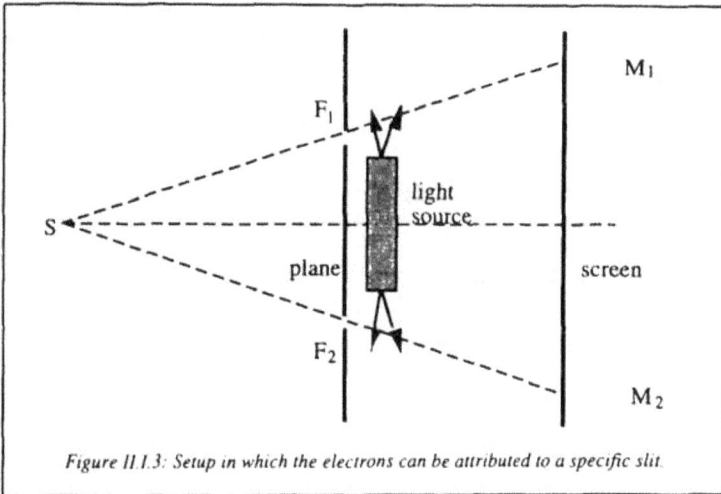

Figure II.1.3: Setup in which the electrons can be attributed to a specific slit.

– It is clear that the measurement performed on each electron changes its state since the interference pattern disappears completely during the measurement. Whatever care is taken to improve the experiment, measurement still disturbs the quantum system.

This experiment introduces the idea that the concept of measurement has to be taken into account in the description of a microscopic system. In the macroscopic domain, measurements can be made without affecting the system. This is no longer true in the microscopic domain, and the effect of measurements must be taken into account by the postulates.

1.4 Energy quantization

Let us leave the thought experiment on electron diffraction, and turn to atomic spectroscopy. We have known since the 19th century that atoms (or molecules) do not emit or absorb radiation energy in a random way. Each molecule only absorbs certain, well-defined frequencies. In particular, the hydrogen spectrum in visible light was studied in 1885 by Balmer who suggested an empirical relation yielding the positions of all absorbed or emitted frequencies

$$v = R c \left(\frac{1}{2^2} - \frac{1}{n^2} \right)$$ (II.1.5)

with n = 3, 4, 5, ... ; c = the velocity of light ; R = Rydberg's constant.

These results cannot be explained by the classical laws of physics; Rutherford's planetary model of the atom in which electrons gravitate around a nucleus did not account for any discontinuity of the energy. Indeed, according to this model and following the classical laws of electrodynamics, an electron gravitating around a nucleus should radiate energy continuously and spiral in towards the nucleus. This would definitely not be a stable system, and thus classical electrodynamics is unable to provide a correct model for the hydrogen atom.

In 1913, Niels Bohr suggested that the hydrogen atom can have only certain discrete, well-defined energies (see Fig. II.1.4), and that the emitted radiations observed corresponded to transitions of the atom from one energy level to another. The frequency of the photons emitted would then be given by the relation $E_n - E_0 = hv_n$. Only the discrete frequencies corresponding to these transitions can be absorbed or emitted.

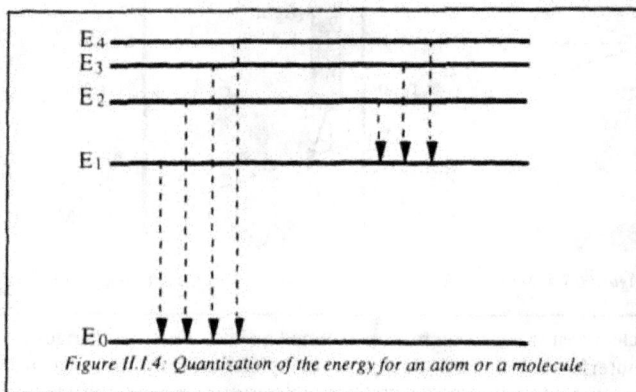

Figure II.1.4: *Quantization of the energy for an atom or a molecule.*

Bohr managed to account for these discrete energy values using Rutherford's model and classical laws, by assuming that the electron trajectory is circular and that the modulus of the angular momentum of the atom can take on only discrete values $|l| = n\hbar$, where n is a whole number. Yet the discrete nature of this angular momentum cannot be explained by classical mechanics.

All these considerations point to a need to elaborate new laws of physics for the microscopic domain. As quantum physics was gradually developed, it had to meet the following criteria :

It had to abandon the deterministic approach of classical mechanics in favor of a probabilistic approach, where the notion of a probability of finding the particle at any point in space replaced the concept of trajectory.

- It had to apply to both matter and radiation, thus unifying the laws of physics concerning these two domains.

- It had to allow for the effect of measurements, as measurements necessarily disturb the state of the system under consideration.

It had to explain the fact that, in general, the energy of atoms or molecules can take only discrete values.

- Finally, at the limit between microscopic and macroscopic systems (i.e., for instance, when the linear momentum of the system under consideration is very large compared to $\hbar$) quantum physics had to be consistent with classical physics.

It is worth mentioning that the fact that the quantum effects revealed by the above experiments have not been observed in the macroscopic domain does not mean that the laws governing the microscopic domain (quantum physics) do not apply to macroscopic systems. For instance, let consider the diffraction experiment again, using billiard balls. The slits diffracting these particles, and the distance separating the slits, must be larger than the dimension of the particles. On the other hand, the wavelength $\lambda = h/p$ of the "associated wave" is extremely small for reasonable velocities of the particles. This means that the interference fringes on the screen are so narrow that no existing device is able to detect them.

In the same way, the quantization of the energy levels of a macroscopic particle is impossible to observe since the levels are so close to each other that their "width" is very much greater than the distance between neighboring levels. Therefore, the energy continuum observed is coherent with classical observations.

2. Lagrange and Hamilton formalisms of classical mechanics

As previously mentioned, the quantum description of a microscopic system must be induced from a classical model. We must now try to find the classical description which is best adapted to this transposition. We already know that, in the quantum description, a wave function $\Psi(\mathbf{r},t)$ will correspond to the state of the system to be described. Fortunately, one description of the laws of classical mechanics uses the same approach: The Lagrangian function provides a complete description of the classical system under consideration. Quantum mechanics is based on the Lagrange formalism, as well as the closely-related Hamilton formalism.

We will start by stating the major features of this way of presenting the laws of physics.

Conservative systems

Let us assume that, from now, on all the systems studied are subjected to a set of forces deriving from a potential V which does not explicitly depend on time. In this case, it is easy to show, on the base of the fundamental principle of dynamics, that the sum of kinetic and potential energy is conserved (i.e., it does not change over time). We can write $E = T + V = $ constant. These systems are called conservative systems.

The Lagrangian equation of a conservative system of N particles,
expressed in rectangular coordinates.

The following function is called the Lagrangian of a system

$$L = T - V = \sum_{i=1}^{N} \frac{1}{2} m_i \left(\dot{x}_i^2 + \dot{y}_i^2 + \dot{z}_i^2 \right) - V(x_i, y_i, z_i) \qquad \text{(II.I.6)}$$

With this function of 6N variables we can express the fundamental law of dynamics in another way:

$$\left.\begin{array}{l} \dfrac{d}{dt}\left(\dfrac{\partial L}{\partial \dot{x}_i}\right) - \dfrac{\partial L}{\partial x_i} = 0 \\[2mm] \dfrac{d}{dt}\left(\dfrac{\partial L}{\partial \dot{y}_i}\right) - \dfrac{\partial L}{\partial y_i} = 0 \\[2mm] \dfrac{d}{dt}\left(\dfrac{\partial L}{\partial \dot{z}_i}\right) - \dfrac{\partial L}{\partial z_i} = 0 \end{array}\right\} \Leftrightarrow -\nabla_i V_i = F_i = m_i \ddot{r}_i \qquad \text{(II.I.7)}$$

In this way, 3N equations are generated. These 3N equations are called the Lagrange equations and are equivalent to the 3N equations of the fundamental law.

Lagrange equations expressed as a function of generalized coordinates

Rectangular coordinates are not always those best adapted to a problem, and sometimes restraining conditions reduce the number of coordinates to be considered. So, taking into account the internal constraints of the problem, it is necessary to determine a number n of independent parameters q_i. These are the generalized coordinates of the system. These coordinates are time-dependent, since they describe the movement of the system. Their first derivatives with respect to time, $\dot{q}_i = dq_i/dt$, are the generalized velocities. In a conservative system, the Lagrangian, which is dependent on the 2N coordinates $q_i(t)$ and $\dot{q}_i(t)$, is expressed as

$$L(q_i, \dot{q}_i, t) = T - V \qquad \text{(II.I.8)}$$

Time appears explicitly only if the reference frame moves with time.

The n Lagrange equations are then expressed as

$$\frac{d}{dt}\left(\frac{\partial L}{\partial \dot{q}_i}\right) - \frac{\partial L}{\partial q_i} = 0 \qquad \text{(II.I.9)}$$

As soon as the system under consideration reaches a certain degree of complexity, generalized coordinates must be used. Newton's formalism quickly becomes unwieldy whereas the equivalent Lagrangian formalism provides a much simpler way of solving problems. The formalism can be derived from Hamilton's principle of least action, which may be used instead of the fundamental law: $F = m\ddot{r}$ (where F is the force and $\ddot{r}$ the acceleration of the particle). We shall not discuss this principle here, although it is in fact very close to the postulates of quantum mechanics. The Lagrange formalism immediately leads to Hamilton's formalism, which in turn serves as a stepping stone to the quantization of microscopic systems. We shall now describe this formalism in more detail.

Conjugated momentums of generalized coordinates

Let us assume that the Lagrangian of a system, $L(q_i, \dot{q}_i, t)$, is known. It completely defines the movement of the system. In the case of a conservative system, it is usually easy to determine L, which does not explicitly depend on time unless the axes of the reference frame move. In a very general way, it is possible to define the conjugated momentum p_i corresponding to the generalized coordinate q_i by the relation

$$p_i = \partial L / \partial \dot{q}_i \qquad\qquad (\mathrm{II.I.10})$$

p_i and q_i are called canonical variables.

In the special case of a conservative system, expressed in rectangular coordinates, the Lagrangian is as follows:

$$L = \sum_i \frac{1}{2} m_i (x_i^2 + y_i^2 + z_i^2) - V(x_1, y_1, z_1)$$

from which we conclude

$$p_i = \partial L / \partial x_i = m_i x_i \qquad\qquad (\mathrm{II.I.11})$$

and similar relations for y_1 and z_1.

In this case, the conjugated momentums of the position coordinates $r_i(x_i, y_i, z_i)$ are simply the linear momentums.

Hamiltonians

In the case of systems where the Lagrangian is not explicitly time-dependent ($\partial L / \partial t = 0$), the following expression is a constant of the motion:

$$- L + \sum_i p_i \dot{q}_i \qquad\qquad (\mathrm{II.I.12})$$

As p_i is equal to $\partial L / \partial \dot{q}_i$, it can be expressed as a function of q_i, $\dot{q}_i$ and t, so that the above expression, which depends explicitly on q_i, $\dot{q}_i$, p_i and t only, in fact, depends on $(2n+1)$ independent variables. It is therefore possible to eliminate n of the aforementioned variables. The function obtained by eliminating the generalized velocities $\dot{q}_i$ is known as the Hamiltonian $H(q_i, p_i, t)$

$$H(p_i, q_i, t) = - L + \sum_i p_i \dot{q}_i \qquad\qquad (\mathrm{II.I.13})$$

In the case of conservative systems, if the change of variable leading to the generalized coordinates is not time-dependent, it can be shown that

$$H = T + V = E \qquad\qquad (\mathrm{II.I.14})$$

In other words, the Hamiltonian is equal to the function representing the total energy of the system and is therefore a constant of the motion.

Canonical equations or Hamitonian equations

Starting from the expression $H = \sum_i p_i \dot{q}_i - L$, it is possible to demonstrate that the Lagrange equations are equivalent to the following set of 2n first-order partial differential equations

$$\frac{\partial H}{\partial q_i} = -\dot{p}_i \qquad \frac{\partial H}{\partial p_i} = \dot{q}_i \qquad \frac{\partial H}{\partial t} = -\frac{\partial L}{\partial t} \qquad\qquad (II.I.15)$$

These equations are called the Hamilton equations or canonical equations of motion.

Remark: It is easy to derive the Hamilton equations (Eq.II.I.15) from the Lagrange equations by writing the differential dH in two different ways:

$$dH = \sum_i \left(\frac{\partial H}{\partial q_i} dq_i + \frac{\partial H}{\partial p_i} dp_i \right) + \frac{\partial H}{\partial t} dt = \sum_i (\dot{p}_i \, d\dot{q}_i + \dot{q}_i \, dp_i) - dL$$

then developing dL

$$dH = \sum_i \left(\dot{p}_i \, d\dot{q}_i + \dot{q}_i \, dp_i - \frac{\partial L}{\partial q_i} dq_i - \frac{\partial L}{\partial \dot{q}_i} d\dot{q}_i \right) - \frac{\partial L}{\partial t} dt$$

Then, taking Eqs. II.I.9 and II.I.10 into account, and identifying the corresponding factors, we obtain Eq. II.I.15.

3. Elementary tools: Operators and their properties

As suggested in the introduction, we shall represent the state of a system by a wave function $\Psi(r_i,t)$. The electronic spin will be added later, in Sec. II.I.5. The set of states of this system constitutes the vectorial space of states F. One way of representing an element of F is $\Psi(r_i,t)$.

These concepts will be explained in more detail in the following section. In order to deal with the evolution of the system, we shall have to look at the transformation of our function by mathematical entities called *operators*. The main properties of these operators are described below.

Definition and examples of linear operators

Operator A, which acts on vectorial space F, is a mathematical entity that maps each vector Ψ of space F onto another vector Ψ' of the same space

$$\psi(r_i, t) \xrightarrow{A} \psi'(r_i, t)$$

and is expressed as

$$\Psi'(r_i, t) = A \, \Psi(r_i, t) \qquad\qquad (II.I.16)$$

Operator A is said to be linear if, for any vectors ψ_1 and ψ_2 of space F and for any two numbers λ_1 and λ_2, the following relation is true

$$A \left(\lambda_1 \Psi_1 + \lambda_2 \Psi_2 \right) = \lambda_1 A\Psi_1 + \lambda_2 A\Psi_2 \qquad\qquad (II.I.17)$$

Here are some examples:

Linear operators	Multiplication by x: $A = x$ $\Psi'(\mathbf{r}, t) = A\Psi(\mathbf{r}, t) = x\Psi(\mathbf{r}, t)$ Derivative with respect to x: $A = \dfrac{\partial}{\partial x}$ $\Psi'(x, y, t) = A\Psi(x, y, t) = \dfrac{\partial}{\partial x}\Psi(x, y, t)$ Parity operator: $A = \pi$ $\Psi'(\mathbf{r}, t) = \pi\Psi(\mathbf{r}, t) = \Psi(-\mathbf{r}, t)$
Nonlinear operators	Logarithmic operator Ln : $A = \mathrm{Ln}$ $\psi'(\mathbf{r}, t) = A\,\psi(\mathbf{r}, t) = \mathrm{Ln}\,\psi(\mathbf{r}, t)$ $\mathrm{Ln}(\lambda_1\psi_1 + \lambda_2\psi_2) \neq \lambda_1\,\mathrm{Ln}\,\psi_1 + \lambda_2\,\mathrm{Ln}\,\psi_2$

Eigenvectors, eigenvalues

It is possible to associate a set of functions u_n belonging to F and a set of complex numbers a_n with each linear operator acting on F, so that

$$A u_n (x_1, y_1, z_1, t) = a_n u_n(x_1, y_1, z_1, t) \tag{II.I.18}$$

The u_n functions are called the eigenfunctions of operator A and the numbers a_n their associated eigenvalues. The set of eigenvalues is also called the spectrum of A. An eigenvalue a_n may be associated with several independent eigenfunctions. The number of linearly independent eigenfunctions corresponding to an eigenvalue is called its degeneracy. It is easy to see that, because of the linearity of the operator, all functions of the form $\lambda u_n(x_1, y_1, z_1, t)$, where λ is a complex number, are also eigenfunctions associated with the same eigenvalue a_n, but since they are proportional to u_n, they do not add to the degeneracy of the eigenvalue, so they are not considered as different eigenfunctions.

Example: Find the eigenvalues of the parity operator π
The equation yielding the eigenvalues is
$$\pi u_n(x, y, z, t) = a_n u_n(x, y, z, t)$$
Using the definition of the parity operator, we obtain
$$\pi u_n(x, y, z, t) = u_n(-x, -y, -z, t) = a_n u_n(x, y, z, t)$$
And by applying the parity operator again, we obtain
$$\pi[\pi u_n(x, y, z, t)] = u_n(x, y, z, t) = a_n \pi u_n(x, y, z, t) = a_n^2 u_n(x, y, z, t)$$
Obviously $a_n^2 = 1$ implies two eigenvalues $a_n = +1$ and $a_n = -1$
$a_n = +1$ implies $u_n(x, y, z, t) = u_n(-x, -y, -z, t)$ and this is the definition of the set of spatially even functions.
$a_n = -1$ implies $u_n(x, y, z, t) = -u_n(-x, -y, -z, t)$; this defines the set of spatially odd functions.
In both cases, degeneracy is infinite.

Hermitian scalar product in F

Let us associate a complex number, usually written (f,g), with any two functions of F, $f(\mathbf{r},t)$ and $g(\mathbf{r},t)$, taken in this order. (f,g) is called the Hermitian scalar product of g by f. By definition, it is equal to

$$(f, g) = \int f^* g \, dv \qquad (\text{II.I.19})$$

In this expression, the star, *, represents the conjugate of the function: f^* is the function obtained by changing i into $-$ i in the expression of f. dv represents the volume element of the space in which f and g are defined. Integration is extended to the whole space in which the variable v is defined.

(For example, for N particles defined in all space, using rectangular coordinates, $dv = dx_1 dy_1 dz_1 ... dx_N dy_N dz_N$), we obtain

$$\int f^* g \, dv = \int_{x_1 = -\infty}^{x_1 = +\infty} \int_{y_1 = -\infty}^{y_1 = +\infty} \int_{z_1 = -\infty}^{z_1 = +\infty} \cdots$$

$$\cdots \int_{x_N = -\infty}^{x_N = +\infty} \int_{y_N = -\infty}^{y_N = +\infty} \int_{z_N = -\infty}^{z_N = +\infty} f^* g \, dx_1 \, dy_1 \, dz_1 \cdots dx_N \, dy_N \, dz_N$$

The following is a brief reminder of the properties of the Hermitian scalar product

$$(f, g) = (g, f)^*$$

$$(f, \lambda_1 g_1 + \lambda_2 g_2) = \lambda_1 (f, g_1) + \lambda_2 (f, g_2)$$

$$(\lambda_1 f_1 + \lambda_2 f_2, g) = \lambda_1^* (f_1, g) + \lambda_2^* (f_2, g)$$

where λ_1 and λ_2 represent any number. The definition of the norm of a function is

$$N_f = \sqrt{(f, f)} = \sqrt{\int f^* f \, dv} \qquad (\text{II.I.20})$$

f is called a normalized function if $N_f = 1$, which implies that

$$\int |f|^2 dv = 1 \qquad (\text{II.I.21})$$

Hermitian operators

Hermitian operators are very important in quantum mechanics. By definition, they must satisfy the following relations:

$$(f, Ag) = (g, Af)^* = (Af, g) \qquad (\text{II.I.22})$$

These relations must hold true for any functions f and g of the vectorial space F of functions we are studying, namely those describing the states of systems.

Let us now state and prove two fundamental properties of Hermitian operators:

The eigenvalues of a hermitian operator are real.

Where A is a Hermitian operator and $A u_n = a_n u_n$ its eigenvalue equation, we obtain

$$(u_n, A u_n) = (u_n, a_n u_n) = a_n (u_n, u_n)$$

However, as A is Hermitian

$$(u_n, A u_n) = a_n (u_n, u_n) = (u_n, A u_n)^* = a_n^* (u_n, u_n)^*$$

Due to the definition of the scalar product

$$(u_n, u_n) = (u_n, u_n)^*$$

we obtain $a_n = a_n^*$ and therefore a_n is a real number.

If two eigenfunctions u_n and $u_{n'}$ of a hermitian operator correspond to two different eigenvalues a_n and $a_{n'}$, they are orthogonal.

Let us assume $\qquad a_n \neq a_{n'}$ (II.I.23)

The eigenvalue equations may be expressed as

$$A u_n = a_n u_n \qquad\qquad A u_{n'} = a_{n'} u_{n'}$$

As A is hermitian, we can write

$$(u_n, A u_{n'}) = a_{n'} (u_n, u_{n'}) = (u_{n'}, A u_n)^* = a_n^* (u_{n'}, u_n)^* = a_n (u_{n'}, u_n)^*$$

From which we conclude that

$$(a_{n'} - a_n) (u_n, u_{n'}) = 0$$

And thus

$$(u_n, u_{n'}) = 0$$

On the other hand, when several linearly independent eigenfunctions correspond to the same degenerate eigenstate, they can always be chosen so as to be orthogonal.

By definition, the set of eigenfunctions of an operator A is said to be orthonormal if $(u_n, u_n) = 1$ and $(u_n, u_{n'}) = 0$ for all $n \neq n'$. In this case, we write $(u_n, u_{n'}) = \delta_{nn'}$. The Kronecker delta $\delta_{nn'}$ is equal to 1 if $n = n'$ and to 0 if $n \neq n'$.

Commutators of two operators - Properties of commuting operators

A and B are two operators, applied to f in turn; first B followed by A; then A followed by B

$$A B f = g \qquad \text{and} \qquad B A f = g'$$

if we use the notation $\qquad f_1 = Bf \qquad$ and $\qquad f'_1 = Af$

we obtain $\qquad g = A f_1 \qquad$ and $\qquad g' = B f'_1$

In general, g' is different from g. The commutator of A and B is the operator $[A,B]$, defined as follows:

$$[A, B] = AB - BA \tag{II.I.24}$$

Two operators are said to commute if their commutator is zero. In that case we obtain

$$g - g' = (AB - BA) f = [A, B] f = 0 \text{ for all } f \tag{II.I.25}$$

Let us now prove two important properties related to commutation.

If two operators A and B are hermitian, the product AB is hermitian provided that $[A,B] = 0$.

Because A is hermitian, we can write $(f, ABg) = (Bg, Af)^*$

and, according to the definition of the scalar product $(Bg, Af)^* = (Af, Bg)$

As B is also Hermitian, we can write $(Af, Bg) = (g, BAf)^*$
And if $[A,B] = 0$, then $(g, BAf)^* = (g, ABf)^*$
So therefore $(f, ABg) = (g, ABf)^*$
which proves the hermiticity of AB (see Eq. II.1.22).

If u_n is an eigenfunction of A associated to the nondegenerate eigenvalue a_n, then u_n is an eigenfunction of all operators B which commute with A.

Indeed, defining the eigenfunctions, then the commutativity of A and B yields

$$BA\, u_n = a_n\, B\, u_n = AB\, u_n$$

The second equality shows that Bu_n is an eigenfunction of A with eigenvalue a_n. This means that Bu_n is proportional to u_n and therefore that u_n is an eigenfunction of B

$$Bu_n = b_n\, u_n$$

If a_n is a degenerate eigenvalue of A, the property is modified as follows: If q linearly independent eigenfunctions u_n are associated with the same eigenvalue a_n (whose degree of degeneracy is therefore q) it is always possible to find q linearly independent combinations of these functions u_n which are eigenfunctions of any operator B commuting with A. We can conclude from this that two commuting operators always have at least one common set of eigenfunctions.

4. The first seven postulates

4.1 - Statements

We wish to lay the foundation of a set of laws of physics governing the microscopic domain, valid not only for matter, but also for radiation. We are about to state some postulates. However, these postulates are not completely universal, as they apply only to particles with speeds very much lower than the velocity of light. These are the postulates of nonrelativistic quantum mechanics. The various theories of mechanics can be organized in a table according to the domains in which they apply. If we are concerned with lower velocities, the nonrelativistic theories apply, whereas at velocities on the order of the speed of light, relativistic concepts come into play. If we are concerned with angular momentums on the order of Planck's constant, only quantum theories are applicable, whereas at much greater values, classical theories can be used.

Even though the notion of velocity is not defined for quantum particles, we can extract a "velocity" from the mean value of the angular momentum, as this mean value can always be calculated.

In this book, we are concerned only with the fundamental notions of non-relativistic quantum mechanics. The relativistic corrections given by Dirac's relativistic quantum mechanics are of little consequence in understanding the basic concepts of chemical bonding.

TABLE II.I.1 — **The various theories of mechanics.**
$(c = 3 \times 10^8 \text{ m/s}, h = 6.6 \times 10^{-34} \text{ J s}, \hbar = h/2\pi)$

	velocity	$v \ll c$	$v \approx c$
angular momentum	$1 \gg \hbar$	classical mechanics	relativistic classical mechanics
	$1 \approx \hbar$	nonrelativistic quantum mechanics	relativistic quantum mechanics

Nevertheless, whereas the effect of relativity (except for spin) is indeed negligible for light atoms, this does not hold true for heavy atoms.

Let us first recall the main characteristics of quantum physics as determined in Section II.I.1:

– It provides a probabilistic representation of the state of a microscopic system.

– The effect of measurements must be included in the theory.

– Measuring certain physical quantities, or observables, has to lead to a discrete set of values.

– Quantum physics must be consistent with classical physics in the case of macroscopic systems.

Description of the state of a system

Quantum physics postulates that the state of a system can be described by a function Ψ (from which a probability density can be derived). This function belongs to a space of functions F, called the space of states. First, let us look at the case of quantum systems described by functions depending exclusively on the spatial coordinates r_i of i particles, and time t (later we shall see that this definition can be extended to descriptions of particles with spin, a nonspatial intrinsic, property).

As we already suggested, function $\Psi(r_i,t)$ belongs to the space of functions which maps R^{3N} space onto C. However, as $|\Psi(r_i, t)|^2$ is also intended to represent the probability density of finding one of the particles in question within the element of volume dv_i, the wave function must satisfy a certain number of criteria:

— $\Psi(r_i,t)$ must be defined, continuous and derivable everywhere.

— $\Psi(r_i,t)$ must be such that its square can be normalized, in other words, such that the following scalar product is always equal to a finite number

$$(\psi(r_i, t), \psi(r_i, t)) = \int | \psi(r_i, t)|^2 dv \qquad (\text{II.1.26})$$

This restricted portion of the function space constitutes a subspace, called the space of states F. It is easy to show that space F is a vectorial space in which scalar product is the hermitian scalar product defined in Section 3.

1st postulate:

At any time t, the state of a system can be described by a wave function $\Psi(r_i, t)$ belonging to vectorial space F, subspace of the space of continuous, derivable functions defined everywhere.

Description of measurable physical quantities, also called observables

We would like to provide a theoretical explanation for the experimentally observed fact that, when measured, some physical quantities, such as energy and momentum, can give rise to a discrete distribution of values.

The fact that mathematical entities called operators usually have a discrete spectrum of eigenvalues, gave rise to the idea of setting up a correspondence between measurable physical quantities, or observables, and operators. If it is then postulated that the only possible values measured on observables are the eigenvalues of its associated operator, a discrete distribution of observable values is obtained.

2nd postulate:

Any observable A can be described by an operator A acting on the vectorial space of states F, and the orthonormalized set of the eigenfunctions u_n of this operator constitutes a base of F.

Postulates concerning the measurement of an observable

3rd postulate:

When measuring an observable A, the only possible results are the eigenvalues of the corresponding operator A.

If operator A has a discrete spectrum of eigenvalues, measuring observable A will yield a discrete set of measured values. However, as the eigenvalues of A must correspond to measurement results, they must be real. For this reason, operator A must be hermitian. Let u_n represent the set of eigenfunctions of A associated with eigenvalues a_n. We assume that all values a_n have a degeneracy of one. Any state $\Psi(r_i, t)$ of the system can be expanded on the infinite base of eigenfunctions u_n in the following way:

$$\Psi(r_i, t) = \sum_n \alpha_n u_n (r_i, t)$$

Then, we need a postulate stating the probability of finding one of the values a_n as a result of a measurement of observable A on the system in state $\Psi(r_i, t)$.

4th postulate: *(in the case of a discrete, nondegenerate eigenvalue)*

When measuring observable A on a system in a state represented by function $\Psi(r_i,t)$ *(assumed to be normalized), the probability* $P(a_n)$ *of finding* a_n *as the result of the measurement is*

$$P(a_n) = |(u_n, \Psi)|^2 = \left| \int u_n^* \Psi \, dv \right|^2 \qquad (II.I.27)$$

where $u_n(r_i,t)$ is the normalized eigenfunction of A associated to a_n.

In fact, we do not need to calculate the integral in $P(a_n)$ provided we know the expansion of Ψ on the base of u_n, as

$$(u_n, \Psi) = \left(u_n, \sum_m \alpha_m u_m \right) = \sum_m \alpha_m (u_n, u_m) = \sum_m \alpha_m \delta_{nm} = \alpha_n$$

and thus

$$P(a_n) = |(u_n, \Psi)|^2 = |\alpha_n|^2 \qquad (II.I.28)$$

It is not very usual to make just a single measurement of A on a system in state Ψ. In general, either a set of measurements are performed on several identical systems, all in state Ψ, or a series of measurements are carried out on the same system, which is returned to state Ψ after each measurement. What then would be the mean value $\bar{a}_\psi$ for this set of measurements of observable A on the system in state Ψ? To calculate this mean value, add all the measured values, and divide this sum by the total number N of measurements. As there will be $NP(a_n)$ measurements, each producing a result of a_n, the mean value of all measurements will be

$$\bar{a}_\psi = \frac{1}{N} \sum_n NP (a_n) a_n = \sum_n P (a_n) a_n = \sum_n |(u_n, \Psi)|^2 a_n = \sum_n |\alpha_n|^2 a_n \qquad (II.I.29)$$

or

$$A\Psi = A \sum_n \alpha_n a_n u_n \quad \text{et} \quad (\Psi, A\Psi) = \sum_n |\alpha_n|^2 a_n \qquad (II.I.30)$$

For a normalized state Ψ, the 4th postulate may therefore also be expressed as

$$\bar{a}_\psi = (\psi, A \psi) = \int \psi^* A \psi d\tau \qquad (II.I.31)$$

This mean value of a is called the expectation value of observable A for state Ψ. If the system happens to be in an eigenstate u_n of operator A, the result of the measurement is known with certainty:

$P(an) = 1$ and $P(am) = 0$ if $m \neq n$ and clearly $\bar{a}_\psi = a_n$.

Earlier, we supposed that operator A had a discrete nondegenerate spectrum with values a_n, but the 4th postulate can be extended to the case of degenerate eigenvalues and also to the case of a continuous spectrum.

Case of a degenerate discrete spectrum

Let $u_n{}^i$ (i = 1, ...g) *represent the g eigenvectors of A corresponding to eigenvalue* a_n *which is g times degenerate. The probability of finding value* a_n *as the result of a measurement on a system in state* Ψ *is given by the formula:*

$$P(a_n) = \sum_{i=1}^{g} |(u_n^i, \Psi)|^2 = \sum_{i=1}^{g} \left| \int u_n^i {}^*\Psi \, dv \right|^2 \qquad (\text{II.I.32})$$

Case of a continuous spectrum

In case of a continuous spectrum of eigenvalues, it is no longer possible to define the probability of obtaining value a_n, *but only the elementary probability* $dP(a_n)$ *of finding a value between* a_n *and* $a_n + da_n$.

$$dP(a_n) = |(u_n, \Psi)|^2 da_n = \left| \int u_n^* \Psi dv \right|^2 da_n \qquad (\text{II.I.33})$$

The postulate can also be extended to the case of degenerate continuous spectra and to partially continuous spectra, but these cases are of little use in the study of chemical bonding.

We shall now describe the effect of performing a measurement. We experimentally observed that measuring an observable A disturbs the state of the system concerned. This disturbance is also defined in a postulate.

5th postulate:

If the measurement of an observable A on a system in any state Ψ *results in the nondegenerate eigenvalue* a_n *of A, then immediately after measurement the system is in the state represented by the eigenfunction* u_n *associated with* a_n.

Of course, this postulate also extends to the degenerate case: If state Ψ is expanded on the base of eigenfunctions u_m of operator A, and if u_n^i are g eigenfunctions of A associated with the same eigenvalue a_n, then Ψ can be expanded as follows:

$$\Psi = \alpha_1 u_1 \cdots + (\alpha_n^1 u_n^1 ... + \alpha_n^g u_n^g) + \alpha_{n+1} u_{n+1} + \cdots \qquad (\text{II.I.34})$$

and if we call

$$\alpha_n^1 u_n^1 + \cdots + \alpha_n^g u_n^g = \Psi_n \qquad (\text{II.I.35})$$

then ψ_n is the projection of Ψ on the subspace formed by the eigenfunctions u_n^i associated with a_n. If a measurement of A yields eigenvalue a_n, the system goes from state Ψ to state Ψ_n.

Postulate governing the evolution of a system

Using these first five postulates, we are now in a position to represent a system and its observables. We also know the results of measurements performed on this system. We still do not know how Ψ (r_i, t) evolves with time. Another postulate tells us that the evolution of Ψ (r_i, t) is determined by the Schrödinger equation. It is possible to show how this equation is derived from classical mechanics, though we shall not go into this here. However, since by definition a postulate is not proven, the equation is justified only because the conclusions to which it leads are consistent with observations.

6th postulate:

If $\Psi(r_i,t)$ *represents the wave function of a quantum system, the evolution of* $\Psi(r_i,t)$ *is governed by the following equation:*

$$i\hbar \frac{\partial \Psi(r_i, t)}{\partial t} = H(t)\,\Psi(r_i, t) \qquad (\text{II.1.36})$$

$H(t)$ is the operator associated with the hamitonian of the system. This is called the *Schrödinger equation* of the system. H is called the hamiltonian operator, or simply the Hamiltonian of the system.

Postulate concerning the relation between observables and operators

In order to use the equation governing the evolution of the system, we need some practical rules for constructing the operator A associated with observable A. First, we must express $A(r_i, p_i, t)$ using the hamilton formalism. p_i is the momentum associated with particle i at point r_i. If there is no external electromagnetic field, the total momentum is equal to the linear momentum:

$$p_i = m_i \dot{r}_i$$

7th postulate:

The operator A associated with observable A is obtained by taking the classical expression of A and substituting the following operators for the quantities x_i, y_i, z_i, p_{xi}, p_{yi}, p_{zi} *(expressed in rectangular coordinates) as indicated:*

$$
\begin{array}{lll}
x_i = x_i & p_{xi} = -i\hbar \dfrac{\partial}{\partial x_i} & \\[2mm]
y_i = y_i & p_{yi} = -i\hbar \dfrac{\partial}{\partial y_i} & r_i = r_i \qquad p_i = -i\hbar \left\{ \begin{array}{l} \dfrac{\partial}{\partial x_i} \\[2mm] \dfrac{\partial}{\partial y_i} \\[2mm] \dfrac{\partial}{\partial z_i} \end{array} \right. = -i\hbar \nabla_i \\[2mm]
z_i = z_i & p_{zi} = -i\hbar \dfrac{\partial}{\partial z_i} &
\end{array}
$$

With these substitutions, it is easy to show that $[x_i, p_{xi}] = i\hbar\,I$ and that $[x_i, x_j] = 0 = [x_i, p_{xj}]$ for $i \neq j$. It should be borne in mind that there are several possible rules for converting observables into operators. Those given here correspond to the Schrödinger representation.

4.2 Physical content of the first seven postulates

Let us take a conservative system consisting of N noninteracting particles, placed in a force field deriving from a potential $V(\mathbf{r})$. The state of the system is described by a wave function $\Psi(\mathbf{r}_1, ..., \mathbf{r}_N, t) = \Psi(\mathbf{r}_i, t)$. This state satisfies the Schrödinger equation (see Eq. II.I.36).

To use this equation, we must determine the expression of operator H, and so we must first find the classical expression of $H(\mathbf{r}_i, \mathbf{p}_i)$, as follows :

$$H = T + V = \sum_{i=1}^{N} \left(\frac{|\mathbf{p}_i|^2}{2m_i} + V(\mathbf{r}_i) \right) \tag{II.I.37}$$

It is then possible to use the correspondence postulate, remembering that

$$|\mathbf{p}_i|^2 = p_{x_i}^2 + p_{y_i}^2 + p_{z_i}^2 \tag{II.I.38}$$

and that operator $p_{x_i}^2$, for example, is obtained by applying p_{xi} twice. Thus

$$p_{x_i} = -i\hbar \frac{\partial}{\partial x_i} \left(-i\hbar \frac{\partial}{\partial x_i} \right) = -\hbar^2 \frac{\partial^2}{\partial x_i^2} \tag{II.I.39}$$

$$H = \sum_{i=1}^{N} \left[-\frac{\hbar^2}{2m_i} \Delta_i + V(\mathbf{r}_i) \right] \tag{II.I.40}$$

with

$$\Delta_i = \frac{\partial^2}{\partial x_i^2} + \frac{\partial^2}{\partial y_i^2} + \frac{\partial^2}{\partial z_i^2} \text{ (Laplacian)} \tag{II.I.41}$$

Of course, having obtained this expression for H using the correspondence postulate rules, it may be expressed in any other reference system used to localize particles i by writing Δ_i and $\mathbf{r}_i$ in the new coordinate system.

The evolution of the state of the system is governed by the Schrödinger equation, so we have

$$i\hbar \frac{\partial \Psi(\mathbf{r}_1, \cdots, \mathbf{r}_N, t)}{\partial t} = \left[\sum_{i=1}^{N} \left(-\frac{\hbar^2}{2m_i} \Delta_i + V(\mathbf{r}_i) \right) \right] \Psi(\mathbf{r}_1, \cdots, \mathbf{r}_N, t) \tag{II.I.42}$$

If we solve this equation, we shall be able to find the state $\Psi(\mathbf{r}_i, t)$ of the system at time t, provided we know its initial state $\Psi(\mathbf{r}_i, 0)$ at time $t = 0$.

4.3 Physical interpretation of the postulates restricted to their consequences on chemical bonding

In this section, we shall examine a few points which are especially important in their effect on chemical bonding.

Wave function and the probability density of finding an electron

We have to check that the postulates we just listed indeed result in the probabilistic mechanics we set out to construct. Let us take the state of a particle described by Ψ (r,t). What is the probability of finding this particle in a volume dv_0 surrounding point r_0? If we know the eigenfunctions $u_{r_0}(x,y,z,t)$ of operator r (found using the Schrödinger representation) , postulate 4 gives us this probability. The equation

$$r\, u_{r_0}\,(x,\,y,\,z,\,t) = r_0 u_{r_0}\,(x,\,y,\,z,\,t) \qquad \qquad (\text{II.I.43})$$

is equivalent to the three "scalar" equations

$$\begin{aligned} x u_{r_0}\,(x,\,y,\,z,\,t) &= x_0 u_{r_0}\,(x,\,y,\,z,\,t) \\ y u_{r_0}\,(x,\,y,\,z,\,t) &= y_0 u_{r_0}\,(x,\,y,\,z,\,t) \\ z u_{r_0}\,(x,\,y,\,z,\,t) &= z_0 u_{r_0}\,(x,\,y,\,z,\,t) \end{aligned} \qquad (\text{II.I.44})$$

The solution of the equation $(x - x_0)\, u_{r_0}\,(x,y,z,t) = 0$ is well known in mathematics. It is a function such that $u_{r_0}\,(x,y,z,t)$ is equal to 0 for all values of $x \neq x_0$, and may take any value at $x = x_0$. This function may be written symbolically as $u_{r_0}\,(x,y,z,t) = \lambda_1(x,y,z,t)\delta(x-x_0)$ where $\delta(x-x_0)$ represents the delta — or Dirac— function defined by

$$\delta(x) = 0 \text{ for all } x \neq 0 \text{ and such that } \int_{-\infty}^{\infty} \delta(x)\, dx = 1$$

Solving the three preceding eigenvalue equations in turn yields

$$u_{r_0}\,(x,\,y,\,z,\,t) = \lambda\ \delta(x - x_0)\,\delta(y - y_0)\,\delta(z - z_0) = \lambda\delta(r - r_0)\ (\text{II.I.45})$$

Let us choose $\lambda = 1$. Then we can write in a more compact way $u_{r_0}\,(x,y,z,t) = \delta(r - r_0)$. This choice of λ is arbitrary , as in fact $\delta(r - r_0)$ is not a normalizable function. We shall return to this important point later. Thus the "function" $\delta(r - r_0)$ does not belong to the space of states F, and cannot therefore represent a real state of the particle. In fact, the physical quantity "position r" is not really an observable. In fact, the real observable is the quantity $r \pm \Delta r$, a position measured with a certain degree of uncertainty. This observable is associated with an eigenfunction of the corresponding operator, that does belong to space F. However, for practical reasons, we want to keep the "function" $\delta(r - r_0)$, so that each position r_0 may be associated with a function $\delta(r - r_0)$. The spectrum of the observable "position of the particle" is continuous and all positions are possible. The probability $dP(r_0)$ of finding the particle inside a volume dv_0 around position r_0 is given by Postulate 4, namely:

$$dP(r_0) = |(\delta(r - r_0),\ \Psi(r,\,t))|^2 dv_0 \qquad (\text{II.I.46})$$

Now the specific properties of the delta (or Dirac) function are such that

$$\delta(r - r_0),\ \psi(r,\,t) = \int \delta^*(r - r_0)\psi(r,\,t)\, dv = \psi(r_0,\,t) \qquad (\text{II.I.47})$$

So that finally we find

$$dP(r_0) = |\Psi(r_0,\,t)|^2\, dv_0 \qquad |\Psi(r_0,\,t)|^2 = \frac{dP(r_0)}{dv_0} \qquad (\text{II.I.48})$$

The volume density of the probability of the particle's presence at point r_0 at time t is given by $|\Psi(r_0, t)|^2$. Let us now consider a set of N particles whose collective state is represented by $\Psi(r_1, ..., r_N, t)$. The interpretation is then generalized as follows: $|\Psi(r_{10}, ..., r_{N0}, t)|^2 dv_{10}, ..., dv_{N0}$ represents the probability of finding particle 1 in volume element dv_{10} located at r_{10}, particle i in volume element dv_{i0} located at r_{i0}, ..., and particle N in volume element dv_{N0} located at r_{N0}, simultaneously at time t. Clearly, in the general case, $\Psi(r_i, t)$ does not give the probability of finding a specific particle. In Section III.I.2 we shall see how to calculate the probability of finding any one of the N electrons in a given volume element.

Let us assume that the N particles are independent, which means that the Hamiltonian can be written as $H = H_1 + ... + H_N$ where H_i is the Hamiltonian relative to particle i (this would be the case for the example described above). Let us also assume that the wave functions $\Psi_i(r_i, t)$ are known, where $\Psi_i(r_i, t)$ is the solution of the Schrödinger equation relative to the ith particle in isolation

$$H_i \, \Psi_i \, (r_i, t) = i\hbar \frac{\partial \Psi_i \, (r_i, t)}{\partial t} \qquad (\text{II.I.49})$$

$\Psi_i(r_i, t)$ is assumed to be normalized. We want to show that the wave function which is the product of all the individual wave functions is a solution of the Schrödinger equation corresponding to the system of N particles

$$\Psi(r_1 ..., r_N, t) = \Psi_1(r_1, t) \, \Psi_2(r_2, t) \, ... \, \Psi_N(r_N, t)$$

Here, we also assume that the particles are all different from each other so as to avoid dealing with the problems raised by indistinguishable particles (see Eq. II.I.5). The calculation is as follows :

$$H \Psi(r_1, ..., r_N, t) = \sum_i H_i \Psi \, (r_i, t) =$$

$$\sum_i i\hbar \frac{\delta \Psi_i(r_i, t)}{\partial t} \Psi_1 \Psi_2 .. \Psi_{i-1} \Psi_{i+1} .. \Psi_N = i\hbar \frac{\delta}{\partial t} \Psi(r_1, ..., r_N, t) \qquad (\text{II.I.50})$$

If the system is in this specific state Ψ, it is once again possible to define the probability of finding particle i at r_{i0}. Indeed, the eigenfunction corresponding to position r_i is as follows:

$$u_{r_{i0}}(r_1, ..., r_i, ..., r_N, t)$$

$$= \Psi_1 \, (r_1, t) \, ... \, \Psi_{i-1} \, (r_{i-1}, t) \, \delta \, (r_i - r_{i0}) \, \Psi_{i+1} \, (r_{i+1}, t) \, ... \, \Psi_N \, (r_N, t) \qquad (\text{II.I.51})$$

and as $\qquad (u_{r_{i0}} \, ... \, (r_i, t), \, \Psi(r_1, ..., r_N, t)) = \Psi_i \, (r_{i0}, t) \qquad (\text{II.I.52})$

we can see that $|\Psi_i(r_{i0}, t)|^2$ actually represents the probability density of finding particle i at position r_{i0} at time t.

Stationary states (H does not explicitly depend on time)

In studying chemical bonding, we are mainly concerned with the stable states of molecules, i.e., states whose energies do not change with time. The third postulate tells us that stable energy values of this type, found with certainty each time the energy is measured on the molecule, can only be eigenvalues of the Hamiltonian H operator associated with the observable "energy". Therefore, the state of a stable molecule must be an eigenstate of H. Let us call this eigenstate $u_n(\mathbf{r}_i, t)$, and E_n the corresponding associated eigenvalue. Of course this state satisfies the equation

$$Hu_n(\mathbf{r}_i, t) = E_n u_n(\mathbf{r}_i, t) \qquad (\text{II.I.53})$$

This energy eigenvalue equation is known by the unfortunate name of the "time-independent Schrödinger equation". It must not be confused with the time-dependent Schrödinger equation, postulated to govern the time evolution of the state $\Psi(\mathbf{r}_i, t)$ of a system. This new Schrödinger equation is a specific eigenvalue equation, satisfied by states with constant, well defined energy. These states are called *stationary states* of the system. But of course, we can try to find the time dependence of such states, using the Schrödinger equation. Equations II.I.36 and II.I.53 give the following :

$$Hu_n(\mathbf{r}_i, t) = E_n u_n(\mathbf{r}_i, t) = i\hbar \frac{\partial}{\partial t} u_n(\mathbf{r}_i, t) \qquad (\text{II.I.54})$$

yielding

$$\frac{\frac{\partial}{\partial t} u_n(\mathbf{r}_i, t)}{u_n(\mathbf{r}_i, t)} = -i \frac{E_n}{\hbar} \qquad (\text{II.I.55})$$

The solution of this differential equation is

$$u_n(\mathbf{r}_i, t) = u_n^{(0)}(\mathbf{r}_i) \exp(-iE_n t/\hbar) \qquad (\text{II.I.56})$$

Inversely, it is possible to show that if we are dealing with a conservative system in which H does not depend on t, any state of the system in which the time and spatial dependence factorize is a stationary state. To prove this, let us assume such a factorization:

$$H\Psi(\mathbf{r}_i, t) = \Psi^{(0)}(\mathbf{r}_i) f(t)$$

Applying Eq. II.I.36 gives

$$H\Psi(\mathbf{r}_i, t) = f(t) H \Psi^{(0)}(\mathbf{r}_i) = \Psi^{(0)}(\mathbf{r}_i) i\hbar \frac{df}{dt}$$

which in turn yields a partial differential equation where the first member depends explicitly only on $\mathbf{r}_i$ and the second member depends explicitly only on t

$$\frac{H\Psi^{(0)}(\mathbf{r}_i)}{\Psi^{(0)}(\mathbf{r}_i)} = i\hbar \frac{df(t)/dt}{f(t)}$$

This equation can hold only if both sides are constant. Obviously, the left hand side is equal to the constant energy of the system E_n. Integrating the differential equation in t immediately gives the solution

$$f(t) = A \exp(-iE_n t/\hbar)$$

Evolution of any nonstationary state as a function of time

Let us consider the state $\Psi(r_i, t)$ of a system for which the spatial and temporal dependencies do not factorize. This is therefore a nonstationary state. How does this state evolve over time, assuming it was known to be in state $\Psi(r_i, t_0)$ at time t_0? The eigenstates $u_n^0 \exp(-iE_n t/\hbar)$ of operator H form one of the bases of the space of states F. The state $\Psi(t)$ can be expanded on this base (by using the 2nd postulate):

$$\Psi(r_i, t) = \sum_n c_n(t) u_n(r_i, t) \tag{II.I.57}$$

The coefficients $c_n(t)$ can be interpreted as projections of $\Psi(r_i, t)$ on the basic states $u_n(r_i, t)$

$$c_n(t) = (u_n(r_i, t), \Psi(r_i, t)) \tag{II.I.58}$$

Thus, the evolution of Ψ over time is known, provided the $c_n(t)$ coefficients are known. Let us apply Eq.II.I.36 a second time

$$i\hbar \frac{\partial \Psi}{\partial t} = H \Psi = i\hbar \sum_n \frac{\partial}{\partial t}(c_n u_n(r_i, t)) = H \sum_n c_n u_n(r_i, t) \tag{II.I.59}$$

in other words

$$i\hbar \left[\sum_n \frac{\partial c_n}{\partial t} u_n(r_i, t) + c_n \frac{\partial}{\partial t} u_n(r_i, t) \right] = \sum_n c_n H u_n(r_i, t)$$

with

$$\sum_n \frac{\partial c_n}{\partial t} u_n(r_i, t) = 0 \ \forall \ t \tag{II.I.60}$$

This equation implies that the $c_q(t)$ functions are constants, as the functions $u_n(r_i, t)$ are assumed to be linearly independent.

If the expansion of $\Psi(r_i, t)$ is known at instant t_0

$$\Psi(r_i, t_0) = \sum_n c_n(t_0) u_n^{(0)}(r_i) \exp(-iE_n t_0/\hbar)$$

and if the values $c_n(t_0) = (u_n(r_i, t_0), \Psi(r_i, t_0))$ are also known, then the law governing the evolution over time of the state of the system is given by

$$\Psi(r_i, t) = \sum_n c_n(t_0) u_n^{(0)}(r_i) \exp(-iE_n t/\hbar) \tag{II.I.61}$$

It should be remembered that if we measure the energy of a system which is in this nonstationary state, the result of the measurement is not a definite value. If the energy is measured at time t, the result will be one of the eigenvalues E_n. It is the *probability* $|c_n(t)|^2 = c_n(t_0)|^2$ of finding E_n which is constant.

The expected energy value, or its mean value for a series of measurements, is given by

$$\overline{E} = \sum_n |c_n(t_0)|^2 E_n \qquad (II.1.62)$$

The expected value of the energy is constant, and is conserved throughout the evolution of the nonstationary state $\Psi(r_i, t)$ of the system. The energy is a constant of motion. The law of evolution of the state of the system also implies conservation of the norm (i.e., the modulus) of the wave function over time. Indeed, we have

$$(\Psi(r_i, t_0), \Psi(r_i, t_0)) = (\Psi(r_i, t), \Psi(r_i, t)) = 1$$

as

$$(\Psi(r_i, t), \Psi(r_i, t)) = \sum_n |c_n(t_0) u_n^{(0)}(r_i)|^2 = (\Psi(r_i, t_0), \Psi(r_i, t_0))$$

The variation method

For the description of chemical bonding, we are interested only in the stationary states of molecules, and more especially their ground state, or lowest energy state.

The arguments in the last section show that, in this case, we need only determine the eigenfunctions of H as defined by $HA(r_i) = E_n A(r_i)$. However, the partial differential equation implied by H, depending on $3N$ variables, is usually too complicated to be solved analytically, or even numerically if we do not have access to unlimited computer time. An important part of quantum chemistry is devoted to the development of methods yielding good approximations for eigenfunctions and eigenvalues. Most of these methods derive from the so-called variation method, based on a simple but very powerful mathematical theorem, called the theorem of variations.

Theorem of variations

Let us consider a system in a state represented by function Ψ. The expected energy value is

$$\overline{E}_\Psi = \frac{\int \Psi^* H \Psi dv}{\int \Psi^* \Psi dv} \qquad (II.1.63)$$

Expanding Ψ on the orthonormal base of stationary states, i.e., states u_n such that $Hu_n = E_n u_n$, yields

$$\Psi = \sum_n c_n u_n$$

The mean energy $\bar{E}_\Psi$ may be expressed as a function of the values of c_n and of E_n

$$\bar{E}_\Psi = \frac{\int \left(\sum_n c_n u_n\right)^* H \left(\sum_m c_m u_m\right) dv}{\int \left(\sum_n c_n u_n\right)^* \left(\sum_m c_m u_m\right) dv} = \frac{\sum_n |c_n|^2 E_n}{\sum_n |c_n|^2} \qquad \text{(II.1.64)}$$

If E_0 is the energy of the ground state ($E_0 < E_n$ for all $n \neq 0$), we can minor the above expression for the expected energy value

$$\bar{E}_\Psi \geq \frac{\sum_n |c_n|^2}{\sum_n |c_n|^2} E_0 = E_0 \qquad \text{(II.1.65)}$$

With this in mind, the theorem of variations gives the following :

$$\frac{\int \Psi^* H \Psi dv}{\int \Psi^* \Psi dv} \geq E_0 \qquad \text{(II.1.66)}$$

Equality can occur only if $\Psi \equiv \Psi_0$, that is, if the system is in its groundstate.

Variations method

Equation II.1.66 leads to the idea that if we take a trial function $\Psi(\lambda_i)$ depending on a certain number of parameters λ_i, and define $W(\lambda_i)$ as follows :

$$W(\lambda_i) = \frac{\int \Psi^*(\lambda_i) H \Psi(\lambda_i) dv}{\int \Psi(\lambda_i)\Psi^*(\lambda_i) dv} \geq E_0 \qquad \text{(II.1.67)}$$

then we find the smallest possible value for expression $W(\lambda_i)$ by varying these parameters.

The values $W(\lambda_i^0)$ and $\Psi(\lambda_i^0)$ are the best possible approximations for E_0 and to Ψ_0 which can be obtained starting from trial function $\Psi(\lambda_i)$. If the trial function was pertinent, $W(\lambda_i^0)$ and $\Psi(\lambda_i^0)$ are good approximations for describing the groundstate of the molecule.

The Hartree-Fock method is mainly based on the variations method. This method is extensively developed in Chapter IV of this book.

4.4 Example: The hydrogen atom

Classical model

The hydrogen atom consists of a proton with mass $m_p = 1.7 \times 10^{-27}$ kg and charge $q = 1.6 \times 10^{-19}$ coulomb, and of an electron with mass $m_e = 0.91 \times 10^{-30}$ kg and charge $-q$. These particles interact electrostatically. The particles exert an attractive force on each other. The amplitude of this force is given by Coulomb's law

$$|F| = q^2/4\pi\varepsilon_0 r^2 \qquad (II.1.68)$$

and it lies on the straight line joining one particle to the other. The corresponding interaction potential (see Eq. III.1.3) is equal to

$$V(r) = -q^2/4\pi\varepsilon_0 r = -e^2/r \qquad (II.1.69)$$

with $e^2 = q^2/4\pi\varepsilon_0$ and r is the distance between the two particles.

Separating the movement of the center of gravity

This system, consisting of two particles, has twelve degrees of freedom defined by the positions and velocities of the particles. These can be called r_p, $\dot{r}_p$, r_e, and $\dot{r}_e$ in an arbitrary reference frame with center O (see Fig. II.1.5). However, instead of taking coordinates r_p and r_e to write the equations of the system, we may also use the coordinate R of its center of mass G, and the coordinate r giving the relative position of the two particles. The two sets of coordinates are related through the following equations:

$$\left| \begin{aligned} R &= \frac{m_p}{m_p + m_e} r_p + \frac{m_e}{m_p + m_e} r_e \\ r &= r_e - r_p \end{aligned} \right. \rightarrow \left| \begin{aligned} r_p &= R - \frac{m_e}{m_p + m_e} r \\ r_e &= R + \frac{m_p}{m_p + m_e} r \end{aligned} \right. \qquad (II.1.70)$$

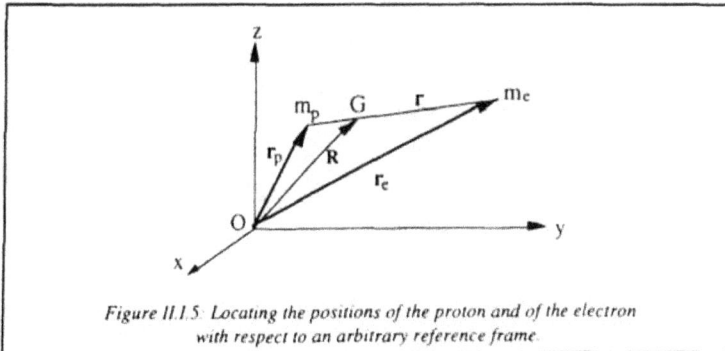

Figure II.1.5: *Locating the positions of the proton and of the electron with respect to an arbitrary reference frame.*

The kinetic energy T can be expressed as a function of **R** and **r**

$$T = \frac{1}{2}\left(m_p\, \dot{\mathbf{r}}_p^2 + m_e\, \dot{\mathbf{r}}_e^2\right) = \frac{1}{2}\left(m_p + m_e\right)\dot{\mathbf{R}}^2 + \frac{1}{2}\frac{m_p\, m_e}{m_p + m_e}\dot{\mathbf{r}}^2$$

$$= \frac{1}{2}M\dot{\mathbf{R}}^2 + \frac{1}{2}\mu\dot{\mathbf{r}}^2 \tag{II.I.71}$$

where $M = m_p + m_e$ (the total mass) and $\mu = (m_p\, m_e)/(m_p + m_e)$ (the reduced mass).

The field of force depends only on $|\mathbf{r}| = r$. The same is true for the potential $V = V(r)$. Therefore the center of mass G has a uniform movement. If we now place ourselves within the reference frame tied to the center of mass G (set **R**=0), we are left with the relative movement of the two particles. This relative movement is equivalent to the movement of a single fictitious particle of mass μ (the reduced mass) placed in a potential $V(|\mathbf{r}|)$ where $|\mathbf{r}|$ represents the distance between the origin and this fictitious particle. Thus the problem is simplified and brought back to the study of a single particle placed in a central potential.

Equivalent particle in a central field

It is easy to show that the motion has two constants:

— The total energy, as we are dealing with a conservative system

$$T + V = \frac{1}{2}\mu\dot{\mathbf{r}}^2 + V(r) = E_0 = \text{constant} \tag{II.I.72}$$

— The angular momentum $\mathbf{l} = \mathbf{r}\times\mathbf{p}$.

This is true as the potential energy $V(r)$ consists of only a radial component so that $F_r = -\partial V/\partial r$ (see Eq. III.I.4) and its momentum with respect to the origin O vanishes ($\mathbf{r}\times\mathbf{F} = 0$). Next, the theorem on angular momentum says that $d\mathbf{l}/dt = \mathbf{r}\times\mathbf{F} = 0$. This means that the angular momentum $\mathbf{l}$ is a constant vector; $\mathbf{l} = \mathbf{l}_0$. This in turn means that vectors **r** and **p** are confined to the plane perpendicular to this vector $\mathbf{l}_0$, and thus the movement itself is plane. Within this plane, the position of the fictitious particle can be described by polar coordinates r and φ. Let us write the constants of the motion with these coordinates

$$E_0 = \frac{1}{2}\mu\left(\dot{r}^2 + r^2\dot{\varphi}^2\right) + V(r) \tag{II.I.73}$$

$$|\mathbf{l}_0| = \mu r^2\dot{\varphi} \tag{II.I.74}$$

and the hamiltonian function, which is equal to the energy, can be expressed exclusively as a function of the radial coordinates, i.e., of r, and of the conjugated variable p_r of r, ($p_r = \mu\dot{r}$), as follows:

$$H(r, p_r) = \frac{p_r^2}{2\mu} + \frac{|\mathbf{l}_0|^2}{2\mu r^2} + V(r) = E_0 \tag{II.I.75}$$

A similar property will be conserved for the associated hamiltonian operator H.

The Bohr atom

The trajectories of the equivalent particle depend on the initial conditions (velocity and position), and, therefore, on the values of E_0 and l_0. Let us determine the conditions on E_0 and l_0 giving rise to a circular trajectory. For a circular trajectory, dr/dt must vanish at all times. The Hamilton equation (see Eq. II.1.15) applied to the conjugated variables r and p_r can be expressed

$$\frac{\partial H}{\partial r} = -p_r \text{ or } -\frac{|l_0|^2}{\mu r^3} + \frac{e^2}{r^2} = -\mu \frac{d^2 r}{dt^2} \qquad (\text{II.1.76})$$

This yields a relation between the radius of the trajectory and the angular momentum

$$r = r_0 = |l_0|^2/\mu e^2 \qquad (\text{II.1.77})$$

And as in this case p_r vanishes at all times, E_0 and $|l_0|$ must also satisfy the relation:

$$E_0 = -\mu e^4/2|l_0|^2 \qquad (\text{II.1.78})$$

So a circular trajectory is possible for certain favorable initial conditions. Moreover, in order to account for the experimental spectroscopic observations, Bohr suggested that l_0 could only take on values which were multiples of $\hbar$: $|l_0(n)| = n\hbar$. The immediate consequence of this is that the energy of the hydrogen atom must necessarily be equal to

$$E_0(n) = -\mu e^4/2\hbar^2 n^2 \text{ where } n \text{ is an integer} \qquad (\text{II.1.79})$$

This formula gives the correct spectrum of the hydrogen atom. Another consequence is that the radii of the corresponding trajectories $r_0(n)$ can only take on a set of discrete values:

$$r_0(n) = n^2 \hbar^2/\mu e^2 \qquad (\text{II.1.80})$$

Using Eq. II.1.70, it is easy to derive the trajectories of the proton and the electron. Nevertheless, as the ratio m_p/m_e of the mass of the proton to the mass of the electron is so large — about 1836 — the center of mass of the system is practically identical with the position of the proton. Thus, in a first approximation, we can consider that the electron rotates uniformly around the proton, and we can replace μ by m_e in the above expressions. This yields:

$$r_0(n) = n^2 a_0 \text{ where } a_0 = \hbar^2/m_e e^2$$

$$E_0(n) = -m_e e^4/2\hbar^2 n^2 = -e^2/2a_0 n^2 \qquad (\text{II.1.81})$$

The radius $r_0(1) = a_0$ of the smallest orbital of the Bohr atom, and the absolute value of the corresponding energy $2E_0(1) = m_e e^4/\hbar^2 = e^2/a_0$ are often used as length and energy units when dealing with atoms and molecules. They are defined as

$$a_0 = 0.0529 \text{ nm} = 1 \text{ Bohr}$$
$$e^2/a_0 = 27.21 \text{ eV} = 1 \text{ Hartree} \qquad (\text{II.1.82})$$

These units, known as *atomic units*, simplify the expression of many atomic and molecular calculations.

Nevertheless, it is fundamental to understand that the quantization of the energy levels and orbital radii has been obtained after *arbitrarily* quantifying the angular momentum. This quantization cannot be justified by the postulates of classical mechanics.

Quantization

If we want a quantum description of the system, we have to discard the idea of describing the hydrogen atom in terms of two classical particles with well-defined trajectories. The whole system of two interacting quantum particles is described by a wave function $\Psi(r_p, r_e, t)$ which completely determines the state of the system. The first step, using the Schrödinger equation, is to write the partial differential equation which must be satisfed by Ψ. The reasoning used in the classical model may be used here to simplify the equations. Indeed, it can be shown in nonrelativistic quantum mechanics that the frame of reference may be placed at the center of mass of the system. As in the classical model, the center of mass travels at constant speed (its translation has a constant linear momentum $\mathbf{P_G}$ and the corresponding energy $p^2_G/2(m_e+m_p)$ can take on any value; the movement is not quantized. Therefore, the system is equivalent to a single particle of mass $\mu = m_p m_e/(m_p+m_e)$ placed in a potential $V(r)$. Once again, we shall make use of the approximation resulting from the great difference in mass between the proton and the electron. As the center of mass practically merges with that of the proton, the positions of the equivalent particle and that of the electron are considered to be identical.

Stationary states of the hydrogen atom

Only those states Ψ_n corresponding to a well-defined energy value are of interest for the study of chemical bonding. Like all other states, these satisfy the Schrödinger equation:

$$H\Psi_n (\mathbf{r}, t) = i\hbar \frac{\partial \Psi_n}{\partial t} (\mathbf{r}, t) \qquad (\text{II.1.83})$$

but in this case, the time and the spatial dependencies separate

$$\Psi_n (\mathbf{r}, t) = \chi_n(\mathbf{r}) \exp[-iE_n t/\hbar] \quad \text{with} \ H\chi_n(\mathbf{r}) = E_n \chi_n (\mathbf{r}) \qquad (\text{II.1.84})$$

The Hamitonian operator H which must be used to derive the $\chi_n(\mathbf{r})$ functions is found by applying Eq. II.1.41 to the case of a single particle. The symmetry of the system suggests, however, that we solve this problem using spherical coordinates r, θ, φ as defined in Fig. II.1.6. Expressing the delta operator Δ in these coordinates yields the following eigenvalue equation:

$$\left\{ -\frac{\hbar^2}{2m_e} \left[\frac{1}{r} \frac{\partial^2}{\partial r^2} r + \frac{1}{r^2} \left(\frac{\partial^2}{\partial \theta^2} + \frac{1}{\mathrm{tg}\,\theta} \frac{\partial}{\partial \theta} + \frac{1}{\sin^2\theta} \frac{\partial^2}{\partial \varphi^2} \right) \right] - \frac{e^2}{r} \right\} \chi_n (r, \theta, \varphi)$$

$$= E_n \chi_n(r, \theta, \varphi) \qquad (\text{II.1.85})$$

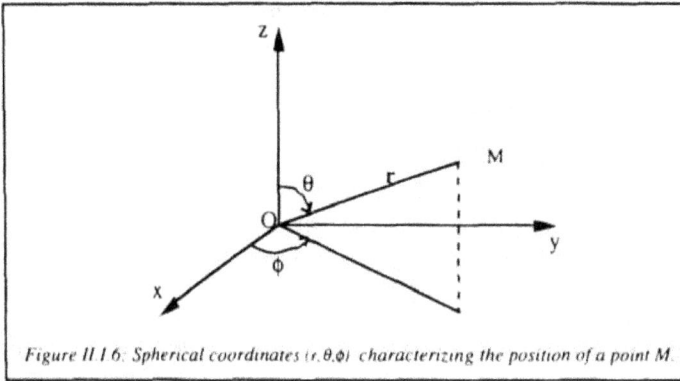

Figure II.1.6: Spherical coordinates (r, θ, ϕ) characterizing the position of a point M.

To solve this eigenvalue equation of the Hamiltonian operator, it is convenient to use the property described earlier (see Eq. II.1.3), that two commuting operators share a set of eigenfunctions.

Starting with the definition of the angular momentum ($\mathbf{l} = \mathbf{r} \times \mathbf{p}$) the projection of $\mathbf{l}$ on the z-axis is expressed by

$$l_z = -i\hbar \left[x \frac{\partial}{\partial y} - y \frac{\partial}{\partial x} \right] \tag{II.1.86}$$

Expressed in spherical coordinates, it simply reduces to

$$l_z = -i\hbar \partial/\partial\varphi \tag{II.1.87}$$

Similarly, we can express the modulus of the angular momentum $|\mathbf{l}|^2 = l_x^2 + l_y^2 + l_z^2$ in spherical coordinates, yielding

$$l^2 = \hbar^2 \left[\frac{\partial^2}{\partial\theta^2} + \frac{1}{\text{tg }\theta} \frac{\partial}{\partial\theta} + \frac{1}{\sin^2\theta} \frac{\partial^2}{\partial\varphi^2} \right] \tag{II.1.88}$$

Using these expressions, it is easy to show that the operators H, l_z, et l^2 commute

$$[H, l_z] = [H, l^2] = [l_z, l^2] = 0 \tag{II.1.89}$$

As a result, these three operators have a common set of eigenfunctions. On the other hand, it is also easy to express H simply in terms of l^2 and r as follows:

$$H = -\frac{\hbar^2}{2m_e} \frac{1}{r} \frac{\partial^2}{\partial r^2} r + \frac{l^2}{2m_e r^2} - \frac{e^2}{r} \tag{II.1.90}$$

Since H and l^2 have a common set of eigenfunctions, this equation shows that, if we can find the eigenfunctions of l^2, we shall only need to solve a radial differential equation in order to find the eigenvalues E_n of H. In fact, this expression of H as a function of l^2 and r is not surprising. It is comparable to the classical expression (Eq. II.1.73) of the hamiltonian function.

Eigenfunctions common to l_z and l^2: Spherical harmonics

To find the eigenfunctions of l^2, we use the fact that it commutes with l_z; the eigenfunctions of l_z are also eigenfunctions of l^2. As operator l_z is especially simple, its eigenfunctions are easy to find

$$l_z \chi(r, \theta, \varphi) = a\chi(r, \theta, \varphi) \to -i\hbar\, \partial\chi/\partial\varphi = a\chi$$

$$\chi = A(r, \theta) \exp(ia\varphi/\hbar) \tag{II.I.91}$$

If χ is to belong to the space of states F, it must be "well-behaved". Among others, it must be defined everywhere, meaning that it must be single-valued at all points in space.

This implies $\chi(r, \theta, \varphi + 2\pi) \equiv \chi(r, \theta, \varphi)$, i.e.,

$$\exp[i\, a\varphi/\hbar] = \exp[ia(\varphi + 2\pi)/\hbar] = \exp[i\, a\varphi/\hbar]\exp[ia\, 2\pi/\hbar] \tag{II.I.92}$$

and this is possible only if $a/\hbar$ is equal to an integer m, which may be positive, negative or zero.

Thus, physically acceptable eigenfunctions of operator l_z take the form

$$\chi_m = A(r, \theta) \exp[im\,\varphi] \tag{II.I.93}$$

and their associated eigenvalues are

$$a = m\hbar \qquad \text{with} \quad m = 0, \pm 1, \pm 2, ... \tag{II.I.94}$$

they are integral multiples of h. Thus, quantum mechanics justifies the intuitive hypothesis of Bohr.

If these functions are also to be eigenfunctions of l^2, they must satisfy

$$l^2 \chi_m = b\chi_m$$

so:

$$-\hbar^2 \left[\frac{\partial^2}{\partial\theta^2} + \frac{1}{tg\,\theta}\frac{\partial}{\partial\theta} + \frac{1}{\sin^2\theta}\frac{\partial^2}{\partial\varphi^2} \right] A(r, \theta)e^{im\varphi} = bA(r, \theta)e^{im\varphi} \tag{II.I.95}$$

As the variable r does not intervene in operator l^2, $A(r, \theta)$ may be expressed as $A(r, \theta) = R(r) P(\theta)$, where $P(\theta)$ satisfies the differential equation

$$-\hbar^2 \left[\frac{\partial^2}{\partial\theta^2} + \frac{1}{tg\,\theta}\frac{\partial}{\partial\theta} - \frac{m^2}{\sin^2\theta} + \frac{b}{\hbar^2} \right] P(\theta) = 0 \tag{II.I.96}$$

We shall not give the solutions of this differential equation in this book. Let us just say that the change of variable $\alpha = \cos\theta$ transforms this equation into the associated Legendre equation, which has well known solutions. We finally obtain that

– For $P(\theta)$ to be "well-behaved" [$\Psi(r, \theta, \varphi)$ must be defined everywhere, and also a quadratically integrable function], b must take on values of the form

$$b = l(l + 1)\hbar^2 \qquad \text{where l is an integer such that } l \geq |m| \tag{II.I.97}$$

– The corresponding eigenfunctions, written $P_{l,m}$ are the associated Legendre functions. A few examples of these functions will be given later in this book.

Therefore, the eigenfunctions of l^2 associated with eigenvalue $l(l + 1)\hbar^2$ are

$$\chi_{l,m}(r, \theta, \varphi) = R(r) \, Y_l^m (\theta, \varphi) \qquad (II.I.98)$$

where $R(r)$ is an arbitrary function of r, and $Y_l^m(\theta, \varphi) = P_{lm}(\theta)e^{im\varphi}$ are functions known as spherical harmonics. The eigenvalue $l(l + 1)\hbar^2$ is degenerate, as all the eigenfunctions $\chi_{l,m}$ with the same l but with different values of m correspond to this same eigenvalue. As $|m| \leq l$, there are $(2l+1)$ values of m for each value of l. The degeneracy of eigenvalue $l(l+1)\hbar^2$ is therefore equal to $(2l+1)$.

Examples of spherical harmonics

(The coefficients in front of the functions are normalization coefficients.)

$$l = 0 \quad Y_{0,0} = 1/\sqrt{4\pi} \qquad\qquad m = 0$$

$$l = 1 \quad \begin{cases} Y_{1, \pm 1}(\theta, \varphi) = (\sqrt{3}/\sqrt{8\pi}) \sin \theta \exp (\pm i\varphi) & m = \pm 1 \\[2mm] Y_{1,0}(\theta, \varphi) = (\sqrt{3}/\sqrt{4\pi}) \cos \theta & m = 0 \end{cases}$$

$$(II.I.99)$$

$$l = 2 \quad \begin{cases} Y_{2, \pm 2}(\theta, \varphi) = (\sqrt{15}/\sqrt{32\pi}) \sin^2 \theta \exp (\pm 2i\varphi) & m = \pm 2 \\[2mm] Y_{2, \pm 1}(\theta, \varphi) = (\sqrt{15}/\sqrt{8\pi}) \sin \theta \cos \theta \, \exp (\pm i\varphi) & m = \pm 1 \\[2mm] Y_{2,0}(\theta, \varphi) = (\sqrt{5}/\sqrt{16\pi})(3\cos^2 \theta -1) & m = 0 \end{cases}$$

Solving the radial equation

We now know that the eigenfunctions of the Hamiltonian are of the form

$$\chi_{k, l, m}(r, \theta, \varphi) = R_{k, l, m}(r) \, Y_l^m(\theta, \varphi) \qquad (II.I.100)$$

where the radial functions are characterized by the indices l and m of the associated spherical harmonic, and by index k, characterizing the state. These functions must satisfy the Hamiltonian eigenvalue equation:

$$H \, [R_{k,l,m}(r) \, Y_l^m(\theta, \varphi)] = E_{k,l,m} \, [R_{k,l,m}(r) \, Y_l^m(\theta,\varphi)] \qquad (II.I.101)$$

writing H explicitly gives

$$\left(- \frac{\hbar^2}{2m_e} \frac{1}{r} \frac{\partial^2}{\partial r^2} r + \frac{l^2}{2\mu r^2} - \frac{e^2}{r} \right) [R_{k,l,m}(r) \, Y_l^m(\theta, \varphi)]$$

$$= E_{k,l,m} \, [R_{k,l,m}(r) \, Y_l^m(\theta, \varphi)] \qquad (II.I.102)$$

This may be simplified by remembering that

$$l^2 \, [R_{k,l,m}(r) \, Y_l^m(\theta, \varphi)] = l(l + 1) \, \hbar^2 \, [R_{k,l,m}(r) \, Y_l^m(\theta, \varphi)] \qquad (II.I.103)$$

yielding, finally, a simple radial equation:

$$\left[-\frac{\hbar^2}{2m_e} \frac{1}{r} \frac{\partial^2}{\partial r^2} r + \frac{l(l+1)\hbar^2}{2m_e r^2} - \frac{e^2}{r} \right] R_{k,l,m}(r) = E_{k,l,m} R_{k,l,m}(r) \qquad \text{(II.I.104)}$$

The energies $E_{k,l,m}$ of a hydrogen atom are the eigenvalues of the operator in square brackets. This operator depends on l, but not on m, so, clearly, neither $E_{k,l,m}$ nor $R_{k,l,m}(r)$ can depend on m. We can therefore simply call these energies $E_{k,l}$ and these functions $R_{k,l}(r)$.

Setting $R_{k,l}(r) = u_{k,l}(r)/r$ makes it possible to transform the operator, yielding the following equation:

$$\left[-\frac{\hbar^2}{2m_e} \frac{d^2}{dr^2} + \frac{l(l+1)\hbar^2}{2 m_e r^2} - \frac{e^2}{r} \right] u_{k,l}(r) = E_{k,l} u_{k,l}(r) \qquad \text{(II.I.105)}$$

This eigenvalue equation is formally equivalent to the one associated with a particle of mass m_e in a effective potential equal to

$$V(r) = -\frac{e^2}{r} + \frac{l(l+1)\hbar^2}{2 m_e r^2}$$

in a one-dimensional problem (i.e., a particle moving on an axis where r is the distance from the origin), but with the restriction that $r > 0$. There are two terms in this potential $V(r)$: An attractive Coulombian term in $-e^2/r$, and a term due to the rotation of the system which tends to push the particle away from the origin; the centrifugal potential

$$(l+1)\hbar^2 / (2 m_e r^2)$$

Fig. II.I.7 shows the $V(r)$ potential for a few values of l, as a function of r/a_0 (in other words, r is shown in atomic units).

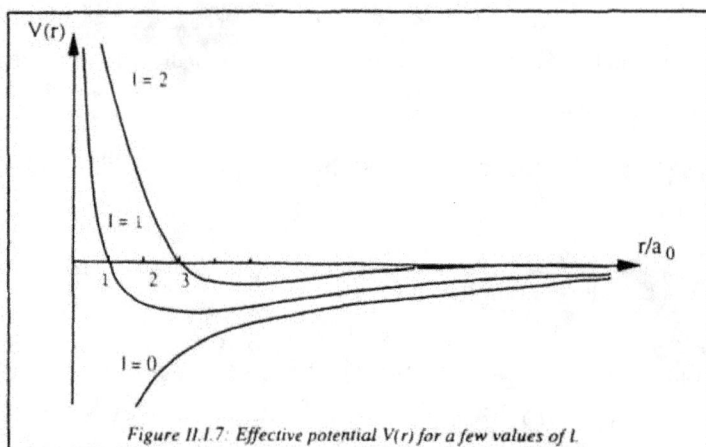

Figure II.I.7: *Effective potential V(r) for a few values of l.*

We shall not give a detailed solution of the radial equation here. Let us simply say that we are only interested in the negative values of $E_{k,l}$. As, for all values of l, the potential $V(r)$ vanishes for very large values of r, it can be shown that only negative values of $E_{k,l}$ correspond to bonded states of the electron. The positive values of $E_{k,l}$ (not quantized) correspond to nonbonded states, i.e., states where the electron escapes to infinity, overcoming the attraction of the proton.

The full solution of the equation also gives the following result: For $u_{k,l}(r)$ to stay finite as r tends to infinity, the only acceptable energy values $E_{k,l}$ are

$$E_{k,l} = -\frac{1}{2}\frac{m_e e^4}{\hbar^2}\frac{1}{(1+k)^2} \qquad (\text{II.I.106})$$

where $k = 1, 2, 3 \ldots$ is a positive nonvanishing integer. If we set $n = k + l$, and taking into account Eq. II.I.97, we find that n must be a positive integer. Thus, we derive the energies of the Bohr atom

$$E_n = -\frac{1}{2}\frac{m_e e^4}{\hbar^2}\frac{1}{n^2} = -\frac{1}{2}\frac{e^2}{a_0}\frac{1}{n^2} \qquad (\text{II.I.107})$$

The functions associated with these energies take the form

$$\chi_{n,l,m}(r, \theta, \varphi) = R_{n,l}(r)\, Y_l^m(\theta, \varphi) \qquad (\text{II.I.108})$$

To solve this equation it is necessary to introduce three whole numbers. These are called *quantum numbers* and must satisfy the relation

$$n > l \geq |m| \qquad (\text{II.I.109})$$

n, which determines the energy, is the *principal quantum number*.

l, which gives the modulus of the angular momentum, is the *secondary*, or *azimutal, quantum number*.

m, which determines the z-component of the angular momentum, is the *magnetic quantum number*.

If n is greater than one, each level E_n is degenerate for two reasons:

– For each pair of quantum numbers (n,l) there are $(2l + 1)$ eigenfunctions, corresponding to the $(2l + 1)$ values of m. We shall see later that this degeneracy is due to symmetry of the wave function (see Sec. II.II.4.1).

– For each value of $n = l + k$, there are n possible values of l: $l = 0, 1, 2, \ldots$, $(n - 1)$. This degeneracy is called "accidental"

As a result, the degeneracy of a level characterized by energy E_n is

$$\sum_{l=0}^{n-1} (2l + 1) = n^2 \qquad (\text{II.I.110})$$

For historical reasons, concerning the spectra of alkali metals, the states corresponding to $l = 0, 1, 2$, and 3 are referred to as s, p, d, and f states. For higher values of l, the letters follow in alphabetical order (g for $l = 4$, etc.) except that the letter j is not used. The value of n is placed in front of the letter (e.g., 1s ; 2s ; 2p ; 3s ; ...). This method of naming energy levels does not take into account the value of m, and each level is degenerated $(2l + 1)$ times. The value of m may be added if needed, as a lower right hand subscript. Fig. II.I.8 gives a schematic representation of the lowest energy levels of the hydrogen atom.

The following are a few examples of wave functions for these lower energy levels. The coefficients in front of the functions are normalization coefficients. In calculating these coefficients, it should be remembered that the volume element in spherical coordinates takes the form

$$dv = r^2 \sin \theta \, dr \, d\theta \, d\varphi \qquad\qquad (II.I.111)$$

and that the integration limits are $(0, \infty)$ for r ; $(0,\pi)$ for θ ; $(0, 2\pi)$ for φ. This yields the following integral:

$$\int \chi^*\chi \, dv = \int_0^{2\pi} \int_0^{\pi} \int_0^{\infty} \chi^*\chi \, r^2 \sin \theta \, dr \, d\theta \, d\varphi \qquad (II.I.112)$$

$$n=1 \left[\begin{array}{l} l = 0 \quad ; \quad m = 0 \text{ (ground state)} \\[2mm] \text{state 1s} \\[2mm] \chi_{1,0,0} = R_{10}(r) \, Y_0^0(\theta, \varphi) = \dfrac{1}{\sqrt{\pi a_0^3}} \, e^{-r/a_0} \end{array} \right.$$

$$n=2 \left[\begin{array}{l} l = 0 \quad ; \quad m = 0 \\[2mm] \text{state 2s :} \\[2mm] \chi_{2,0,0} = R_{20}(r) \, Y_0^0(\theta, \varphi) = \dfrac{1}{2\sqrt{2\pi a_0^3}} \, \left(1-\dfrac{r}{2a_0}\right) e^{-r/2a_0} \\[4mm] l = 1 \quad ; \quad m = 0 \qquad\qquad\qquad\qquad (II.I.113) \\[2mm] \text{state } 2p_0 : \\[2mm] \chi_{2,1,0} = R_{21}(r) \, Y_1^0(\theta, \varphi) = \dfrac{1}{4\sqrt{2\pi a_0^3}} \, \dfrac{r}{a_0} \, e^{-r/2a_0} \cos\theta \\[4mm] l = 1 \quad ; \quad m = \pm 1 \\[2mm] \text{state } 2p_{\pm 1} : \\[2mm] \chi_{2,1,\pm 1} = R_{21}(r) \, Y_1^{\pm 1}(\theta, \varphi) = \dfrac{1}{\sqrt{8\pi a_0^3}} \, \dfrac{r}{a_0} \, e^{-r/2a_0} \sin\theta \, e^{\pm i\varphi} \end{array} \right.$$

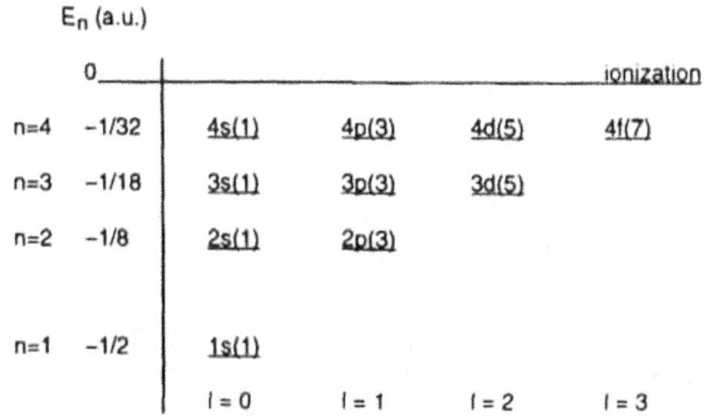

E_n (a.u.)

		$l = 0$	$l = 1$	$l = 2$	$l = 3$
0					ionization
n=4	−1/32	4s(1)	4p(3)	4d(5)	4f(7)
n=3	−1/18	3s(1)	3p(3)	3d(5)	
n=2	−1/8	2s(1)	2p(3)		
n=1	−1/2	1s(1)			

Figure II.1.8: First states of the hydrogen atom, ordered with respect to the values of l and n. The degeneracies are shown in parentheses.

We can calculate a certain number of characteristics for these stationary states, if we know their wave functions explicitly. For example:

– The mean distance between the proton and the electron :

$$\bar{r} = (\chi, r\chi) \int \chi^* r\chi \; dv = \int_0^{2\pi} \int_0^{\pi} \int_0^{\infty} \chi^* \chi \; r^3 \sin\theta \; dr \; d\theta \; d\varphi \qquad \text{(II.I.114)}$$

(for the groundstate $\bar{r} = 3a_0/2$).

– The expression:
$$\sqrt{\overline{r^2}} = \sqrt{(\chi, r^2\chi)}$$

(for the groundstate $\sqrt{\overline{r^2}} = \sqrt{3} \; a_0$)

– The probability of finding the electron at a distance from the proton between r and r + dr:

$$P(r) \; dr = \int_0^{2\pi} d\varphi \int_0^{\pi} \sin\theta \; d\theta \; \chi\chi^* \; r^2 \; dr \qquad \text{(II.I.115)}$$

For the groundstate, this probability is

$$P(r) \; dr = (4/a_0^3) \; r^2 \exp(-2r/a_0) \; dr$$

which reaches a maximum at $r = a_0$, i.e., exactly the value corresponding to the radius of the first Bohr orbital.

– The expected kinetic energy value $\bar{T} = \overline{p^2/2m_e}$.
As $H\chi_n = (T + V)\chi_n = E_n\chi_n$, the expected kinetic energy value may be expressed as

$$\bar{T} = (\chi_n, T\chi_n) = -(\chi_n, V\chi_n) + (\chi_n, E_n\chi_n) = E_n - \bar{V} = E_n + e^2 \; \overline{1/r}$$

In the groundstate, $\overline{1/r} = 1/a_0$, as $E_1 = - e^2/2a_0$ this yields for $\overline{T}$:

$$\overline{T} = - E_1 = e^2/2a_0 \qquad (II.I.116)$$

Although velocity does not convey the same meaning in quantum mechanics, we can calculate the classical velocity v giving the same kinetic energy value for the electron :

$$v = \sqrt{2\overline{T} / m_e} = 2.2 \times 10^6 \ ms^{-1} \qquad (II.I.117)$$

This result justifies the use of nonrelativistic calculations. It also indicates, however, that relativistic corrections would not be totally negligible.

Atomic orbitals

The function $\Psi_{n, l, m} (\mathbf{r}, t) = R_{n, l} (r) Y_l^m (\theta, \varphi) \exp (- iE_n t/\hbar)$ represents the wave function corresponding to a stationary state of energy $E_n = - 1/2 \ n^2$ (a.u) of the hydrogen atom.

The physical interpretation of this function is as follows: The probability of finding the electron in a volume element $dv = r^2 \sin\theta dr d\theta d\varphi$ located around position r, θ, φ is given by

$$P(r, \theta, \varphi) \ dv = \Psi^*_{n,l,m} (\mathbf{r}, t) \ \Psi_{n,l,m} (\mathbf{r}, t) \ dv = \chi^*_{n,l,m} (\mathbf{r}) \chi_{n,l,m} (\mathbf{r}) \ dv$$

$$= |R_{nl} (\mathbf{r})|^2 \ |Y_l^m (\theta,\varphi)|^2 \ dv \qquad (II.I.118)$$

We are dealing with a stationary state, so P does not depend on time. The distribution of P in space is constant and characteristic of the state. The spatial functions $c_{nlm} = R_{nl} (r) Y_l^m (\theta, \varphi)$, which depend on the coordinates of a single electron are called atomic orbitals.

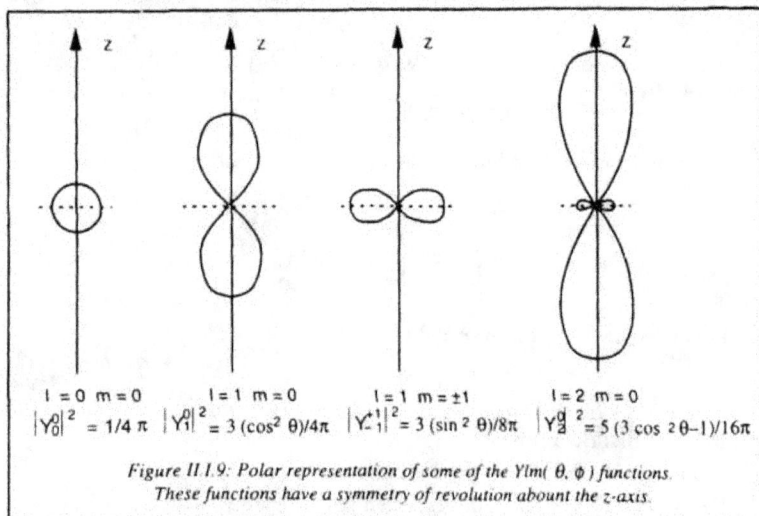

$$\begin{array}{cccc} l = 0 \ m = 0 & l = 1 \ m = 0 & l = 1 \ m = \pm 1 & l = 2 \ m = 0 \\ |Y_0^0|^2 = 1/4 \pi & |Y_1^0|^2 = 3 (\cos^2 \theta)/4\pi & |Y_{-1}^{+1}|^2 = 3 (\sin^2 \theta)/8\pi & |Y_2^0|^2 = 5 (3 \cos 2\theta - 1)/16\pi \end{array}$$

Figure II.1.9: Polar representation of some of the Ylm(θ, ϕ) functions.
These functions have a symmetry of revolution about the z-axis.

They present many similarities with other monoelectronic functions used to build approximate wave functions, also called atomic orbitals, for atoms with more than one electron, as well as molecules. It is interesting to look at some of the details of these probability distributions.

Why more than one graphical representation is needed to represent atomic orbitals

A proper representation of the values of a function defined in three-dimensional space really necessitates a graph in four-dimensional space: Three dimensions to indicate the point at which the function is defined, and the fourth to plot the corresponding value of the function, in this case $\chi_{n,l,m}$ (r,θ,φ). These values would constitute a "surface" in this four-dimensional space. However, as we do not have a four-dimensional space at our disposal, we give several complementary graphic representations of the atomic orbitals in two or three dimensions.

— First, the angular dependency of the atomic orbitals is obtained by associating to each direction in space a vector pointing in this direction with a length of $|Y_l^m(\theta, \varphi)|$ or of $|Y_l^m(\theta, \varphi)|^2$. If we look at the expression for Y_l^m, we see that $|Y_l^m(\theta, \varphi)|^2$ does not depend on φ, so the corresponding surface has a symmetry of revolution around the z-axis. Figure II.1.9 above shows a few examples.

— Second, the radial dependency of the orbitals is given by the functions $R_{nl}(r)$ (see Fig. II.1.10). Only the states corresponding to $l = 0$ do not vanish at the origin.

— $\chi_{n,l,m}$ can also be represented in three-dimensional space by the surfaces formed by the set of points where $|\chi_{n,l,m}$ $(r, \theta, \varphi)|$ is equal to a given value. This representation yields a set of contour-surfaces from which we can define regions of space where there is a high probability of finding the electron. Examples of these kinds of surfaces are given below.

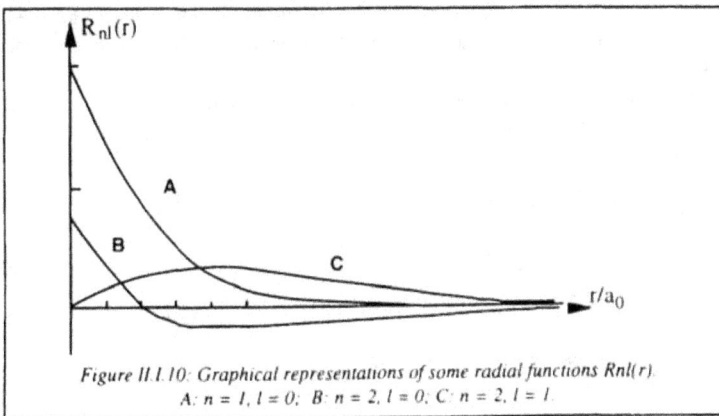

Figure II.1.10: Graphical representations of some radial functions Rnl(r).
A: $n = 1, l = 0$; B: $n = 2, l = 0$; C: $n = 2, l = 1$.

Real atomic orbitals: the s, p, and d orbitals.
Finding real-valued orbitals

The stationary states $\Psi_{n,l,m}(\mathbf{r}, t) = R_{nl}(r)\, Y_l^m(\theta,\varphi)\, \exp(-iE_n t/\hbar)$ are not the only possibilities for the hydrogen atom. Indeed, as was pointed out earlier, energy E_n depends exclusively on the principal quantum number n so that the degeneracy of this energy level is n^2. So for a given n and E_n, there are n^2 eigenfunctions $R_{nl}(r)\, Y_l^m(\theta,\varphi)$ with $l < n$ and $|m| \le l$. However, as any linear combination of these n^2 eigenfunctions of energy E_n is also an eigenfunction of H associated with E_n, we have

$$\left.\begin{aligned} H\, R_{nl_1}\, Y_{l_1}^{m_1} &= E_n\, R_{nl_1}\, Y_{l_1}^{m_1} \\ H\, R_{nl_2}\, Y_{l_2}^{m_2} &= E_n\, R_{nl_2}\, Y_{l_2}^{m_2} \end{aligned}\right\} \rightarrow \begin{aligned} & H\,(\lambda_1 R_{nl_1}\, Y_{l_1}^{m_1} + \lambda_2 R_{nl_2}\, Y_{l_2}^{m_2}) \\ & = E_n\,(\lambda_1 R_{nl_1}\, Y_{l_1}^{m_1} + \lambda_2 R_{nl_2}\, Y_{l_2}^{m_2}) \end{aligned} \qquad \text{(II.I.119)}$$

With this property in mind we can find a large number of different stationary eigenstates, all corresponding to the same value E_n but for which the quantum numbers l and m are not defined. For these states, the probability of finding an electron at $\mathbf{r}$, defined by $|\chi(r)|^2$, is different from that already calculated for the stationary states $R_{nl}Y_l^m$. The $R_{nl}Y_l^m$ orbitals are complex-valued functions since $Y_l^m(\theta,\varphi) = P_l^m(\theta)\, e^{im\varphi}$. Therefore, the graph showing the modulus of these orbitals does not give an exact idea of their angular dependency, as the dependency in φ disappears. This gives the surfaces of revolution around the z-axis shown on Fig. II.I.9 which show only a partial view of the $\chi_{n,l,m}$ orbital. If, by making linear combinations of these $\chi_{n,l,m}$ orbitals, we can construct stationary states represented by real-valued functions, with a value equal to their modulus, we obtain a much more significant representation. These orbitals are easier to handle when building chemical bonds. In the following paragraphs, we give a few of the most classical examples of real orbitals.

Definition and representation of the s and p orbitals

For s orbitals, $l = 0$ and $\chi = \chi_{n,0,0} = R_{n,0}(r)/\sqrt{4\pi}$. As $m = 0$, the real orbital is identical with $\chi_{n,0,0}$. So, in this case, the representations are those we gave earlier for $l = 0$ and $m = 0$. In particular, the angular dependency shows a sphere centered at O.

The following three linear combinations may be built for the p orbitals ($l = 1$):

$$\chi_{p_z} = \chi_{n,1,0} = (\sqrt{3}/\sqrt{4\pi})\, R_{n,1}(r)\, \frac{z}{r} = (\sqrt{3}/\sqrt{4\pi})\, R_{n,1}(r)\, \cos\theta$$

$$\chi_{p_x} = -(\sqrt{1}/\sqrt{2})(\chi_{n,1,1} + \chi_{n,1,-1}) = (\sqrt{3}/\sqrt{4\pi})\, R_{n,1}(r)\, \frac{x}{r}$$

$$= (\sqrt{3}/\sqrt{4\pi})\, R_{n,1}(r)\, \sin\theta\, \cos\varphi \qquad \text{(II.I.120)}$$

$$\chi_{p_y} = (i/\sqrt{2})(\chi_{n,1,-1} - \chi_{n,1,+1}) = (\sqrt{3}/\sqrt{4\pi})\, R_{n,1}(r)\, \frac{y}{r}$$

$$= (\sqrt{3}/\sqrt{4\pi})\, R_{n,1}(r)\, \sin\theta\, \sin\varphi$$

These expressions are such that all the orbital representations of X_{p_x} and of X_{p_y} can be found by taking those of X_{p_z} and exchanging the x-axis or the y-axis with the z-axis. Their angular dependency is shown in Fig. II.I.11.

The equivalence of the three axes is also clearly seen on a representation showing the set of contour surfaces where $|\chi|$ is equal to a given constant value (see Fig. II.I.12). We can simplify this representation by deciding to show only one arbitrarily chosen surface. For instance, we can choose the closed surface corresponding to $|\chi| = \alpha_0 \, |\chi|_{max}$ *defined in such a way that the probability of finding the electron in the enclosed volume is equal to 0.9.* The representation of this surface for orbital χ_{pz} is shown in Fig. II.I.13. The representations for p_x and for p_y are obtained by swapping the axes. It should be kept in mind that these representations define a volume with a high probability of presence, and must not be confused with the angular dependency graph which gives no information about distances.

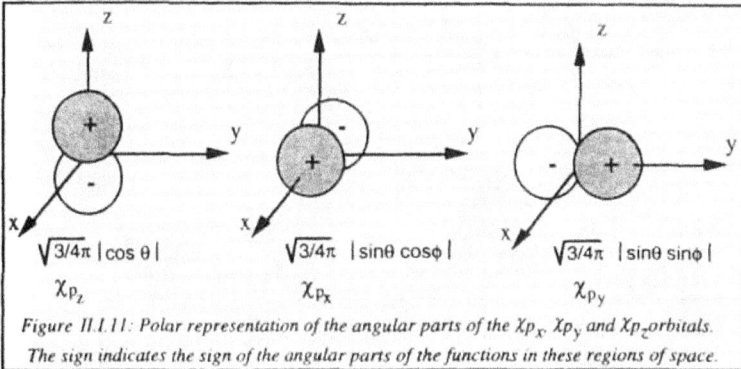

$\sqrt{3/4\pi}\ |\cos\theta|$ $\sqrt{3/4\pi}\ |\sin\theta\cos\phi|$ $\sqrt{3/4\pi}\ |\sin\theta\sin\phi|$

χ_{p_z} χ_{p_x} χ_{p_y}

Figure II.I.11: Polar representation of the angular parts of the χ_{p_x}, χ_{p_y} and χ_{p_z} orbitals. The sign indicates the sign of the angular parts of the functions in these regions of space.

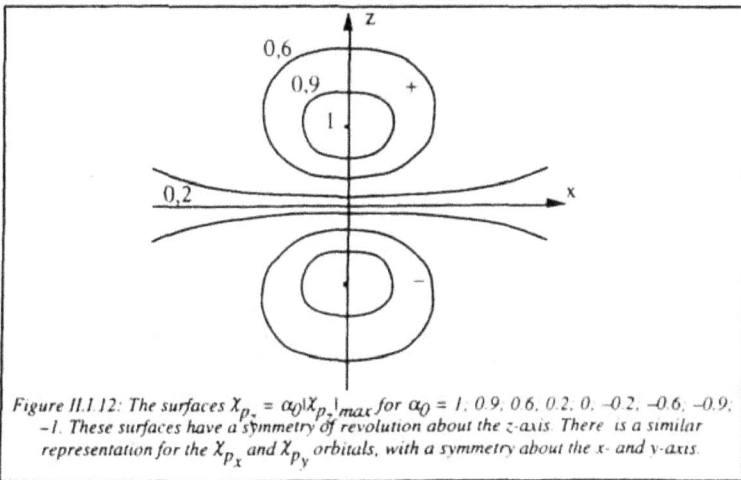

Figure II.I.12: The surfaces $\chi_{p_z} = \alpha_0 |\chi_{p_z}|_{max}$ for $\alpha_0 = 1, 0.9, 0.6, 0.2, 0, -0.2, -0.6, -0.9, -1$. These surfaces have a symmetry of revolution about the z-axis. There is a similar representation for the χ_{p_x} and χ_{p_y} orbitals, with a symmetry about the x- and y-axis.

Hybrid orbitals

As E_n depends only on n, it is also possible to make linear combinations of the $\chi_{n,l,m}$ with the same n but different values of l. Thus, sp orbitals are obtained by linear combinations of χ_s and χ_{p_z} orbitals:

$$\chi_{n,sp_z} = (1/\sqrt{2})\,[\chi_{n,s} + \chi_{n,p_z}] = (1/\sqrt{2})\,[\chi_{n,0,0} + \chi_{n,1,0}]$$

$$\chi'_{n,sp_z} = (1/\sqrt{2})\,[\chi_{n,s} - \chi_{n,p_z}] = (1/\sqrt{2})\,[\chi_{n,0,0} - \chi_{n,1,0}] \qquad (II.I.121)$$

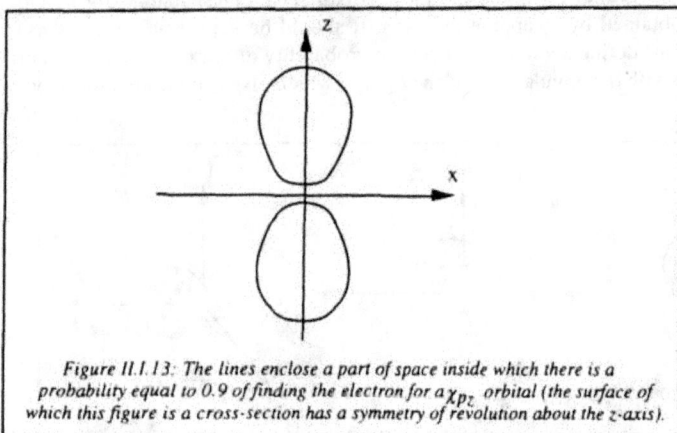

Figure II.1.13: *The lines enclose a part of space inside which there is a probability equal to 0.9 of finding the electron for a* χ_{p_z} *orbital (the surface of which this figure is a cross-section has a symmetry of revolution about the z-axis).*

On the graph showing the angular dependencies of the sp orbitals, we see that there is a higher probability of finding the electron around either direction of the z-axis. sp² orbitals are obtained in a similar fashion by linear combinations of χ_s, χ_{p_x} and χ_{p_y} orbitals:

$$\chi_{n,sp_xp_y} = (1/\sqrt{3})\,\chi_{n,s} + (\sqrt{2}/\sqrt{3})\,\chi_{n,p_x}$$

$$\chi'_{n,sp_xp_y} = (1/\sqrt{3})\,\chi_{n,s} - (1/\sqrt{6})\,\chi_{n,p_x} + (1/\sqrt{2})\,\chi_{n,p_y} \qquad (II.I.122)$$

$$\chi''_{n,sp_xp_y} = (1/\sqrt{3})\,\chi_{n,s} - (1/\sqrt{6})\,\chi_{n,p_x} - (1/\sqrt{2})\,\chi_{n,p_y}$$

The coefficients are chosen so that the orthonormal orbitals proceed from one another by a 120° rotation about the z-axis. The first orbital has a symmetry axis (the x-axis), and a plane of symmetry (the xOz plane). The polar representation of angular dependency (Fig. II.I.15) in the xOy plane shows the existence of a privileged direction with a high probability of finding the electron.

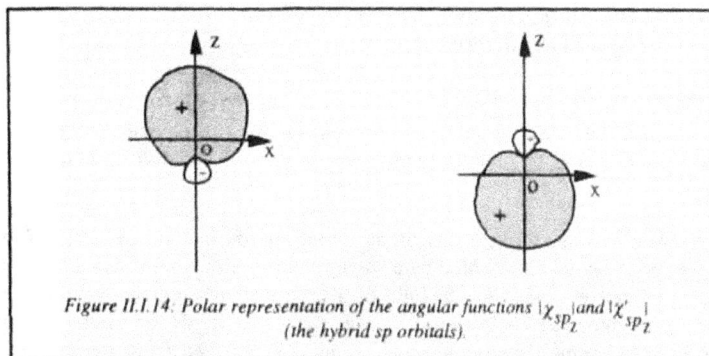

Figure II.1.14: Polar representation of the angular functions $|\chi_{sp_z}|$ and $|\chi'_{sp_z}|$ (the hybrid sp orbitals).

The sp^3 orbitals are obtained by linear combinations of χ_s, χ_{p_z}, χ_{p_x}, and χ_{p_y} orbitals. The resulting orbitals fit together like the segments joining the center of a regular tetrahedron to its four corners (with angles of 109°28'). They result from the following linear combinations:

$$\chi_{n,sp_xp_yp_z} = \frac{1}{2}[\chi_{n,s} + \chi_{n,p_x} + \chi_{n,p_y} + \chi_{np_z}]$$

$$\chi'_{n,sp_xp_yp_z} = \frac{1}{2}[\chi_{n,s} - \chi_{n,p_x} - \chi_{n,p_y} + \chi_{np_z}]$$

$$\chi''_{n,sp_xp_yp_z} = \frac{1}{2}[\chi_{n,s} - \chi_{n,p_x} + \chi_{n,p_y} - \chi_{np_z}]$$

$$\chi'''_{n,sp_xp_yp_z} = \frac{1}{2}[\chi_{n,s} + \chi_{n,p_x} - \chi_{n,p_y} - \chi_{np_z}]$$

(II.1.123)

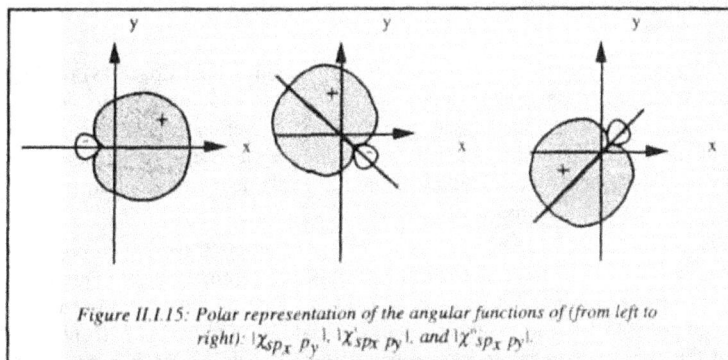

Figure II.1.15: Polar representation of the angular functions of (from left to right): $|\chi_{sp_x\,p_y}|$, $|\chi'_{sp_x\,p_y}|$ and $|\chi''_{sp_x\,p_y}|$.

Other real-valued orbitals can be found, for instance, by looking for real-valued combinations of orbitals with larger values of l (d orbitals ...).

5. The 8th and 9th postulates: How an electronic wave function can be represented by a Slater determinant

5.1 The spin of the electron

In which the position variables do not suffice to characterize the state of a hydrogen atom electron , leading to the need for spin in the relativistic quantum theory.

In the last section, we saw that, in an initial approximation, the state of the hydrogen atom can be represented by a wave function $\Psi(x, y, z, t)$ describing the state of an electron placed in the field of a proton. However, the Schrödinger equation governing the evolution of Ψ over time only holds for small electron kinetic energies (much smaller than m_ec^2) as it is derived from the classical Hamiltonian H established in nonrelativistic mechanics. Fortunately, the mean kinetic energy obtained with this first model is small for bonded states $(\overline{T}/m_ec^2 \approx 10^{-4})$ but it is not small enough to entirely forego relativistic corrections. The task then became to find an equation governing the evolution of Ψ which would be invariant under the Lorentz transforms, unlike the Schrödinger equation. The search for this equation led to the conclusion that it is impossible to find a correct relativistic equation (like the Dirac equation) if the electron is characterized solely by its mass at rest. In particular, the angular momentum of the orbital of the electron, l, no longer commutes with the relativistic Hamiltonian H. The whole quantum construction had been based on the fact that H and the total angular momentum of the system could commute. The solution was to associate an intrinsic angular momentum with the electron, called the *spin*, s. When spin is integrated, the operator corresponding to the total angular momentum $j = l + s$ of the electron commutes with H.

Three kinds of experimental manifestations of electron spin are of historical importance:

— *The fine structure of spectral lines*: If the Lyman α line $(2p \Leftarrow 1s)$ of the hydrogen spectrum is analyzed in detail, it splits up into two components which cannot be accounted for by the simple, spinless theory given in the last section. If we admit the existence of spin, then it can be shown that the 2p level has two components. This fine structure, which cannot be explained without spin, shows up in most atomic spectral lines.

— *The Stern-Gerlach experiment*: A jet of silver atoms traveling through a magnetic gradient forms two spots on the screen intercepting the jet. This can only be explained by admitting that the silver atoms have an angular momentum component equal to $\pm \hbar/2$ in the direction of the magnetic field. This half-integer value of the angular momentum cannot be attributed to the electron orbitals as these always give rise to values of l_z equal to integer multiples $m\hbar$ of $\hbar$. By attributing spin to the electrons, it becomes possible to account for these half-integer values of the angular momentum component.

— *The anomalous Zeeman effect.* If atomic levels with vanishing orbital angular momentum (1 = 0) are immerged in a uniform magnetic field B, the energy should not change when the value of B changes as the magnetic moment associated with the atom should vanish too. However, an experimentally observed displacement occurs, proportional to B (the anomalous Zeeman effect). This shows that the electron has an intrinsic magnetic momentum, which in turn is compatible with the existence of an intrinsic angular momentum.

Clearly, the state of the electron cannot be defined only by the spatial coordinates x, y, z in the space of orbital states F which we used up to now. The wave function corresponding to the state of a particle must also depend on the state of its spin.

We shall not derive relativistic quantum mechanics (the Dirac equation) in this book as it requires the use of complex, difficult mathematics. We shall simply introduce the existence of spin and its properties by way of a postulate, using a method developed by Pauli. Thereby the axiomatics of nonrelativistic quantum mechanics is extended to particles with spin, provided the kinetic energies are small compared to $m_e c^2$

8th postulate

The electron is a particle endowed with intrinsic angular momentum s called spin.

The operator s associated with spin has some properties similar to those of the orbital angular momentum l, namely that its components commute with the operator s^2 If we call $\chi_{m_s}(\sigma)$ the eigenfunctions common to s^2 and s_z, the corresponding eigenvalue equations can be expressed as

$$s^2 \chi_{m_s}(\sigma) = s(s+1) \hbar^2 \chi_{m_s}(\sigma) \qquad \text{with} \qquad s = \frac{1}{2}$$

$$s_z \chi_{m_s}(\sigma) = m_s \hbar \, \chi_{m_s}(\sigma) \qquad \text{with} \qquad m_s = \pm \frac{1}{2}$$

$$(\text{II.1.124})$$

Two extra quantum numbers s and m_s, correspond, for electronic spin s , to quantum numbers 1 and m for orbital angular momentum l, except that they take on half-integer values. Electrons are called "spin 1/2" particles, referring to the value of s. As a result, the space of states of spin F_s is a two-dimensional space. The following two eigenfunctions of operator s_z constitute the base of this space of states:

$$\chi_{1/2}(\sigma) = \alpha(\sigma) \quad \text{et} \quad \chi_{-1/2}(\sigma) = \beta(\sigma) \qquad (\text{II.1.125})$$

Therefore, any state with arbitrary spin can be described by a linear combination of these two functions:

$$\chi(\sigma) = c_+ \alpha(\sigma) + c_- \beta(\sigma) \qquad (\text{II.1.126})$$

As $\alpha(\sigma)$ and $\beta(\sigma)$ are eigenfunctions of s^2, as is $\chi(\sigma)$, associated with the same double degenerate eigenvalue $s(s+1)\hbar^2 = 3/4 \, \hbar^2$. This shows that s^2 is proportional to the identity operator I: $s^2 = (3/4) \hbar^2 I$.

It is practical to use the fact that the spin variable can take on only two values, $\sigma = 1/2$ or $\sigma = -1/2$. Then the functions $\alpha(\sigma)$ and $\beta(\sigma)$ are defined by the following relations:

$$\alpha(1/2) = \beta(-1/2) = 1 \quad \text{and} \quad \alpha(-1/2) = \beta(1/2) = 0$$

Using the definition for the scalar product of two spin functions $\chi(\sigma)$ and $\chi'(\sigma)$

$$(\chi(\sigma), \chi'(\sigma)) = \sum_{\sigma} \chi^*(\sigma)\,\chi'(\sigma)$$

it is easy to check that $\alpha(\sigma)$ and $\beta(\sigma)$ are orthonormal functions:

$$(\alpha(\sigma), \alpha(\sigma)) = (\beta(\sigma)), \beta(\sigma)) = 1$$
$$(\alpha(\sigma), \beta(\sigma)) = (\beta(\sigma)), \alpha(\sigma)) = 0 \tag{II.I.127}$$

Taking Eq. II.I.127 into account, we find

$$(\chi(\sigma), \chi'(\sigma))$$
$$= (c_+\,\alpha(\sigma) + c_-\,\beta(\sigma),\ c'_+\,\alpha(\sigma) + c'_-\,\beta(\sigma) = c^*_+ c'_+ + c^*_- c'_- \tag{II.I.128}$$

In a very general way, an operator A acting on a given space is well-defined if we know how it transforms the basic vectors of this space. In the case of space F_s, the basic vectors can always be transformed as follows :

$$A\,\alpha(\sigma) = A_{++}\,\alpha(\sigma) + A_{-+}\beta(\sigma)$$
$$A\,\beta(\sigma) = A_{+-}\,\alpha(\sigma) + A_{--}\beta(\sigma) \tag{II.I.129}$$

So operator A is defined if we know the 2 x 2 matrix

$$(A) = \begin{pmatrix} A_{++} & A_{+-} \\ A_{-+} & A_{--} \end{pmatrix} \tag{II.I.130}$$

This matrix is called the representation of A in base $[\alpha, \beta]$.

s^2 and s_z can also be defined by matrices, instead of defining them by their action on the basic vectors.

The matrices representing s^2 and s_z are easily found to be

$$(s^2) = \frac{3}{4}\hbar^2 \begin{pmatrix} 1 & 0 \\ 0 & 1 \end{pmatrix} \quad \text{and} \quad (s_z) = \frac{1}{2}\hbar \begin{pmatrix} 1 & 0 \\ 0 & -1 \end{pmatrix} \tag{II.I.131}$$

Using the commutation relations between the components of s, it can be shown that the matrices representing s_x and s_y are as follows:

$$(s_x) = \frac{1}{2}\hbar \begin{pmatrix} 0 & 1 \\ 1 & 0 \end{pmatrix} \quad \text{and} \quad (s_y) = \frac{1}{2}\hbar \begin{pmatrix} 0 & -i \\ i & 0 \end{pmatrix} \tag{II.I.132}$$

As we know the matrix representation of the three components of s, s itself can be represented as a particular vector, whose components are matrices

$$(s) = \frac{1}{2}\hbar\,(\sigma) \tag{II.I.133}$$

where (σ_x), (σ_y) and (σ_z) are dimensionless matrices called Pauli matrices:

$$(\sigma_x) = \begin{pmatrix} 0 & 1 \\ 1 & 0 \end{pmatrix} \qquad (\sigma_y) = \begin{pmatrix} 0 & -i \\ i & 0 \end{pmatrix} \qquad \sigma_z = \begin{pmatrix} 1 & 0 \\ 1 & -1 \end{pmatrix} \qquad \text{(II.I.134)}$$

With all these definitions in mind, the state of a hydrogen atom electron can be represented by a function of space E, a product of spaces F and F_s. To build a base of space E, we take the base of F consisting of the eigenfunctions common to H, l^2 and l_z, i.e., the $\chi_{n,l,m}(\mathbf{r})$, and the base of F_s, consisting of the eigenfunctions common to s^2 and s_z. We obtain a base of E by multiplying each basic function of F by all the basic functions of F_s. This yields the base $[\chi_{n,l,m,m_s}(\mathbf{r},s)]$ defined as

$$\chi_{n,l,m,1/2}(\mathbf{r},\sigma) = \chi_{n,l,m}(\mathbf{r})\,\alpha(\sigma) \text{ and } \chi_{n,l,m,-1/2}(\mathbf{r},\sigma) = \chi_{n,l,m}(\mathbf{r})\,\beta(\sigma)$$

These basic functions are eigenfunctions of the operators H, l^2, l_z, s^2, and s_z as an operator acts only upon the functions of the space where it is defined, leaving the other functions unchanged. The following are a few examples:

$$l^2\,\chi_{n,l,m}(\mathbf{r})\,\alpha(\sigma) = l(l+1)\,\hbar^2\,\chi_{n,l,m}(\mathbf{r})\,\alpha(\sigma)$$

and

$$s_z\,\chi_{n,l,m}(\mathbf{r})\,\alpha(\sigma) = (1/2)\,\hbar\,\chi_{n,l,m}(\mathbf{r})\,\alpha(\sigma) \qquad \text{(II.I.135)}$$

Equation II.I.19 defining the hermitian scalar product of two functions f and g must be modified to take into account the spin variable. Let

$$(f, g) = \sum_\sigma \int f^*g\,dv$$

which can be symbolized as

$$(f, g) = \int f^*g\,d\tau$$

where the volume element dv is replaced by $d\tau$, and we shall assume that the integration is also taken over the spin variable.

Similarly, for a system containing more than one electron, we define

$$(f, g) = \sum_{\sigma_1} \sum_{\sigma_2} \cdots \int f^*g\,dv = \int f^*g\,d\tau$$

These definitions warranty that the basic functions we just defined are orthonormal in E. Indeed

$$(\chi_{n,l,m,m_s}(\mathbf{r},\sigma), \chi_{n',l',m',m'_s}(\mathbf{r},\sigma)) = (\chi_{n,l,m}(\mathbf{r}), (\chi_{n',l',m'}(\mathbf{r})](\chi_{m_s}(\sigma) \cdot \chi_{m'_s}(\sigma))$$

$$= \delta_{nlm,n'l'm'}\,\delta_{m_s\,m'_s} \qquad \text{(II.I.136)}$$

which is true as the $\chi_{n,l,m}$ orbitals are orthonormal in space F and the spin functions are orthonormal in F_s.

These functions facilitate a more complete description of the state of a hydrogen atom and are called atomic spin-orbitals. Any bonded state of the hydrogen atom can always be expressed as

$$\chi = \chi^+ (\mathbf{r}) \, \alpha(\sigma) + \chi^- (\mathbf{r}) \, \beta(\sigma) \qquad \text{(II.I.137)}$$

where

$$\chi^+ = \sum_{n,l,m} c_{n,l,m}^+ \chi_{n,l,m} (\mathbf{r}) \quad \text{and} \quad \chi^- = \sum_{n,l,m} c_{n,l,m}^- \chi_{n,l,m} (\mathbf{r})$$

The set of two functions $\begin{pmatrix} \chi^+ (\mathbf{r}) \\ \chi^- (\mathbf{r}) \end{pmatrix}$, characterizing χ, constitutes what is called

a two-component "spinner". The probability of finding an electron with a spin component equal to $s_z = (1/2)\hbar$, or $s_z = (-1/2)\hbar$, in an volume element dv located at point $\mathbf{r}_0$ is (see Eq. II.I.43)

$$dP^+ (\mathbf{r}_0) = \chi^+ (\mathbf{r}_0) {}^* \chi^+ (\mathbf{r}_0) \, dv$$
$$\qquad \qquad \qquad \qquad \text{(II.I.138)}$$
$$dP^- (\mathbf{r}_0) = \chi^- (\mathbf{r}_0) {}^* \chi^- (\mathbf{r}_0) \, dv$$

and the probability of finding the electron at $\mathbf{r}_0$ whatever the value of its spin is

$$dP (\mathbf{r}_0) = dP^+ (\mathbf{r}_0) + dP^- (\mathbf{r}_0) \qquad \text{(II.I.139)}$$

5.2 Identical particles and the antisymmetry principle

Until now, we have restricted ourselves to the study of the hydrogen atom, an atom with a single electron. If we study atoms with several electrons, or molecules, we must tackle the subtle problem of identical particles. The electrons of atoms or molecules are indiscernible, and, exchanging their symbols in the wave function does not change the state of the system under consideration. We shall start by showing that the set of postulates we already know is unable to resolve the ambiguities raised by swapping electrons, especially in terms of physical measurements. We shall then state a new postulate, called the *antisymmetry postulate*, which solves this problem.

The study of a system of two identical particles is simple enough to reveal the problem and will enable us to illustrate the solution. Two particles are said to be identical if all their intrinsic properties (mass, charge, spin, etc.) are exactly the same. No experiment can differentiate between them. All electrons are identical particles, as are all protons, and all hydrogen atoms. If, in a wave function representing a system consisting of two identical particles, the variables of the two particles are exchanged, there should be no change in the physical properties of the system. The description of the state considered must account for this invariance-by-permutation (exchange degeneracy). The state of a two-particle system is represented by a vector of space $E = E^{(1)} \times E^{(2)}$ which is the product of $E^{(1)}$, the space of states of particle (1), and $E^{(2)}$, the space of states of particle $(2)^{(1)}$. Let $\varphi(1, 2)$ be such a normalized function of E. As the two particles are identical, $\varphi(2,1)$ must necessarily also be a normalized function representing the same state, as it results from the permutation of the two particles. Now, remembering what was said in Sec. II.I.3, any normalized linear combination of these two states

$$N[a \; \varphi(1, 2) + b \; \varphi(2, 1)] \tag{II.I.140}$$

must also describe the same state, and therefore produce the same calculated values of observables. As an example, let us look at the spin state of two electrons. Let $\varphi(1, 2) = \alpha(\sigma_1) \; \beta(\sigma_2)$ and $\varphi(2, 1) = \beta(\sigma_1) \; \alpha(\sigma_2)$. Let us then calculate the probability that the S_x component of the total spin of the function of equation (Eq. II.I.140) is equal to $\hbar$.

This probability is obtained (see Sec. II.I.4) by calculating the square of the modulus of the scalar product of the function representing the state with the eigenfunction $S_x = s_x(1) + s_x(2)$ corresponding to eigenvalue $\hbar$. By noticing that $s_x \alpha = (1/2) \hbar \beta$ et $s_y \beta = (1/2) \hbar \alpha$, and by using the Pauli matrices, this eigenvector can be expressed as

$$\psi(1, 2) = \frac{1}{\sqrt{2}} \left[\alpha(\sigma_1) + \beta(\sigma_1) \right] \frac{1}{\sqrt{2}} \left[\alpha(\sigma_2) + \beta(\sigma_2) \right] \tag{II.I.141}$$

Finally, remembering that the normalization coefficient is equal to $1/\sqrt{|a|^2 + |b|^2}$, the probability is equal to

$$\frac{|(a + b)/2|^2}{|a^2| + |b|^2} \tag{II.I.142}$$

However, the value of this expression depends on the values chosen for a and b. This means that the exchange degeneracy is incompatible with the fact that, when the system is in a given state, it must have well-defined physical properties which do not depend on the coefficients a and b used to describe the state. Therefore we are forced to introduce yet another postulate in order to decide which linear combination to choose.

Antisymmetry postulate for two identical particles (fermions):

If a system contains two identical particles with half-integer spin (fermions), only those vectors $\varphi(1, 2)$ of space $E = E^{(1)} \times E^{(2)}$ which are antisymmetrical with respect to a permutation of the two particles correctly represent the physical states of this system.

For example, we have

$$\varphi(1, 2) = \phi_1(1) \; \phi_2(2) - \phi_1(2) \; \phi_2(1) = -\varphi(2,1)$$

This postulate grew out of experimental evidence obtained for all fermions, i.e., particles with half-integer spin, which all have similar behavior. Electrons are fermions, and so are protons. Bosons, particles with integer spin values (e.g., photons, mesons, etc.) behave very differently, but their behavior is of little consequence for the study of chemical bonding.

These methods then had to be generalized to a system of N identical particles. The difficulty arises from the large number of possible permutations between N particles. Let us take N particles, 1, 2, 3, ..., N. Each particle has a specific number and they are ordered as written. Any operation which changes the initial order is called a particle permutation. For example:

$$(1, 2, 3, 4, 5) \overset{P}{\rightarrow} (3, 5, 4, 1, 2).$$

P is the permutation operator. It is easy to define the product of two permutations PP':

$$(1, 2, 3, 4, 5) \overset{P}{\rightarrow} (3, 5, 4, 1, 2) \overset{P'}{\rightarrow} (5, 3, 4, 1, 2)$$

Clearly, $P'P$ is different from PP':

$$(1, 2, 3, 4, 5) \overset{P'}{\rightarrow} (2, 1, 3, 4, 5) \overset{P}{\rightarrow} (3, 5, 4, 2, 1)$$

The specific permutation P' consisting merely of an exchange of two particles, is called a transposition and is expressed as (n,n'). In this case, for instance $P' = (1,2)$ as P' exchanges the first two particles on the list. Any permutation P may be decomposed as a product of transpositions [e.g., $P = (1,3)\ (3,4)\ (2,5)$]. Although this product of transpositions is not unique, the number ω of transpositions used to decompose P must always have the same parity. If ω is an even number, P is said to be even. If ω is an odd number, P is said to be odd. For N particles, the number of possible permutations is equal to N!. For example if N = 3, the 3! = 6 possible ways of ordering the particles are shown in Table II.1.2.

TABLE II.1.2 — **The six possible ways of ordering three particles.**

P	P_1	P_2	P_3	P_4	P_5	P_6
$P_i(1, 2, 3)$	(1, 2, 3)	(1, 3, 2)	(2, 1, 3)	(2, 3, 1)	(3, 1, 2)	(3, 2, 1)

Permutation P_1 is in fact the identity operator (no permutation takes place). P_1 is therefore even, as are $P_4 = (1,2)\ (2,3)$ and $P_5 = (1,3)\ (2,3)$. On the other hand, $P_2 = (2,3)$, $P_3 = (1,2)$, and $P_6 = (1,3)$ are odd permutations.

The state of a system consisting of N particles is represented by a function of the vectorial space $E = E^{(1)} \times E^{(2)} \times E^{(3)} \times \dots \times E^{(N)}$. Let $\varphi(1, 2, \dots, N)$ be a vector of E representing a state of the system. Any permutation P of the N particles must leave the state unchanged. Thus, the set of all the functions corresponding to the N! permutations, $\varphi_p = P\varphi\ (1, 2, \dots, N)$, as well as all their linear combinations, must describe the same state. Such a linear combination can be expressed by

$$\varphi = \sum_p c_p\, P\varphi\, (1, 2 \dots N) = \sum_p c_p\, \varphi_p \qquad (\text{II.1.143})$$

where the sum over p indicates the sum extended over all the N! possible vectors φ_p. [φ_p is vector of space E obtained by taking $\varphi(1, \dots, N)$ and carrying out permutation P on the particles $(1,2,\dots, N)$.] As in the case of two particles, these linear combinations do not all result in the same physical properties for the system. Therefore they are not all equally well-adapted to describing the state in question. This leads to a new postulate.

9th postulate: *Antisymmetry postulate for N identical particles.*

For a system consisting of N identical particles with a half-integer spin (fermions), only those vectors of space $E = E^{(1)} \times E^{(2)} \times \ldots \times E^{(N)}$ *which are antisymmetrical with respect to the transposition of any two particles can correctly describe a physical state of the system.*

Starting with a state $\varphi\,(1, 2, \ldots, N)$, we can build a state Ψ_a which is antisymmetrical with respect to the transposition of any two particles by writing the following linear combination:

$$\Psi_a = \frac{1}{\sqrt{N!}} \sum_p (-1)^{\omega_p} P\varphi\,(1, 2 \ldots N) \qquad (\text{II.I.144})$$

where ω_p characterizes the parity of the permutation: $\omega_p = 1$ or $\omega_p = -1$. Indeed

$$\Psi_a\,(2, 1, 3, \ldots, N) = -\Psi_a\,(1, 2, 3, \ldots, N)$$

and more generally speaking

$$P\Psi_a\,(1, 2, 3, \ldots, N) = \varepsilon\Psi_a\,(1, 2, 3, \ldots, N)$$

with

$$\varepsilon = +1 \ \text{if } P \ \text{is even}$$

$$\varepsilon = -1 \ \text{if } P \ \text{is odd}$$

The vector $\varphi(1, 2, \ldots, N)$ may be obtained by combining the vectors $\varphi_{(i)}$ representing the states of individual particles i, $\phi_{(i)}$ in $E_{(i)}$:

$$\varphi(1, 2, \ldots, N) = \phi_{(1)}\,(1)\,\phi_{(2)}\,(2) \ldots \phi_{(N)}\,(N) \qquad (\text{II.I.145})$$

For example, in the case of three electrons

$$\Psi_a = \frac{1}{\sqrt{N!}} \sum_p (-1)^{\omega_p} P \mid \phi_{(1)}\,(1)\,\phi_{(2)}\,(2)\,\phi_{(3)}\,(3) \mid$$

$$= \frac{1}{\sqrt{6}} [\phi_{(1)}\,(1)\,\phi_{(2)}\,(2)\,\phi_{(3)}\,(3) - \phi_{(1)}\,(1)\,\phi_{(2)}\,(3)\,\phi_{(3)}\,(2) - \phi_{(1)}\,(2)\,\phi_{(2)}\,(1)\,\phi_{(3)}\,(3)$$

$$+ \phi_{(1)}\,(2)\,\phi_{(2)}\,(3)\,\phi_{(3)}\,(1) - \phi_{(1)}\,(3)\,\phi_{(2)}\,(2)\,\phi_{(3)}\,(1) + \phi_{(1)}\,(3)\,\phi_{(2)}\,(1)\,\phi_{(3)}\,(2)]$$

As N, the number of particles, increases, the construction of Ψ_a becomes more and more tedious. Fortunately, we can see that the terms of the construction follow the same pattern as those for developing a determinant. In the case of three particles, we thus have

$$\Psi_a = \frac{1}{\sqrt{6}} \begin{vmatrix} \phi_{(1)}\,(1) & \phi_{(2)}\,(1) & \phi_{(3)}\,(1) \\ \phi_{(1)}\,(2) & \phi_{(2)}\,(2) & \phi_{(3)}\,(2) \\ \phi_{(1)}\,(3) & \phi_{(2)}\,(3) & \phi_{(3)}\,(3) \end{vmatrix} \qquad (\text{II.I.146})$$

And for N functions, it is possible to construct an antisymmetrical state Ψ_a, starting from the states of the individual particles $\phi_{(i)}$ and using them to write what is called the *Slater determinant*:

$$\Psi_a = \frac{1}{\sqrt{N!}} \begin{vmatrix} \phi_{(1)}(1) & \phi_{(2)}(1) & \ldots & \phi_{(N)}(1) \\ \phi_{(1)}(2) & \phi_{(2)}(2) & \ldots & \phi_{(N)}(2) \\ \vdots & \vdots & & \vdots \\ \phi_{(1)}(N) & \phi_{(2)}(N) & \ldots & \phi_{(N)}(N) \end{vmatrix} \qquad (\text{II.I.147})$$

The coefficients $1/\sqrt{6}$ and $1/\sqrt{N!}$ in Eqs. II.I.146 and II.I.147 are the normalization factors when using orthonormal $\phi_{(i)}$ functions.

Nevertheless, we must keep in mind that $\varphi(1, 2, ..., N)$ does not necessarily take the form given by Eq. II.I.145: For interacting particles, the wave function cannot usually be expressed in the form of a single Slater determinant.

Consequences of the antisymmetry postulate. Pauli's exclusion principle.

For noninteracting particles, the functions $\phi_{(i)}$ represent the individual states of the particles. In this case, the wave function of the system can indeed be expressed as a Slater determinant. When two rows of a determinant are identical, the determinant vanishes and cannot, therefore, represent an acceptable state of the system. This means that we have to exclude all states of the system for which two of the fermions would be in the same state [i.e., represented by the same wave function $\phi_{(i)}$]. This, in fact, is Pauli's exclusion principle, which appears here as a consequence of the 9th postulate of quantum mec........' s.

In a noninteracting system, two identical fermions cannot be represented by the same function $\phi_{(i)}$.

This property is maintained for a system of many interacting electrons if its wave function is approximated by a Slater determinant.

6. Application to many-electron systems

In describing an atom with several electrons, it is very important to be able to determine whether or not its wave function can be represented by a Slater determinant, and if so, what functions $\phi_{(i)}$ must be chosen to give a satisfying approximation to this atom's states. These functions $\phi_{(i)}$ are called atomic orbitals.

For a stationary state of the atom, its wave function Ψ must satisfy

$$H \, \Psi(1, 2, ..., N) = E \, \Psi(1, 2, ... N) \qquad (\text{II.I.148})$$

the Hamiltonian being

$$H = \sum_{i=1}^{N} \left(-\frac{\hbar^2}{2m_e} \Delta_i - \frac{Ze^2}{r_i} \right) + \sum_{i=1}^{N} \sum_{j=i+1}^{N} \frac{e^2}{r_{ij}} = \sum_{i=1}^{N} \left(T_i - V_{in} + \sum_{j>i}^{N} V_{ij} \right) \quad \text{(II.I.149)}$$

T_i represents the kinetic energy of electron i, V_{in} is the attraction between the electron and the nucleus, and V_{ij} is the mutual repulsion between two electrons. This last term is the reason that the eigenvalue equation cannot be solved exactly. Indeed, without this term, it would have been possible to factorize Ψ as a product of functions of the different variables r_i (the positions of the electrons) which would then provide a solution to the eigenfunction equation for the Hamiltonian

$$\sum_i (T_i + V_{in})$$

but the repulsion term disturbs this nice separability.

To solve this problem, Hartree and Slater suggested using the *central field approximation*, which can be explained as follows: Each electron is viewed as if it experiences an effective potential energy equal to the sum of the attractive electron-nucleus potential and the average repulsion due to the (N −1) other electrons. This repulsion term represents the screening effect the other electrons have on the nucleus-electron attraction. As a large part of this repulsion has spherical symmetry, we can make the approximation:

$$V_{ij} \approx \sum_i S(r_i)$$

where $S(r_i)$ represents the average repulsion of each of the other electrons. In this approximation, each electron experiences an effective potential $V(r_i)$ which can be expressed as

$$V(r_i) = -\frac{Ze^2}{r_i} + S(r_i) \quad \text{(II.I.150)}$$

The Hamiltonian of this approximation is, therefore, a sum of monoelectronic Hamiltonians

$$H = \sum_{i=1}^{N} \left(\frac{\hbar^2}{2m_e} \Delta_i + V(r_i) \right) = \sum_{i=1}^{N} h_i = H_0 \quad \text{(II.I.151)}$$

The eigenfunctions of H_0 can be factorized as a product of N functions $\phi_i(r_i)$ depending on the variables r_i

$$H_0 \Psi(r_i) = \left(\sum_i h_i \right) \phi_1(r_1) \dots \phi_N(r_N) = (E_1 + \dots + E_N) \phi_1(r_1) \dots \phi_N(r_N)$$

and each function $\phi_i(r_i)$ must be an eigenfunction of the monoelectronic Hamiltonian h_i

$$h_i \phi_i(r_i) = E_i \phi_i(r_i) \quad \text{(II.I.152)}$$

These eigenfunctions $\phi_i(r_i)$ are not atomic orbitals $\chi_{n,l,m}(r_i)$ as they were defined for the hydrogen atom as the potential energy. $V(r)$ is different from the potential varying as $1/r$. However, as we are still dealing with a spherical potential, it is also possible to separate the variables r_i, θ_i, φ_i and factorize $\phi_i(r_i)$ in the following manner:

$$\phi_i(\mathbf{r}_i) = R'_{n_i,l_i}(r_i)\, Y^{m_i}_{l_i}(\theta_i, \varphi_i) = \chi'_{nlm}(\mathbf{r}_i) \qquad \text{(II.I.153)}$$

where Y^m_l is a spherical harmonic and $R'_{n,l}$ is a radial function, different from that of the hydrogen atom. Although the energy E_{n_i,l_i} is still independent of the value of m_i it now depends not only on n_i but also on l_i , due to the fact that the potential is no longer proportional to $1/r$.

These new functions are called central field orbitals. In this case, the total energy of the many-electron atom is the sum of the individual electronic energies

$$E = \sum_{i=1}^{N} E_{n_i,l_i}$$

To construct the anti-symmetrical wave function Ψ_a , we must attribute a spin function χ_i (equal to α or to β) to each ϕ_i function, and express the Slater determinant with these monoelectronic functions

$$\Psi_a(\mathbf{r}_i) = \frac{1}{\sqrt{N}} \begin{vmatrix} \phi_1(1)\chi_1(1) & \phi_2(1)\chi_2(1) & \cdots & \phi_N(1)\chi_N(1) \\ \phi_1(2)\chi_1(2) & \phi_2(2)\chi_2(2) & \cdots & \phi_N(2)\chi_N(2) \\ \vdots & \vdots & & \vdots \\ \phi_1(N)\chi_1(N) & \cdots & \cdots & \phi_N(N)\chi_N(N) \end{vmatrix} \qquad \text{(II.I.154)}$$

The next problem is to decide which functions to choose to represent a given state, in particular the ground state. If we wish to describe the ground state of an atom with a specific number of electrons N, we must know the relative positions of the energies E_{nl}, and also the degeneracies of each energy E_{nl}. Once we have this information, we can attribute the N electrons to distinct orbitals ϕ_i corresponding to the lowest energies E_{nl}, specifying whether the electron has spin α or β. In fact, we proceed as if we filled the orbitals one after the other, taking the level of lowest energy not yet filled.

To obtain an excited state, we must put one or more of the N electrons in a higher energy orbital. A set of simple rules, the *Kleschkowsky rules*, governs finding the succession of orbitals in order of increasing energy. The value of $E_{n,l}$ mostly depends on the value of $(n+l)$. It increases in function of $(n+l)$. For a given value of $(n+l)$, the energy of $E_{n,l}$ increases by the value of n. The arrows in Table II.I.3. indicate the orbitals in order of increasing energy, for the first central-field orbitals.

The degeneracy of each level has two distinct origins. The value of $E_{n,l}$ does not depend on the magnetic quantum number m, and can take $(2l+1)$ different values for a given pair of numbers (n,l). This means that there are $(2l+1)$ linearly independent functions ϕ_i associated with $E_{n,l}$. Moreover, as each function ϕ_i can have two distinct spin states, the degeneracy of level $E_{n,l}$ is equal to $g_{nl} = 2(2l+1)$. Degeneracy does not depend on n.

TABLE II.I.3 : **Energies of the first few central-field orbitals (order indicated by the arrows).**

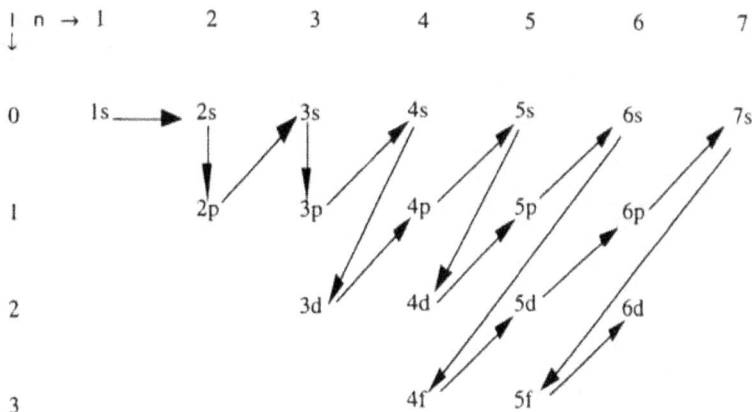

The set of all states associated with the same energy $E_{n,l}$ makes up a "subshell" of the atomic structure, to which we refer by the two numbers (n,l) (e.g., subshell 3p corresponds to n = 3, l = 1). Each s subshell has two associated orbitals and can therefore contain two electrons ($g_{n0} = 2$). Likewise, p subshells can contain six electrons ($g_{n1} = 6$), d subshells can contain ten electrons ($g_{n2} = 10$), and f subshells can contain fourteen electrons ($g_{n3} = 14$). Taking all this into consideration, it is not difficult to find the central-field orbitals occupied by the electrons of an atom in its ground state: They are the N lowest energy orbitals. The term "electronic shell" is the name of the set of all states associated with a given value of n. Of course, it includes all the subshells corresponding to this value of n. Conventionally, shells n = 1, 2, 3, ... are referred to by the capital letters K, L, M, ..., and so on, in alphabetical order.

Each shell can contain $\sum_{l=0}^{n-1} 2(2l+1) = 2n^2$ electrons. A description of an atom's state on the base of the number of electrons in each subshell is called the electronic configuration. The ground configuration corresponds to the groundstate of an atom in which the lowest energy orbitals are occupied first. To write these configurations, the occupied orbitals are listed in order of increasing energies, with the number of electrons in each subshell in superscript. For example, the ground configuration of the chlorine atom (17 electrons) is

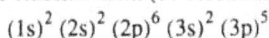

$$(1s)^2 (2s)^2 (2p)^6 (3s)^2 (3p)^5$$

In this example, all the subshells are saturated except the last one which lacks one electron. It is of course this lack of an electron in the 3p subshell which the cause of chlorine's reactivity and electronegativity.

Sometimes, if the shell of an electronic configuration is completely filled, the corresponding subshells are replaced by the symbol for the shell, without superscript. Thus, the ground configuration of the chlorine atom may also be expressed as: $(K) (L) (3s)^2 (3p)^5$. The existence of electronic spin together with the central-field approximation applied to many-electron atoms leads to a simple interpretation of Mendeleev's classification of the elements. It also yields a qualitative explanation for the chemical properties of the various classes of elements, as well as for the various ionization potentials.

Comment: As a result of degeneracy, the choice of spin-orbital with respect to m and m_s, is not unique for a given energy. In this case, it should be remembered that the possible wave functions are linear combinations of Slater determinants. There is no ambiguity in the case of complete shells, as each orbital has an α and β function. In this case, the Slater determinant is as follows:

$$\Psi_a = \frac{1}{\sqrt{N!}} \begin{vmatrix} \phi_1(1)\alpha(1) & \phi_1(1)\beta(1) & \phi_2(1)\alpha(1) & \phi_2(1)\beta(1) & \cdots & \phi_{N/2}(1)\beta(1) \\ \vdots & \vdots & \vdots & \vdots & & \vdots \\ \phi_1(N)\alpha(N) & \phi_1(N)\beta(N) & \phi_2(N)\alpha(N) & \phi_2(N)\beta(N) & \cdots & \phi_{N/2}(N)\beta(N) \end{vmatrix} \qquad (II.I.155)$$

7. Computer exercises and illustrations

7.1 Particle interference through Young slits: Electronic diffraction teaching experiment

We present below the results of three experiments that may be simulated, as we shall see, using software developed by our department. These are examples of the " thought " experiments presented at the very beginning of Chapter I, demonstrating the unusual behavior of subatomic particles and the inadequacy of standard mechanics to describe their movements.

Three thought experiments

Thought experiments are often considered unsuitable for use in teaching, or even criticized as "unscientific" Without going into sterile controversy, it seemed helpful to include a brief reminder of a historical approach illustrating their usefulness. Galileo, considered to be the founder of physics as an experimental science, used thought experiments to convince his audience. We are referring to the famous experiment that consisted of comparing the point of impact of a heavy object dropped from the masthead of a fast-moving ship, with the point of impact of the same object if the ship were stationary. Of course, we now know the answer: The point of impact is just at the base of the mast in both cases. However, at that time, the official scientific answer to this problem was incorrect, as scientists were not capable of dissociating essential effects like inertia from interference such as air resistance. Galileo imagined an ideal experiment, eliminating everything that could complicate the reasoning process and found the correct solution. This type of thought "experiment" may be used to follow the logical development of a theory and, in addition, to asses its coherence[*] We are now going to present a series of three thought experiments, designed as educational tools for distinguishing the essential aspects of each theory without becoming bogged down in superfluous details. These experiments will enable us to understand the unusual behavior of matter on a microscopic level.

Basic particle experiment

Let us consider, first of all, the experimental apparatus described in Fig. II.1.16. There is a source of particles, for example a man shooting a gun in all directions, sending bullets all over space. Opposite this "source" we have set up a plate with two slits in it (labeled F_1 and F_2), then, further away, a screen that detects all the bullets whose initial trajectory passes through F_1 or F_2

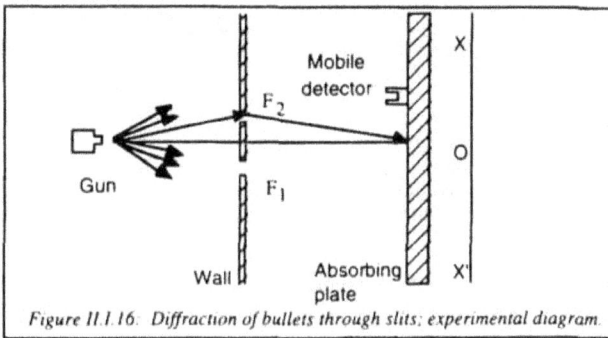

Figure II.1.16: Diffraction of bullets through slits; experimental diagram.

The software shows a display like a movie screen where each bullet impact corresponds to a luminous spot. In the lower part of the screen, we show the graph of the number of impacts in function of the position marked on a straight line (labeled XX') perpendicular to the direction of the slits. We imagined an ideal version of the experiment and supposed that a bullet would go in a totally random direction as it came through the slit (equal probability for every direction). The results are described in Fig. II.1.17.

— Slit F_1 open and slit F_2 closed: The bullets hit the screen rather erratically, one after the other. If we wait long enough, the graph in the lower part of the screen shows the probability of detecting a particle passing through slit F_1, labeled P_1 (a).

---

[*] On this specific point, we would like to mention the rather provocative statement made by Einstein, one of the proponents of a revival of thought experiments in theoretical physics, on the subject of Max Planck. "He was one of the most intelligent people I have ever known ; but during the eclipse in 1919, he stayed up all night to see whether it would confirm the deviation of light in the sun's gravitational field. If he had really understood the theory explaining the equivalence of inert mass and gravitational mass, he would have gone to bed like I did."

— Slit F_1 closed and slit F_2 open: Figure b shows the results. Note that P_2 is the probability corresponding to this experiment.

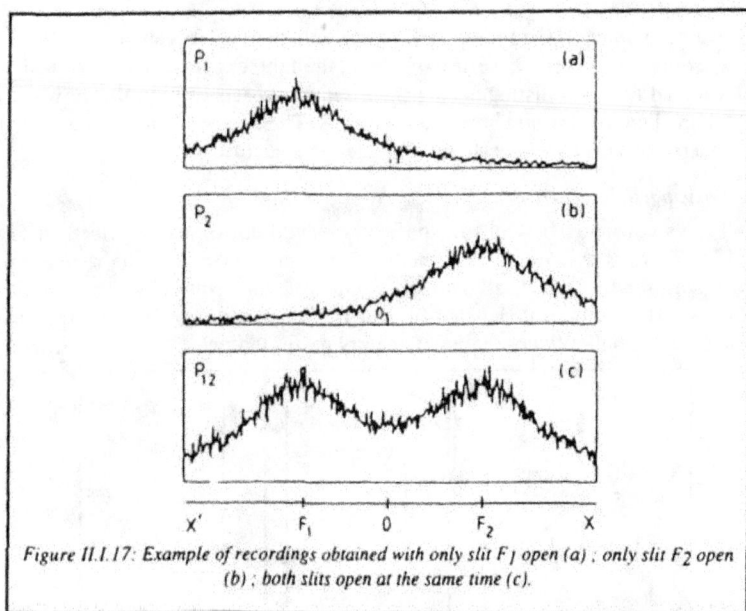

Figure II.I.17: *Example of recordings obtained with only slit F_1 open (a) ; only slit F_2 open (b) ; both slits open at the same time (c).*

— Slits F_1 and F_2 open at the same time: Figure c shows the result corresponding to an acquisition time twice as long as those shown in the previous graphs. This graph corresponds quite closely to the sum of graphs a and b. We therefore conclude that the total probability P_{12} of this process equals the sum of the probabilities of the two previous experiments conducted independently, i.e.,

$$P_{12} = P_1 + P_2$$

This addition illustrates the fact that, in standard mechanics, the probability of observing a phenomenon corresponding to the superposition of several possibilities equals the sum of the probabilities of each of these possibilities being realized. This result is not at all surprising, and coincides with our "intuitive" understanding of natural phenomena.

Standard wave experiment

Now we are going to study a similar arrangement, using waves rather than the particles in the previous experiment. It is simply a matter of replacing the gun (see Fig. II.I.18) by a wave generator (e.g., a loudspeaker in the case of sound waves), and moving a (sound) wave detector with a response proportional to the energy, i.e., to the square of the amplitude $|A|^2$ along the X'X axis. In this case, the idealized experiment consists of assuming that the slits are infinitely fine and the detector is capable of infinite spatial resolution.

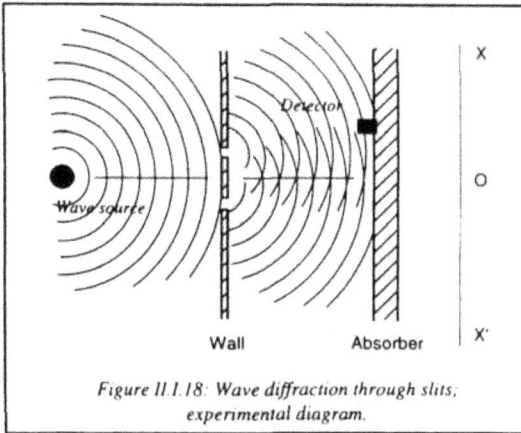

Figure II.1.18: Wave diffraction through slits;
experimental diagram.

We know that diffraction of a wave through an infinitely small hole produces a spherical wave, and an isotropic distribution of energy in the observer's half-space. It is noticeable that, unlike the particle experiment, where all the energy was concentrated in the detected particle, the energy in this case is uniformly spread along the diffracted wave. We expect, therefore, that if F_2 is closed, the graph of the energy $|A_1|^2$ detected along the XX' axis should have a shape similar to that of P_1. In the same way, if slit F_1 is closed, the energy distribution $|A_2|^2$ should be proportional to P_2. A "new" phenomenon appears in this experiment when both slits are open at the same time (see Fig. II.1.19). We know that the total signal amplitude is the sum of the two amplitudes measured separately

$$A_{tot} = A_1 + A_2$$

The energy, proportional to the square of total signal amplitude, is expressed as follows:

$$|A_{tot}|^2 = |A_1|^2 + |A_2|^2 + 2|A_1|\ |A_2|\cos\delta$$

where δ expresses the phase shift between the wave from F_1 and the wave from F_2. Unlike the previous case: $|A_{tot}|^2 \neq |A_1|^2 + |A_2|^2$. Maxima and minima appear in the energy graph. This is the well-known phenomenon of *interference* that we find when we study waves in mechanics or optics.

Electron experiment

We replaced the standard particle or wave source by an electron source (electron gun for example), and the bullet or wave intensity detector by a suitable detector, e.g., a Geiger counter (see Fig. II.1.20). We assume that the source emits monoenergetic electrons and provides wide angular dispersion so that both slits are equally accessible to all the particles. We assume, of course, that the slits are infinitely fine as in both previous experiments.

— Slit 1 open and slit 2 closed: On the observation screen (cf., Fig. II.I.21), each electron detected is visualized by a luminous dot, in the same way as the basic particles. Initially, the points of impact seem to be distributed in a more-or-less random way, and it is only after a long enough time that is becomes possible to obtain the probability of detecting one electron. This law of probability is labeled P_1.

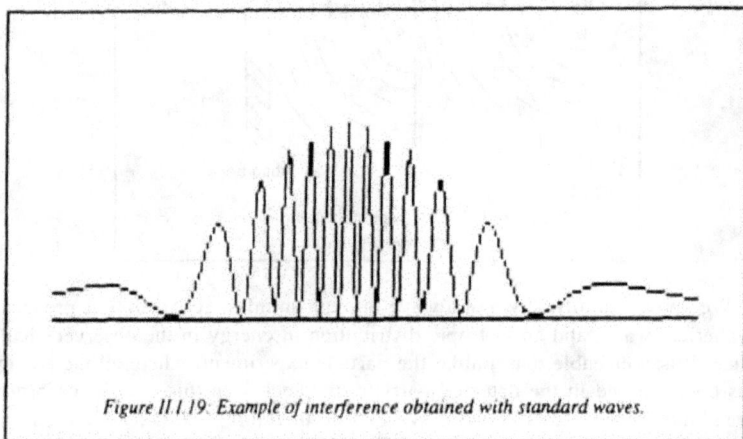

Figure II.I.19: *Example of interference obtained with standard waves.*

We observe that electrons have the same statistical behavior as the rifle bullets in the first experiment. We detect one electron each time. At the moment it is measured, the energy is concentrated at a well-defined point, indicating the impact of the electron on the screen. It should be noted that normal waves do not have this property, as the energy is spread along the entire wave front. It is possible to detect parts of the wave at two different points at the same time, which is impossible with electrons or rifle bullets.

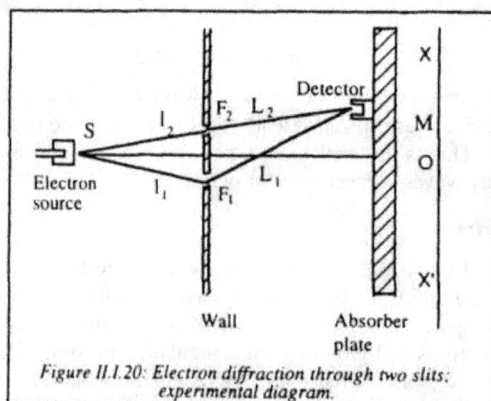

Figure II.1.20: *Electron diffraction through two slits:
experimental diagram.*

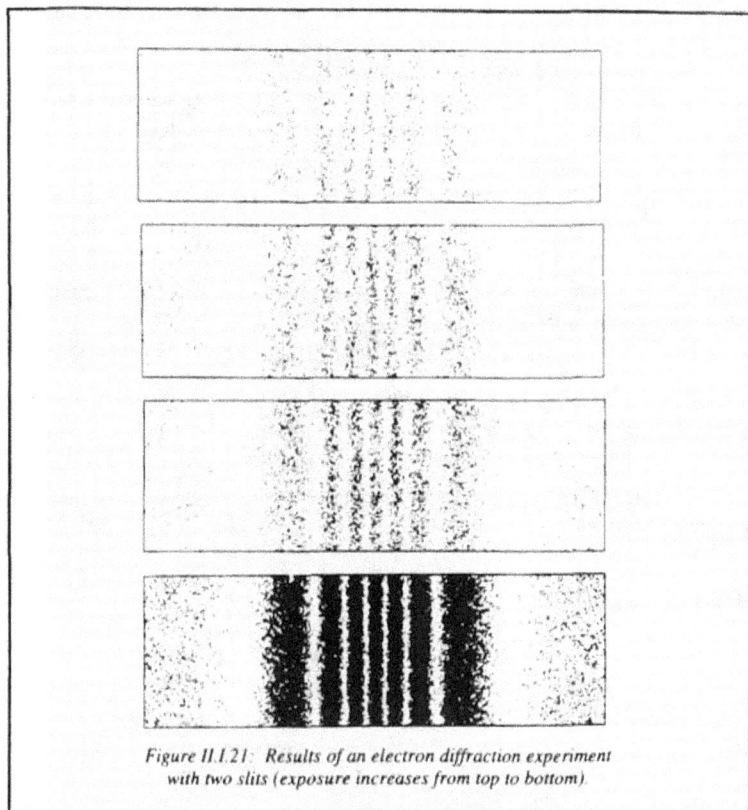

*Figure II.1.21: Results of an electron diffraction experiment
with two slits (exposure increases from top to bottom).*

— Slit 2 open and slit 1 closed: This is symmetrical to the previous situation. We observe similar behavior, with an electron distribution P_2.

— Slits 1 and 2 open: If electrons were really normal objects, like rifle bullets, we would expect that the probability of detecting one electron, when both slits are open, would be as follows:

$$P_{12} = P_1 + P_2$$

In fact, this figure shows the state of the screen at various times during a prolonged exposure. We should first note that electrons hit the screen like particles, as a whole electron is detected each time and not an electron "diluted" across the entire the screen. However, as large enough numbers of electrons are recorded on the surface of the screen, the probability of detecting this type of particle no longer corresponds at all to the standard solution.

We obtain

$$P_{12} \neq P_1 + P_2$$

and we observe an interference pattern similar to that observed with normal waves. We see, consequently, that electrons obey the laws of quantum mechanics, with a corpuscular behavior (each electron detected is indivisible), but a spatial distribution reminiscent of interference patterns obtained with normal waves. An interpretation of these results has already been proposed in Chapter II.I.1: The appearance of fringes is due to interference between the amplitudes of the ψ_1 and ψ_2 functions, representing the electron, associated with trajectories 1 (F_1) and 2 (F_2) (cf., Fig. II.I.20). Probabilities P_1 and P_2 are in fact the corresponding probability volume densities

$$P_1 = |\psi_1|^2 \quad ; \quad P_2 = |\psi_2|^2$$

When both slits are open at the same time, the detector measures $P_{12}|\psi_1 + \psi_2|^2$ given by the ratio (Eq. II.I.4)

$$P_{12}|\psi_1 + \psi_2|^2 = P_1 + P_2 + 2R(\psi_1 \psi_2{}^*)$$

The last term of the above equation (R indicates the real part of the expression in brackets) corresponds to the term

$$2|A_1||A_2|\cos \delta$$

of the expression used in the standard wave experiment, responsible for the interference observed.

7.2 Operator algebra

This section includes four problems and their respective solutions, to familiarize the reader with the use of operators, and particularly their matrix representations, widely used in quantum chemistry.

(I) Show that the eigenvalues of $A \equiv -i\dfrac{d}{dx}$ are nondegenerate.

The eigenvalue equation is as follows:

$$-id[\varphi_n(x)]/dx = a_n\varphi_n(x)$$

The general solution to this equation is $\varphi_n(x) = N\exp(ia_n x)$.

Each value a_n has only one corresponding wave function $\varphi_n(x)$, therefore a_n is nondegenerate.

(II) If A is an operator whose action on a function $f(r,\theta,\varphi)$, expressed in polar coordinates, is to change φ into $\varphi - 2\pi/3$:

$$A f(r,\theta,\varphi,) = f(r,\theta,\varphi - 2\pi/3)$$

Show that $\cos \varphi$ and $\sin \varphi$ form a base for the representation of A.

Using the definition of A, we can write

$A\cos \varphi = \cos(\varphi - 2\pi/3) = \cos \varphi \cos 2\pi/3 + \sin \varphi \sin 2\pi/3 = -1/2 \cos \varphi + \sqrt{3}/2 \sin \varphi$

$A\sin \varphi = \sin(\varphi - 2\pi/3) = \sin \varphi \cos 2\pi/3 - \cos \varphi \sin 2\pi/3 = -\sqrt{3}/2 \cos \varphi - 1/2 \sin \varphi$

As applying operator A to each of the functions of the base (i.e., $\sin \varphi$ and $\cos \varphi$) yields a linear combination of these two functions, they indeed form a base for the representation of A.

Find the matrix associated with this operator.

The matrix consists of the coefficients by which the basic functions were multiplied after A acted upon them, so we can write

$$(M) = \begin{pmatrix} -1/2 & -\sqrt{3}/2 \\ \sqrt{3}/2 & -1/2 \end{pmatrix}$$

Comment: We have just represented operator A by a 2x2 matrix. The matrix representation of operators is fundamental in most of the quantum mechanics applications in chemistry. Please note that the matrix changes if a different base is used. The basic functions must always be specified prior to constructing the matrix representation of an operator.

Find its eigenvalues.

Consider a function u_n expanded on the $\{\cos \varphi, \sin \varphi\}$ base as

$$u_a = c_1 \cos \varphi + c_2 \sin \varphi$$

In a matrix representation, the eigenvalue equation $A u_n = a_n u_n$ takes the form

$$\begin{pmatrix} -\dfrac{1}{2} & -\dfrac{\sqrt{3}}{2} \\ \dfrac{\sqrt{3}}{2} & -\dfrac{1}{2} \end{pmatrix} \begin{bmatrix} c_1 \\ c_2 \end{bmatrix} = a_n \begin{bmatrix} c_1 \\ c_2 \end{bmatrix} \text{ or } \begin{pmatrix} -\dfrac{1}{2} - a_n & -\dfrac{\sqrt{3}}{2} \\ \dfrac{\sqrt{3}}{2} & -\dfrac{1}{2} - a_n \end{pmatrix} \begin{bmatrix} c_1 \\ c_2 \end{bmatrix} = \begin{bmatrix} 0 \\ 0 \end{bmatrix}$$

where $\begin{bmatrix} c_1 \\ c_2 \end{bmatrix}$ is the "row vector" $[u_n]$ representing function u_n in the base consisting of the functions $\cos \varphi$ and $\sin \varphi$. Developing the above equations leads to a system of two homogeneous, linear equations:

$$(-1/2 - a_n)c_1 - \sqrt{3}/2\, c_2 = 0 \qquad\qquad (\text{II.1.156})$$

$$\sqrt{3}/2\, c_1 + (-1/2 - a_n)c_2 = 0$$

Equation II.1.156 is a system containing three unknown quantities (a_n, c_1 et c_2). The obvious solution ($c_1 = c_2 = 0$, a_n undetermined) has no physical meaning. The other solutions are found by specifying that the determinant of the matrix associated with the system must vanish for the system to be compatible. In other words

$$(-1/2 - a_n)^2 + 3/4 = 0$$

Solving this equation gives the two eigenvalues

$$a_{n,1} = -1/2 + i\sqrt{3}/2 = e^{2\pi i/3} \text{ and } a_{n,2} = -1/2 - i\sqrt{3}/2 = e^{-2\pi i/3}$$

Find the eigenfunctions of A.

Inserting these eigenvalues into the first equation of Eq. II.1.156 yields

for $a_{n,1}$: $c_2 = -c_1$

for $a_{n,2}$: $c_2 = c_1$

giving the following eigenfunctions:

$$u_{n,1} = c_1 (\cos \varphi - i \sin \varphi) = c_1 e^{-i\varphi}$$

$$u_{n,2} = c_2 (\cos \varphi + i \sin \varphi) = c_1 e^{i\varphi}$$

Conventionally, the value of coefficient c_1 is determined by normalizing u_n as follows:

$$\int_0^{2\pi} u_{n,i}^* u_{n,i} \, d\varphi = 1$$

so that $c_1 = 1/\sqrt{2\pi}$

(III) Represent operator $A = d/du$ in the base $\{\sin u, \cos u, \sin 2u, \cos 2u\}$. This base is not normalized.

As in Problem (II), we apply operator A to each function of the base:

$$d \sin u / du = \cos u$$
$$d \cos u / du = - \sin u$$
$$d \sin 2u / du = 2 \cos 2u$$
$$d \cos 2u / du = - 2 \sin 2u$$

Using the same procedure as in the second problem, we express the matrix representation of A in this base:

$$(A) = \begin{pmatrix} 0 & -1 & 0 & 0 \\ 1 & 0 & 0 & 0 \\ 0 & 0 & 0 & -2 \\ 0 & 0 & 2 & 0 \end{pmatrix}$$

Calculate the expression $d(\sin u + 3 \sin 2u)/du$, first by conventional derivation, then by using the matrix.

— Conventional method: direct derivation yields

$$d(\sin u + 3 \sin 2u)/du = \cos u + 6 \cos 2u$$

— Matrix method:

Let $\Psi = \sin u + 3 \sin 2u$

The equation $A\psi \equiv \dfrac{d}{du}(\sin u + 3 \sin 2u) = \psi'$ can be expressed using the matrix formalism as

$$(A)[\psi] = \begin{pmatrix} 0 & -1 & 0 & 0 \\ 1 & 0 & 0 & 0 \\ 0 & 0 & 0 & -2 \\ 0 & 0 & 2 & 0 \end{pmatrix} \begin{bmatrix} 1 \\ 0 \\ 3 \\ 0 \end{bmatrix} = [\Psi']$$

The row-vector on the left-hand side of the equation represents the components of Ψ in the chosen base. Multiplying the row vector by the matrix yields

$$[\Psi'] = \begin{bmatrix} 0 \\ 1 \\ 0 \\ 6 \end{bmatrix}$$

These are also components in the above base, therefore

$$\Psi' = \cos u + 6 \cos 2u$$

which is indeed identical to the result obtained by calculating the derivative directly

(IV) Find the matrix associated with operator $A = d/du$ in the orthonormal base

$$\left\{ \frac{1}{\sqrt{\pi}} \cos nu, \quad \frac{1}{\sqrt{\pi}} \sin nu \right\}.$$

Is this operator Hermitian? What about operator $A = i\, d/du$?

Using the same method as in the first problems, we obtain

$$\left| \frac{d}{du} \left(\frac{1}{\sqrt{\pi}} \cos nu \right) = -\frac{n}{\sqrt{\pi}} \sin u \right.$$

$$\left| \frac{d}{du} \left(\frac{1}{\sqrt{\pi}} \sin u \right) = \frac{n}{\sqrt{\pi}} \cos nu \right.$$

In the base given here, the two operators can be expressed as follows

$$\left(\frac{d}{du} \right) \equiv \begin{pmatrix} 0 & n \\ -n & 0 \end{pmatrix} \qquad \left(i\frac{d}{du} \right) \equiv \begin{pmatrix} 0 & in \\ -in & 0 \end{pmatrix}$$

According to the definition of a hermitian operator given in Sec.II.1.3, it is clear that an operator is hermitian if the off-diagonal elements of the associated matrix are each other's complex conjugates and its diagonal elements are real-valued. Thus, by looking at the elements of the matrices, operator (d/du) appears not to be hermitian, while $i(d/du)$ is.

7.3 Particle in one-dimensional box:
Computer calculation of the probability density

Case of an infinite square well

Consider once again the electron of the hydrogen atom. Its potential energy at distance r from the nucleus is (see Fig. II.1.22a)

$$V(r) = -q^2/4\pi\varepsilon_0 r = -e^2/r$$

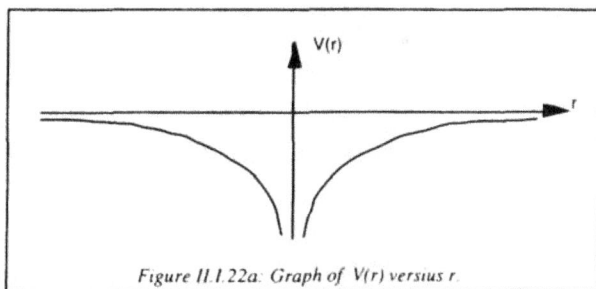

Figure II.1.22a: Graph of V(r) versius r.

The particle is said to be in a one-dimensional box submitted to an infinite square well. Rather complicated calculations are required to solve the equation for the wave functions describing the possible stationary states of the electron in the box. Let us consider, therefore, an infinite one-dimensional (in x) simplified box .

A simplified representation is shown in Fig. II.I.22b:

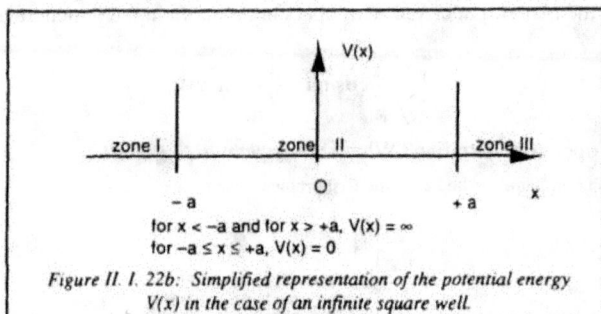

for x < −a and for x > +a, V(x) = ∞
for −a ≤ x ≤ +a, V(x) = 0

Figure II. I. 22b: Simplified representation of the potential energy
V(x) in the case of an infinite square well.

Exercise: Determine the eigenfunctions $\psi(x)$ and associated eigenvalues (energies) of the Hamilton operator H of a particle (mass m_e) in this box. Numerical application: a = 100 pm; m_e = 9.1 10^{-31} kg.

It is possible to prove that, in zones I and III , the eigenfunction is zero. The equation for eigenvalues in zone II is

$$-\frac{\hbar^2}{2m_e}\frac{d^2\psi}{dx^2} = E\,\psi(x) \text{ where E is positive}$$

then

$$\frac{d^2\psi(x)}{dx^2} = -\frac{2m_eE}{\hbar^2}\psi(x)$$

The solutions are of the following type:

$$\psi(x) = A\cos(kx) + B\sin(kx)$$

with

$$k = \sqrt{2m_eE/\hbar^2}$$

These functions must also be continuous to be acceptable in physical terms. This leads to

$$\left\{ \begin{array}{l} A\cos(ka) - B\sin(ka) = 0 \\ A\cos(ka) + B\sin(ka) = 0 \end{array} \right\}$$

then $\cos(ka) = 0$ et $B = 0$ or $\sin(ka) = 0$ and $A = 0$.

There are two types of solutions:

$$\psi(x) = A\cos(kx) \qquad \psi \text{ is an even function.}$$
$$\psi(x) = B\sin(kx) \qquad \psi \text{ is an odd function.}$$

Note that H commutes with the operator space parity Π (see Sec.II.I.3) whose eigenwave functions are even or odd functions of x. The eigenfunctions of H are then even or odd x functions.

The values of k and E are deduced from the previous equations. For the even functions: $\cos(ka) = 0$, so $k = (2n+1)\,\pi/(2a)$, where n is a whole number ≥ 0.

$$k = \sqrt{2m_e E/\hbar^2} \quad \text{leading to} \quad E = (2n+1)^2\,\frac{\hbar^2\pi^2}{8m_e a^2}$$

Consequently, the energy is quantized.
For the odd functions: $\sin(ka) = 0$, so $k = n'\,\pi/a$ where n' is a whole number ≥ 1. We deduce

$$E = n'^2\,\frac{\hbar^2\pi^2}{2m_e a^2}$$

The values of A and B are determined by stating the probability of finding the particle anywhere on the axis x'x is 1. As the functions are real, and by choosing real normalization constants

$$\int_{-\infty}^{+\infty} \psi^2(x)\,dx = 1$$

This equation leads to

$$\int_{-a}^{+a} A^2\cos^2(kx)\,dx = 1 \Rightarrow A = 1/\sqrt{a}$$

or

$$\int_{-a}^{+a} B^2\sin^2(kx)\,dx = 1 \Rightarrow B = 1/\sqrt{a}$$

Numerical application: $\dfrac{\hbar^2\pi^2}{8m_e a^2}$ gives a value of 1.50×10^{-18} J ≈ 9.4 eV.

Expression (in $m^{-\frac{1}{2}}$ unit) of some even functions for $-a \leq x \leq +a$ [$\psi(x)$ is zero everywhere else]:

$$n = 0,\ E = 9.4\ eV;\ \psi(x) = 10^5\cos(\pi x/2a)$$
$$n = 1;\ E = 84\ eV;\ \psi(x) = 10^5\cos(3\pi x/2a)$$
$$n = 2;\ E = 234\ eV;\ \psi(x) = 10^5\cos(5\pi x/2a)$$

Expression (in $m^{-\frac{1}{2}}$ unit) of some odd functions for $-a \leq x \leq +a$ [$\psi(x)$ is zero every where else]:

$$n' = 1;\ E = 37.4\ eV;\ \psi(x) = 10^5\cos(\pi x/a)$$
$$n' = 2;\ E = 149.5\ eV;\ \psi(x) = 10^5\cos(2\pi x/a)$$

Calculate and draw the graph of the one-dimensional probability density $\sigma(x)$ of the particle on the x'x axis.

The probability density is $\sigma(x) = \psi^2(x)$ so, for the even functions:
$\sigma(x) = (1/a)\cos^2(kx)$ for $-a \leq x \leq +a$, $\sigma(x) = 0$ everywhere else
and for the odd functions:
$\sigma(x) = (1/a)\sin^2(kx)$ for $-a \leq x \leq +a$; $\sigma(x) = 0$ everywhere else.

Let us draw the graphs(Figs. II.1.23 and II.1.24) showing the two probability densities (odd and even) corresponding to $n = n' = 1$. The ground state for the particle is the lowest energy state corresponding to the even solution $n=0$. This result will be re-examined in Chapter II.1.7.6.

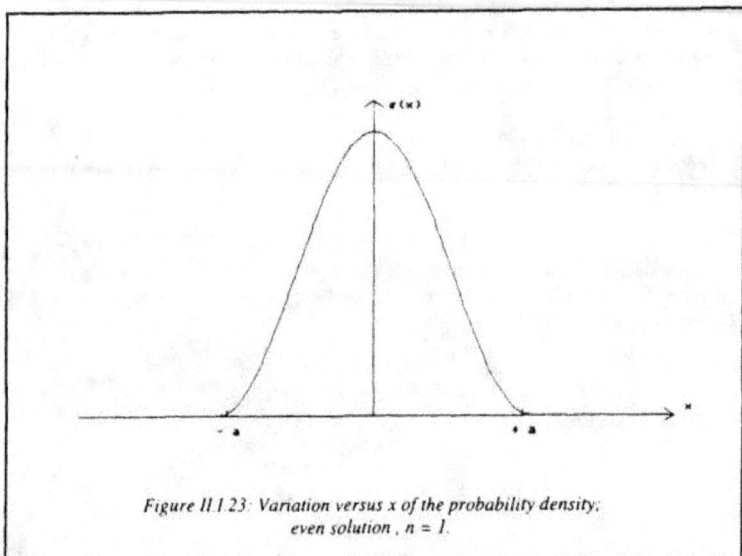

Figure II.1.23: *Variation versus x of the probability density;
even solution , n = 1.*

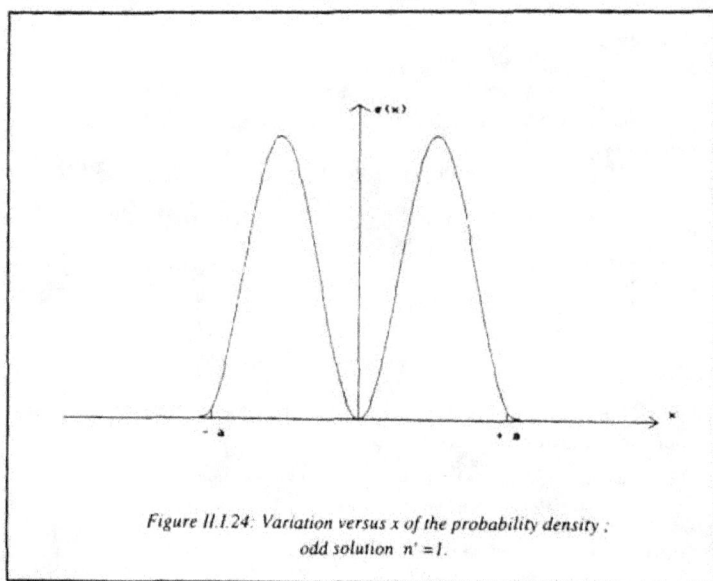

Figure II.1.24: *Variation versus x of the probability density ;
odd solution n' = 1.*

Case of any one-dimensional box

Consider the finite square well described in Fig. II.1.25.

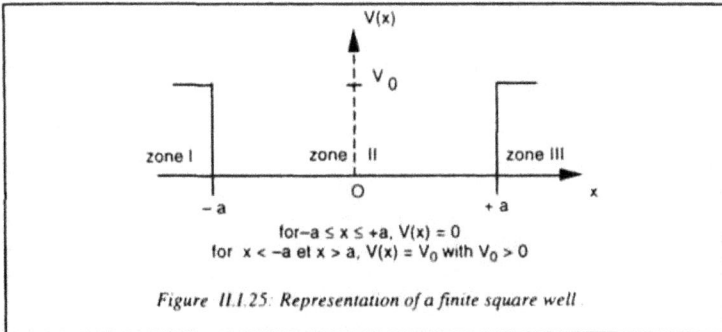

for $-a \leq x \leq +a$, $V(x) = 0$
for $x < -a$ et $x > a$, $V(x) = V_0$ with $V_0 > 0$

Figure II.1.25: Representation of a finite square well

Exercise: Write the eigenvalue equations for a particle (mass m_e) inside this box. Determine the general expression for the eigenfunctions $\psi(x)$.

The equations for eigenvalues (zone II) can be expressed as follows:

$$- \frac{\hbar^2}{2m_e} \frac{d^2\psi(x)}{dx^2} = E\,\psi(x)$$

then

$$\frac{d^2\psi(x)}{dx^2} = - k^2 \psi(x) \text{ with } k^2 = \frac{2m_e E}{\hbar^2}$$

We assume that k is positive. The solutions are of the following type:

$$\psi(x) = A\cos(kx) + B\sin(kx)$$

In zones I et III:

$$- \frac{\hbar^2}{2m_e} \frac{d^2\psi(x)}{dx^2} + V_0\,\psi(x) = E\,\psi(x)$$

This equation can be expressed

$$\frac{d^2\psi(x)}{dx^2} = \alpha^2 \psi(x) \text{ with } \alpha^2 = \frac{-2m_e(E - V_0)}{\hbar^2}$$

Remark: The particle is only described in the box. In this case:
$0 \leq E \leq V_0$. Then, the value $-2m(E-V_0)/\hbar^2$ is positive.

Let us suppose that α is positive. The solutions are of the following type:

$$\psi(x) = C\exp(-\alpha x) + D\exp(\alpha x)$$

These functions must be finite to be acceptable, therefore

zone I: $\Psi(x) = D\exp(\alpha x)$
zone III: $\Psi(x)\,C\exp(-\alpha x)$

Symmetry conditions dictate that these functions must be either even or odd. We can therefore deduce the even functions:

$$\begin{cases} \text{Zone I} & \psi(x) = D\exp(\alpha x) \\ \text{Zone II} & \psi(x) = A\cos(kx) \\ \text{Zone III} : & \psi(x) = D\exp(-\alpha x) \end{cases}$$

and the odd functions:

$$\begin{cases} \text{Zone I} & \psi(x) = F\exp(\alpha x) \\ \text{Zone II} & \psi(x) = B\sin(kx) \\ \text{Zone III} & \psi(x) = -F\exp(-\alpha x) \end{cases}$$

Deduce the corresponding energies from the continuity conditions for $\psi(x)$ and $\psi'(x)$ [the first derivative of $\psi(x)$].

We shall assume that the functions and their first derivatives must be continuous for $x = -a$ and $x = +a$. We find then, for the even functions:

$$D\exp(-\alpha a) = A\cos(-ka) \text{ and } D\,\alpha\exp(-\alpha a) = -Ak\sin(-ka)$$

leading to

$$\tan(ka) = \alpha/k$$

For the odd functions:

$$F\exp(-\alpha a) = B\sin(-ka) \text{ and } F\alpha\exp(-\alpha a) = Bk\cos(-ka)$$

leading to

$$\tan(ka) = -k/\alpha$$

The two previous equations can be graphically solved ; we can also solve them by a numerical method using Newton's algorithm. Let us call: $y(k) = \tan(ka) \pm k/k_0$. We wish to find the values of k where $y(k) = 0$. On the other hand

$$y(k) = a/\cos^2(ka) \pm 1/k_0$$

Newton's algorithm is an iterative process allowing the computation of the successive values of k, using the recurrent relation:

$$k_{n+1} = k_n - \frac{y(k_n)}{y'(k_n)}$$

Let us choose the initial value k_1 and iterate, for example, until we find $|k_{n+1} - k_n| \le |k_n| \times 10^{-3}$ (the algorithm converges towards the desired values of k). We can then calculate the corresponding energy values with the relation:

$$\alpha^2 = -\frac{2m_e(E - V_0)}{\hbar^2}$$

Remark: When integrating the equation for eigenvalues in a very small space around x, we find

$$\int_{x-\epsilon}^{x+\epsilon} \frac{d^2\psi(x)}{dx^2}\, dx + \frac{2m_e}{\hbar^2} \int_{x-\epsilon}^{x+\epsilon} (E - V(x))\; \psi(x)\, dx = 0$$

then

$$\psi'(x+\epsilon) - \psi'(x-\epsilon) + \frac{2m_e}{\hbar^2} \int_{x-\epsilon}^{x+\epsilon} (E - V(x))\; \psi(x)\, dx = 0$$

The integrated equation tends towards zero if ϵ tends towards zero [$V(x)$ and ψ are finite]. Therefore $\psi'(x + \epsilon) - \psi'(x - \epsilon) = 0$ and the first derivative of $\psi(x)$ is continuous

Expression of the eigenfunctions $\psi(x)$

We have to calculate the normalizing constants A, B, D, F for the functions $\psi(x)$. We assume that the probability of finding the particle anywhere on the axis x'x is 1.

For the even functions, we find

$$\int_{-\infty}^{-a} D^2 \exp(2\alpha x)\, dx + \int_{-a}^{+a} A^2 \cos^2(kx)\, dx + \int_{+a}^{+\infty} D^2 \exp(-2\alpha x)\, dx = 1$$

We can deduce

$$\frac{D^2}{\alpha} \exp(-2\alpha a) + A^2 a + \frac{A^2}{2k} \sin(2ka) = 1$$

On the other hand, we had previously found that $D \exp(-\alpha a)$ $A \cos(ka)$
Therefore we can deduce

$$A = \left[\frac{\cos^2(ka)}{\alpha} + \frac{\sin(2ka)}{2k} + a \right]^{-1/2}$$

that is

$$D = A \cos(ka) \exp(\alpha a)$$

and for the odd functions , the same method gives

$$B = \left[\frac{\sin^2 ka}{\alpha} - \frac{\sin(2ka)}{2k} + a \right]^{-1/2}$$

that is

$$F = -B \sin(ka) \exp(\alpha a)$$

The eigenfunctions $\psi(x)$ are now completely determined.

Calculate the one-dimensional probability density $\sigma(x) = \psi(x)^2$

Numerical application: $V_0 = 100\ eV$; $a = 0.1\ nm$; $m_e = 9.1 \times 10^{-31}\ kg$.

Four graphic representations of odd or even functions are given below.

Remark: The probability of finding the particle outside the box is not zero (this is different from classical mechanics) . However, the deeper the well , the smaller the probability.

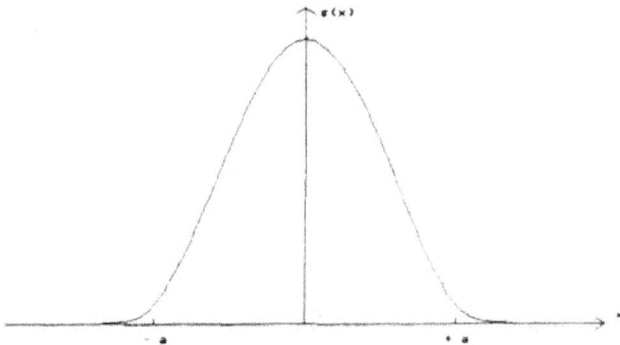

Figure II.1.26: Variation of the one-dimensional probability density versus x
Even function n° 1
$E = 6.6204\ eV$ $k = 1.106 \times 10^{10}\ m^{-1}$ $\sigma_{max} = 8.32 \times 10^9$ $\alpha = 4.9493 \times 10^{10}\ m^{-1}$

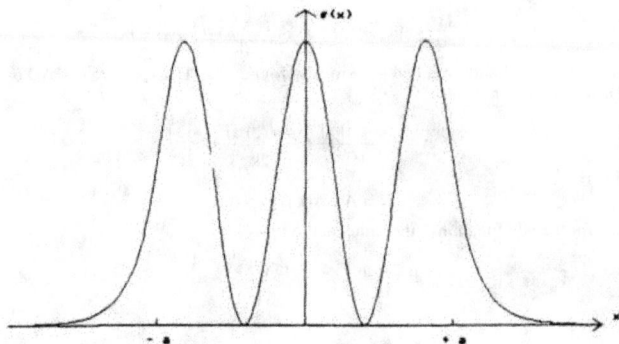

Figure II.1.27: Variation of the one-dimensional probability density versus x.
Even function n° 2:
$$E = 57.2565 \ eV \qquad k = 3.542 \times 10^{10} \ m^{-1}$$
$$\sigma_{max} = 7.696 \times 10^9 \qquad \alpha = 3.3485 \times 10^{10} \ m^{-1}.$$

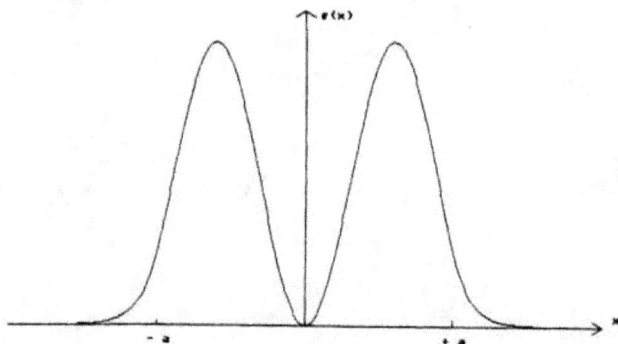

Figure II.1.28: Variation of the one-dimensional probability density versus x.
Odd function n° 1:
$$E = 26.1522 \ eV \qquad k = 2.6048 \times 10^{10} \ m^{-1}$$
$$\sigma_{max} = 7.4822 \times 10^9 \qquad \alpha = 4.4013 \times 10^{10} \ m^{-1}.$$

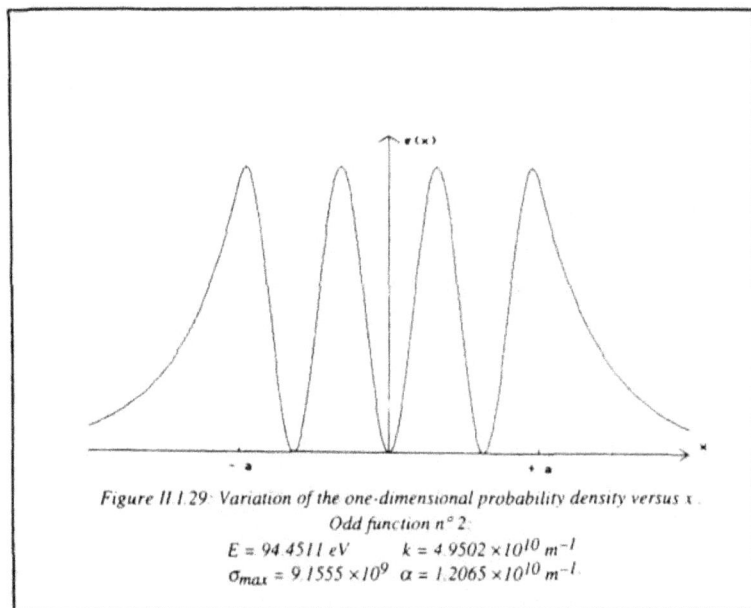

Figure II.1.29: Variation of the one-dimensional probability density versus x.
Odd function n° 2:

$$E = 94.4511\ eV \qquad k = 4.9502 \times 10^{10}\ m^{-1}$$
$$\sigma_{max} = 9.1555 \times 10^{9} \qquad \alpha = 1.2065 \times 10^{10}\ m^{-1}.$$

7.4 Importance of nonstationary states: Wave packets

In this section, we shall examine a problem which is very much inspired by the book of C. Cohen-Tannoudji et al, mentioned at the beginning of this subchapter. We shall illustrate the solutions with some computer simulations. This will provide practice in using the concepts described in the last few paragraphs. It will also demonstrate how to extend them to the concept of nonstationary states (or nonsteady states), which will be more fully developed later, in conjunction with chemical bonding.

A few examples of the use of nonstationary states will be given:

— They can be used to represent the quasi-classical "movement" of an isolated microscopic particle (for instance a particle in an accelerator, or an electron in an oscilloscope).

— They can also be used to account for the "movement" of a macroscopic system (macroscopic harmonic oscillator).

Let us consider a microscopic particle of mass m which is free to move on the x- axis ($V = 0$). We start by assuming that this particle has a constant, perfectly-defined momentum $p_x = p$. Using the postulates, determine a wave function $\Psi_p(x,t)$ representing one of its stationary states. Deduce from Heisenberg's uncertainty principle that, in this state, the particle is completely delocalized (i.e., all positions are equally likely). Show that $\Psi_p(x,t)$ cannot be normalized. What can be said about the state represented by $\Psi_p(x,t)$?

The particle is free and has a perfectly well-determined momentum. This means that $\Psi_p(x,t)$ is an eigenfunction of operator p (according to Postulate 5). Operator p is obtained from the value of p_x using Postulate 7. We obtain

$$p \, \Psi p(x, t) = p \Psi p (x, t) \text{ that is } - i\hbar \frac{\partial}{\partial x} \Psi p (x,t) = p \, \Psi \, p(x,t)$$

Solving this differential equation in x yields

$$\Psi_p(x, t) = A(t) \exp (ipx/\hbar)$$

To find A(t), we can use Postulates 6 and 7

$$i \, \hbar \, \Psi_p(x, t) = H \, \Psi_p (x, t) \text{ with } H = \frac{p^2}{2m} \,\, , \,\, V = 0$$

which, of course, gives

$$A(t) = B \exp (- ip^2 t / 2m\hbar)$$

where B is a constant. Thus the complete expression for the wave function $\Psi_p(x,t)$ is

$$\Psi_p(x, t) \,\, = B \exp (ipx/\hbar) \exp (- ip^2 t/2m\hbar)$$

Heisenberg's uncertainty principle (see reference at the beginning of this subchapter) states that, for a system in state Ψ, if Δx and Δp are the uncertainties concerning the measurements of x and p, then the product of these uncertainties, $\Delta x \Delta p$, must always be greater than a value on the order of $\hbar/2$. As, in this case, $\Delta p = 0$, Δx must necessarily be infinite, which means that the particle is completely delocalized on the x-axis. In fact, this is already shown by the mathematical form of $\Psi_p(x, t)$ as the probability density is independent both of t (it is a steady state) and x.

$$|\Psi_p(x, t)|^2 = B^2 = \text{constant.}$$

If we try to calculate the norm of the wave function (using Eq. II.1.20), or rather its square to simplify the equation, we obtain

$$N^2 = \int_{-}^{+\infty} |\Psi_p(x,t)|^2 \, dx = B^2 \int_{-}^{+\infty} dx$$

Thus N diverges and $\Psi_p(x, t)$ cannot be normalized. Therefore $\Psi_p(x, t)$ does not belong to the space of states and cannot stand for a real state of the particle. As a matter of fact, the concept of a free particle (i.e., with a well-defined value of p) is completely unrealistic as a particle must always be characterized by a distribution of values of p around a mean value p_0.

Let us now try to find a real state of the particle represented by wave function $\Psi(x, t)$. This function can always be expanded as a superposition of stationary states such as $\Psi_p(x, t)$, where the amplitude associated with each component Ψ_p is given by g(p)

$$\Psi(x, t) = \int_{-\infty}^{+\infty} g(p) \, \Psi_p(x, t) \, dp$$

This superposition of stationary waves is called a wave packet.

Show that, at each instant t, $\Psi(x, t)$ can be expressed as the Fourier transform of a function, and find this function.

Using the above definition of $\Psi(x,t)$:

$$\Psi(x, t) = B \int_{-\infty}^{+\infty} g(p) \exp(ipx/\hbar) \exp(-ip^2 t/2m\hbar)dp$$

if we define $p = \hbar k$, this changes to

$$\Psi(x, t) = \int_{-\infty}^{+\infty} f(k, t) e^{ikx}dk$$

where $f(k, t) = \hbar Bg(\hbar k) \exp(-i\hbar k^2 t/2m)$.

$\Psi(x,t)$ indeed appears as the Fourier transform of $f(k)$, and the conjugated variables are x and k. The shape of $f(k)$ is arbitrary, as is that of $g(p)$. It is always possible to choose $g(p)$ in such a way that the Fourier transform $\Psi(x,t)$ of $f(k,t)$ can be normalized.

Figure II.1.30 illustrates how a wave packet can be produced by superposing three discrete sinusoidal base functions.

As t increases, the center of the wave packet moves and the shape of the packet changes. This behavior can be illustrated by taking a Gaussian shape for $g(p)$, to facilitate analytical calculations:

$$g(p) = \alpha \exp[-a^2(p - p_0)^2 / 4\hbar^2]$$

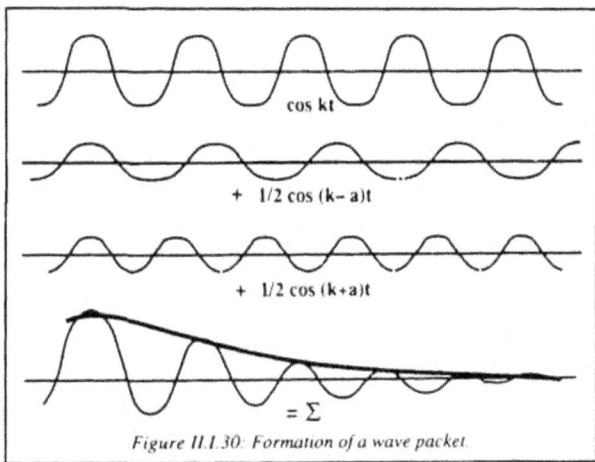

$$\cos kt$$
$$+ \ 1/2 \cos(k-a)t$$
$$+ \ 1/2 \cos(k+a)t$$
$$= \Sigma$$

Figure II.1.30: Formation of a wave packet.

A simple calculation yields

$$|\Psi(x, t)|^2 = \frac{\beta}{\sqrt{1 + 4\hbar^2 t^2 / m^2 a^4}} \exp[-2a^2(x - \hbar k_0 t/m)^2 / (a^4 + 4\hbar^2 t^2 / m^2)]$$

with $k_0 = \dfrac{p_0}{\hbar}$ and $\beta = \dfrac{4\pi\hbar^2 B^2 \alpha^2}{a^2}$ = constant

Show that $\Psi(x,t)$ can be normalized. Calculate the velocity of the center of the wave packet.

The norm of $\Psi(x,t)$ is given by $= N' = \displaystyle\int_{-\infty}^{+\infty} |\Psi(x,t)|^2\, dx$

We know that the integral is taken over a Gaussian function, therefore it converges and is nonvanishing, and $\Psi(x,t)$ can be normalized. The center of the wave packet is localized at $x = \hbar k_0 t/m$. It has a linear movement, and moves at constant speed $v_0 = \hbar k/m$. If we remember that $\hbar k_0 = p_0$, we can see that this speed is equal to the ratio between momentum $p_0 = mv_0$ and particle mass. This result converges with the classical result. Of course, the shape of the packet changes over time.

Figure II.1.31 represents the snap shot of a free wave packet at time $t = 0$, and shows how it is "squashed" as it moves to the right.

Let us now consider a macroscopic particle of mass m placed on the x-axis subjected to a restoring force $F = -kx$ with respect to the origin O (this is a one-dimensional harmonic oscillator). Let $\Psi(x,t)$ be the following function:

$$\Psi(x,t) = \alpha \exp(-\beta x^2)\exp(-iEt/\hbar)$$

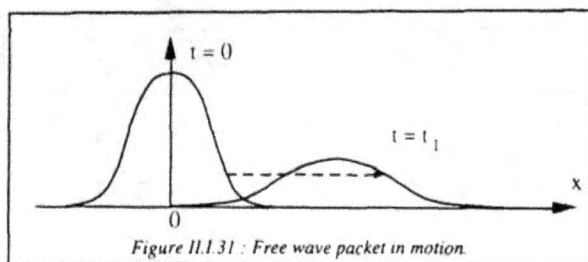

Figure II.1.31 : Free wave packet in motion.

Show that this function represents a stationary state of the particle for certain values of β and of E. Calculate these values (for the ground state of the oscillator). Set $\omega^2 = k/m$.

As shown earlier, when studying the physical consequences of the postulates, if $\Psi(x,t)$ is a stationary state of a system, it can be expressed in the form:

$$\Psi(x,t) = \Psi^{(0)}(x)\exp(-iEt/\hbar)$$

where $\Psi^{(0)}(x)$ is a solution of the equation $H\,\Psi^{(0)}(x) = E\,\Psi^{(0)}(x)$.

The Hamiltonian of the system is as follows : $H = (-\hbar^2/2m)\,d^2/dx^2 + kx^2/2$ and $e^{-\beta x^2}$ must satisfy the equation:

$$-\frac{\hbar^2}{2m}\frac{d^2}{dx^2}(e^{-\beta x^2}) + \frac{1}{2}kx^2 e^{-\beta x^2} = E\,e^{-\beta x^2}$$

This is indeed the case. Moreover, we can easily find E and β by identification:

$$\beta = m\omega/2\hbar \qquad E = \hbar\omega/2$$

Calculate the expected values $\bar{x}$ and $\bar{p}$ in this stationary state. What can we conclude about the representation of a macroscopic particle by a stationary state?

We proved earlier (as a consequence of Postulate 4) that

$$\bar{a}_\psi = \int \Psi^* A\Psi d\tau$$

In this case

$$\bar{x} = \int_{-\infty}^{+\infty} x\,|\Psi(x,t)|^2 dx$$

The integral over a spatially odd function vanishes. A similar calculation shows that $\bar{p}$ also vanishes. These values, which are always zero, are in contradiction with experimental observations of an oscillating phenomenon over time. Therefore, this stationary state cannot represent the macroscopic particle.

In fact, the macroscopic particle is in a specific nonstationary state $\Psi_c(x,t)$ (assumed to be normalized with a norm equal to one), which is an eigenstate of the operator a defined as follows:

$$a = (1/\sqrt{2})\left[\sqrt{m\omega/\hbar}\,x + (1/\sqrt{\hbar m\omega})\,p\right]$$

We shall not try to find the exact mathematical form of $\Psi_c(x,t)$. The square of its norm is assumed to be equal to

$$|\Psi_c(x,t)|^2 = |\Psi[x - A\cos(\omega t + \varphi)]|^2$$

where A and φ are arbitrary real-valued constants.
Calculate x. Comment on the result.

$$\bar{x} = \int_{-\infty}^{+\infty} x\,|\psi[x - A\cos(\omega t + \varphi)]|^2 dx$$

The variable is changed as follows: $x' = x - A\cos(\omega t + \varphi)$ This yields

$$\bar{x} = \int_{-\infty}^{+\infty} [x' + A\cos(\omega t + \varphi)]\,|\Psi(x',t)|^2 dx'$$

The first term, an odd function, yields a vanishing integral, so we are left with

$$\bar{x} = A\int_{-\infty}^{+\infty} \cos(\omega t + \varphi)|\Psi(x',t)|^2 dx' = A\cos(\omega t + \varphi)$$

The expected value for the position of the macroscopic particle corresponds to the experimentally observed result. Figure II.I.32 shows a similar behavior for a macroscopic particle in a square potential well. The wave packet moves alternately in one direction or the other without being "squashed".

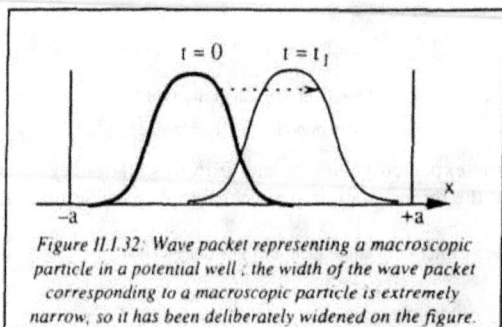

*Figure II.1.32: Wave packet representing a macroscopic
particle in a potential well ; the width of the wave packet
corresponding to a macroscopic particle is extremely
narrow, so it has been deliberately widened on the figure.*

7.5 Problem of measuring the position of a microscopic particle

In this section, we come back to the fundamental problem of measuring the position of a quantum particle. We showed earlier (when discussing the physical interpretation of the postulates) why the Dirac- or delta-function $\delta(x - x_0)$ cannot belong to the space of states and also why the physical quantity "position of a particle" cannot be measured with infinite precision. The problem we are about to study illustrates these discussions.

Let us consider a quantum particle of mass m located inside an infinite potential well defined by $V(x) = 0$ for $|x| < a$ and $V(x) = \infty$ for $|x| > a$. A detector D (detecting particle position) is placed on $x = x_0$; this detector signals a presence at time t if at time t the particle is located between $x_0 - \varepsilon$ and $x_0 + \varepsilon$ (with $\varepsilon > 0$, $\varepsilon \ll a$, and $|x_0 \pm \varepsilon| < a$). At time $t = 0$, the particle is in its ground state $\Psi_1(x)$. Detector D measures the probability $P^{(\varepsilon)}_{x_0}$ of finding the particle between $x_0 - \varepsilon$ and $x_0 + \varepsilon$ at time $t = 0$. (Here, we assume that the measurement is repeated many times on identical systems in the ground state). Determine the function $P^{(\varepsilon)}_{x_0}$. For what value of x_0 does $P^{(\varepsilon)}_{x_0}$ reach a maximum? Would the probability be different if the measurement were made with a value for time t other than zero? What comment can you make about this result? Compare the probability $P^{(\varepsilon)}_{x_0}$ with the probability corresponding to a classical particle placed in the same potential well.

The solution to Problem 7.3 provided the wave function $\Psi_1(x)$ representing the groundstate of the particle:

$$\Psi_1(x) = (1/\sqrt{a}) \cos(\pi x/2a)$$

The probability $P^{(\varepsilon)}_{x_0}$ is thus given by

$$P^{(\varepsilon)}_{x_0} = \int_{x_0-}^{x_0+} |\Xi_1(x)|^2 dx = (\varepsilon/a) + (1/\pi) \sin(\pi\varepsilon/a)\cos(\pi x_0/a)$$

As $\varepsilon/a \ll 1$, $P^{(\varepsilon)}_{x_0}$ can be approximated by the expression

$$P_{x_0}^{(\epsilon)} = (\epsilon/a)\,(1 + \cos \pi\, x_0/a)$$

$P_{x_0}^{(\epsilon)}$ reaches a maximum at $x_0 = 0$ and $P_{x_0}^{(\epsilon)}{}_{max} = 2\epsilon/a$.

The probability does not depend on the time at which the measurement is made as we are dealing with a steady state system. The classical particle travels back and forth in the well and the probability of being detected is equal to ϵ/a, which does not depend on x_0. To conclude, we see that the behavior of the quantum particle in its ground state is fundamentally different from that of a classical particle.

Locating a particle at time $t = 0$ using detector D placed on $x_0 = 0$ means that a signal appeared on D at this time. Once the particle is located at $x_0 = 0$, its wave function is no longer $\Psi_1(x)$, but has changed to $\Psi_\epsilon(x)$, defined as follows:

$$\Psi_\epsilon(x) = A/2\epsilon \ \text{ for } \ |x| < \epsilon \quad \text{and} \ \Psi_\epsilon(x) = 0 \ \text{ for } \ |x| > \epsilon$$

Show that $\Psi_\epsilon(x)$ can be normalized and calculate the normalization constant A (assuming A is a positive real number). Is this change of state after measurement compatible with the postulates of quantum mechanics? Deduce from this a property of the operator corresponding to the observable measured by D, without trying to determine this operator exactly.

Let us calculate the norm N

$$N^2 = \left(A^2/4\epsilon^2\right)\int_{-\epsilon}^{+\epsilon} dx = A^2/2\epsilon$$

So $\Psi_\epsilon(x)$ can be normalized. Setting $A^2/2\epsilon = 1.1$ yields $A = \sqrt{2\epsilon}$

and

$$\Psi_\epsilon(x) = 1/\sqrt{2\epsilon} \quad \text{if } -\epsilon < x < +\epsilon$$
$$\Psi_\epsilon(x) = 0 \quad \text{if } |x| < \epsilon$$

This change of state follows Postulate 5: After a measurement, the system must be in an eigenstate corresponding to the eigenvalue which was the result of the measurement. The observable measured was the value of the position x (with an accuracy of $\pm \epsilon$). The state of the system after measurement is indeed the eigenfunction $\Psi_\epsilon(x)$ associated with the eigenvalue $(0 \pm \epsilon)$. The eigenfunctions of operator $(x_0 \pm \epsilon)$ are

$$\Psi_\epsilon^{(x_0)}(x)$$

$$(x \pm \epsilon)\,\Psi_\epsilon^{(x_0)}(x) = (x_0 \pm \epsilon)\,\Psi_\epsilon^{(x_0)}(x)$$

with
$$\begin{cases} \Psi_\epsilon^{(x_0)} = 1/\sqrt{2\epsilon} \ \text{if } (x_0 - \epsilon) < x < x_0 + \epsilon \\ \Psi_\epsilon^{(x_0)}(x) = 0 \quad \text{if } |x - x_0| > \epsilon \end{cases}$$

Once the particle is in state $\Psi_\epsilon(x)$, calculate the expected value of its position, $\bar{x}$, of the square of its position, $\overline{x^2}$, and the root mean square deviation Δx. Using Heisenberg's uncertainty principle, find the lower boundary of the root mean square deviation of momentum Δp.

As in Problem 7.4, we have $\bar{x} = 0$ as $\Psi_\varepsilon^{(x_0)}(x)$ is an even function.

$$\overline{x^2} = \int_{-\varepsilon}^{+\varepsilon} \left(x^2/2\varepsilon\right) dx = \varepsilon^2/3$$

$$\Delta x = \sqrt{\overline{x^2} - \bar{x}^2} = \varepsilon / \sqrt{3}$$

Remembering that $\Delta p \Delta x \geq \hbar / 2$, we find that $\Delta p \geq \hbar \sqrt{3} / 2\varepsilon$

With the particle still in state $\Psi_\varepsilon(x)$, we measure its energy. What values can we expect to find, and with what probabilities?

The 3rd and 4th postulates tell us that energy E will be found with the probability

$$P(E_n) = \left| \int_{-\varepsilon}^{+\varepsilon} \Psi_n^*(x)\, \Psi_\varepsilon(x)\, dx \right|^2$$

E_n and $\Psi_n(x)$ are the eigenvalues and the eigenfunctions of the Hamiltonian H, respectively. Replacing $\Psi_\varepsilon(x)$ by its value for x between $-\varepsilon$ and $+\varepsilon$ yields

$$P(E_n) = \left(1/2\varepsilon\right) \left| \int_{-\varepsilon}^{+\varepsilon} \Psi_n^*(x) dx \right|^2$$

Clearly, this probability vanished for odd eigenfunctions $\Psi_{odd}(x)$. In Problem 7.3 we proved that the even eigenfunctions may be expressed as follows:

$$\Psi_p(x) = \cos[(2k + 1)\, \pi x/2a]/\sqrt{a} \qquad \text{(k is a whole number)}$$

so that the probability is

$$P(E_n) = \left| \int_{-\varepsilon}^{+\varepsilon} \cos\left[(2k + 1)\, \pi x / 2a\right] dx \right|^2 / 2\, a\varepsilon = 8\, a\, \sin^2\left[(2k + 1)\, \pi\varepsilon / 2a\right] / \varepsilon\pi^2 (2k + 1)^2$$

and Problem 7.3 gave the associated energy E_{even} as equal to

$$E_p = (2k + 1)^2\, \pi^2\, \hbar^2 / 8\, a^2\, m$$

7.6 Wave function of a particle driven by a periodic potential on a circle: A simple simulation of a cyclic molecule

Any object studied in the context of a physics theory requires the construction of more-or-less sophisticated models, depending on the degree of accuracy you wish to achieve when the calculation results are compared with experimental data. Our aim here is to try for a qualitative rather than a quantitative approach, so we deliberately simplify our models as far as possible, without contradicting the essential principles of physics. Starting with a model of a hydrogen atom using a square well with the dimensions described in Sec. II.I.7.3, we extended this approach to the ion molecule H_2^+, then we deduced a phenomenological description of chemical bonding that we applied to the study of cyclic molecules, then one-dimensional crystals, presented as equivalent to infinite cyclic molecules. A justification of the results obtained is given in Chapter IV, using the molecular orbital method best suited to studying chemical bonds (cf., Sec. IV.III.6.3).

Precise calculation. Presentation of a double well

Roughly speaking, an electron of a hydrogen atom, or any other element, may be considered to be held prisoner in a potential well, assumed to be square in order to simplify the calculations (cf., Sec. II.I.7.3). This provides a qualitative explanation as to why the energy levels of atoms and of bonded systems in general, are discrete. We pursue this approach by modeling a hydrogen molecule by combining two wells (both hydrogen atoms) separated by a barrier, whose width and height depend on the distance between the two protons. Simply by varying these two parameters, it is possible to show the energy level disturbances caused by the presence of another atom in the vicinity of the first. Finally, it should be noted that, to simplify the problem still further, we assumed that both of the wells were infinitely deep. An example of potential energy is shown in Fig. II.I.33.

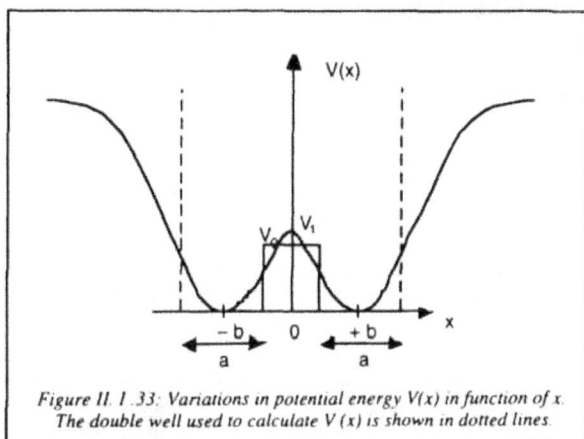

Figure II. 1 .33: *Variations in potential energy V(x) in function of x. The double well used to calculate V (x) is shown in dotted lines*

Evolution over time

We saw in Sec. II.I.7.3 that the solution to the square well problem (specifically Hamiltonian solution) was stationary, i.e., that the probability density function was not time-dependent. On the contrary, all wave functions that are not eigenfunctions of the Hamiltonian may be expressed using a base of time-independent functions, leading to a time dependence of the probability density. The exercise at Sec. II.I.7.4 illustrates this aspect in the building up of a wave packet.

In the case of a double well, it is possible to study changes over time in a state represented at time t = 0 by an eigenfunction of the infinitely deep well (see Fig. II.I.34). Note that, if the barrier is very wide and very high, the probability density varies very little over time. This means that the presence of the second well disturbs the first one very little and that, as a result, the wave function tested is practically an eigenfunction of the problem: The energy levels are not modified. On the other hand, if the barrier is not very wide, a leakage of probability density is observed from one well to the other, with an oscillating movement. A probability current is set up between the two wells and this result is interpreted as a modification in energy levels (Hamiltonian eigenvalues) due to the presence of the second well. The eigenfunctions of the infinitely deep well are no longer eigenfunctions of the double well.

It should be noted that this flow of probability from one well to the other through the barrier, is a typically quantum phenomenon , known as the tunnel effect. Its application to the microscope of the same name earned its inventors the Nobel prize in 1986.

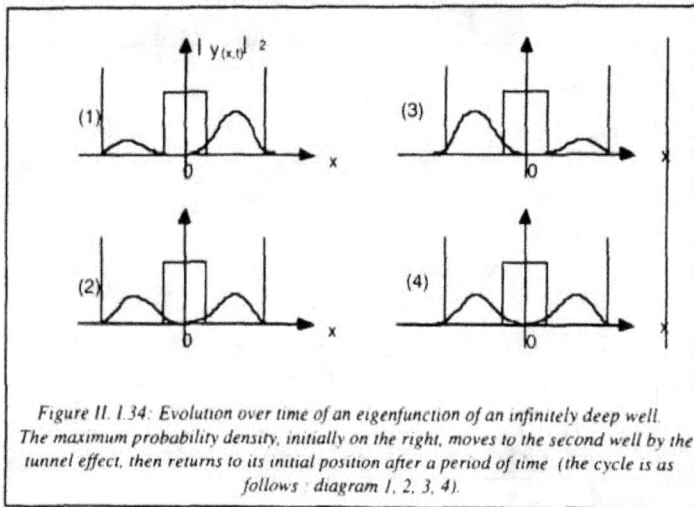

Figure II. 1.34: Evolution over time of an eigenfunction of an infinitely deep well. The maximum probability density, initially on the right, moves to the second well by the tunnel effect, then returns to its initial position after a period of time (the cycle is as follows : diagram 1, 2, 3, 4).

Determining stationary energy levels

From the previous study, we deduced that the eigenfunctions of single wells did not provide a good approximation of stationary wave functions for double wells, except in the rather uninteresting case where the barrier is very wide. We deduced that only a full study, solving the Schrödinger equation in a double well, would provide an answer to our problem.

Note, first of all, that the $V(x)$ potential is an even function. We looked for a base of eigenfunctions with definite parity: either symmetrical, or antisymmetrical (as the Hamiltonian H commutates with the parity operator π). The results of this calculation lead to two transcendental equations:

— For the symmetrical solution:

$$\mathrm{tg}\,(k_s\,a) = -\frac{k_s}{\sqrt{\alpha^2 - k_s^2}}\,\coth\left[\sqrt{\alpha^2 - k_s^2}\left(b - \frac{a}{2}\right)\right]$$

— And for the antisymmetrical solution:

$$\mathrm{tg}\,(k_a\,a) = -\frac{k_a}{\sqrt{\alpha^2 - k_a^2}}\,\mathrm{th}\left[\sqrt{\alpha^2 - k_a^2}\left(b - \frac{a}{2}\right)\right]$$

where $\alpha^2 = 2m_e V_0/\hbar^2$ and $k^2 = 2m_e E/\hbar^2$ the index shows whether it is a symmetrical or antisymmetrical solution (see Fig. II.1.33 for the geometrical explanation of V_0, a and b). These equations may be solved graphically using specialized software.

Measuring the degeneracy of the levels

The solution to these two equations gives two energies E_s and E_a that are different, except where V_0 or the distance between the two protons is infinite. In cases where the two energies are indistinguishable, we find solutions for infinitely deep single wells. On the other hand, when the barrier allows a nonnegligible tunnel effect, we observe a splitting of levels which had previously been indistinguishable, known as a degeneracy break.

Influence of barrier height and width

The split, i.e., the distance between energies E_a and E_s, becomes larger as the thickness of the barrier decreases. In the same way, the number of levels inside each well increases as the height of the barrier increases. However, for a given barrier height, there is less splitting at deep levels than at levels close to the top of the barrier. Our software makes it possible to test these various combinations at will. For coherence, we note that the levels E_a and E_s produced from the same undisturbed level, remain neighbors for thick enough barriers by comparison to the difference between the initial, undisturbed levels.

Phenomenological approach . Presenting the model

We have just shown, in the preceding section, that when two atoms moved towards each other, the energy levels moved apart. A more elaborate process would have achieved the same result, at least from a qualitative point of view, and would also have made it possible to demonstrate that, from a quantitative point of view, the H_2^+ ion molecule, in its basic state, has a favorable energy configuration for a distance between the two protons corresponding to those measured experimentally. This approach to the problem is still too complicated if we wish to deal with molecules consisting of more than two atoms.

We shall therefore consider levels E_a and E_s as resulting from a disturbance on the initial level E_0, and we treat the effect of the barrier in a purely phenomenological way. The problem is thus reduced to processing a two-level system, as the other levels are assumed to be far enough away not to intervene to any significant extent.

Let us start by noting that, if the barrier is infinite, states ϕ_s and ϕ_a have the same energy E_0. The functions

$$\phi_1 = \frac{1}{\sqrt{2}}(\phi_s + \phi_a) \text{ and } \phi_2 = \frac{1}{\sqrt{2}}(\phi_s - \phi_a)$$

are therefore the eigenfunctions of the hamiltonians corresponding to the two infinite square wells respectively, located on the left and right of Fig. II.I.34. They also correspond, therefore, to a position of the electron in each of these two wells. The hamiltonian is expressed as follows in the base $\{\Phi_1, \Phi_2\}$:

$$(H_0) = \begin{pmatrix} E_0 & 0 \\ 0 & E_0 \end{pmatrix}$$

The effect of a noninfinite barrier is then introduced phenomenologically by "disturbing" (H_0) by means of a nondiagonal matrix:

$$(H_1) = \begin{pmatrix} 0 & -A \\ -A & 0 \end{pmatrix}$$

where A is a positive energy. The hamiltonian total is thus represented in the $\{\Phi_1, \Phi_2\}$ base by the matrix

$$(H) = \begin{pmatrix} E_0 & -A \\ -A & E_0 \end{pmatrix}$$

Diagonalizing this matrix produces two energy levels: $(E_0 + A)$ corresponding to the eigenstate ϕ_a and $(E_0 - A)$ corresponding to the eigenstate ϕ_s. This means that the split (amplitude 2A in this case) of the undisturbed level E_0 induced by the barrier has been phenomenologically reproduced. The influence of barrier height and width may be studied at will by modifying the A coupling term.

Extension to more than two wells

It is relatively easy to extend the preceding calculation to the case of a linear molecule consisting of three atoms. As coupling only occurs between adjacent atoms, the following matrix represents the hamiltonian on the base of undisturbed states:

$$\begin{pmatrix} E_0 & -A & 0 \\ -A & E_0 & -A \\ 0 & -A & E_0 \end{pmatrix}$$

When this matrix is diagonalized, three eigenvalues: $E_0 - A\sqrt{2}$, E_0 and $E_0 + A\sqrt{2}$ are obtained. Once again, the initial level splits into several levels, depending on the intensity of the coupling. The calculation may theoretically be continued for any number of wells.

Cyclic case

The process described above, which is complicated for a molecule with a large number of atoms, is simplified, according to the properties of symmetry, in the case of cyclic molecules. Consider, for example, the case of the large number of electrons π in the benzene molecule. We know that each carbon atom is at the summit of a regular hexagon, giving this molecule a property of invariance for a rotation of an angle $2\pi/6$. The reasoning given in Sec. II.II. shows that the ψ_n eigenfunctions of the hamiltonian that may be built up from the undisturbed states of the electrons in each of the wells associated with carbon atoms $(\phi_1, \phi_2, ..., \phi_6)$ is, as a result of symmetry, in the following simple form:

$$\psi_n = \frac{1}{\sqrt{6}} \sum_{q=1}^{6} e^{2\pi i n q/6} \phi_q$$

where n is an integer in the interval [1, 6]. A simple calculation thus gives six energy levels

$$E_n = E_0 - 2A \cos(n\pi/3)$$

A more accurate analysis of these results shows that, if the six π electrons in the molecule are on the three lowest levels (basic state), the total energy is: $E_{total} = 2(E_0 - 2A) + 4(E_0 - A) = 6E_0 - 8A$. This is less than the energy corresponding to the three uncoupled double bonds $[3 \times (2E_0 - 2A)]$ in a formula developed by Kekulé.

The probability P_q that an electron, described by the wave function ψ_n, is located adjacent to a carbon atom q (given by the square of the coefficient module of ϕ_q in ψ_n) is

$$\left| \frac{1}{\sqrt{6}} e^{2\pi i n q/6} \right|^2 = \frac{1}{6}$$

Each site therefore has the same probability of being occupied.

A simple model of a one-dimensional crystal: From molecules to crystals.

The Born-Von Karman cyclic conditions make it possible to extend the preceding developments to a one-dimensional crystal. It is simply necessary to consider that the crystal's wave function, assumed to be infinite, is reproduced identically every N atoms, where N may be of any size. The mathematical problem is thus the same as that of a cyclic molecule with N atoms and the energy levels are as follows:

$$E_n = E_0 - 2A \cos(2\pi n/N) \text{ for } n = 1, 2, ..., N$$

It is important to note that, whatever the number N of atoms, the splitting of the initial state stays between the extreme values $E_0 - 2A$ and $E_0 + 2A$. If we give N a large enough value, in order to model a macroscopic crystal, the energy levels are so close to one another that it is as if there were a continuous energy band of width 4A.

Until now, we have only coupled the states ϕ_q corresponding to the basic state of each well, but, if we consider the higher energy states ϕ_q, we obtain a series of authorized energy bands separated by prohibited energy bands (Fig. II.I.35).

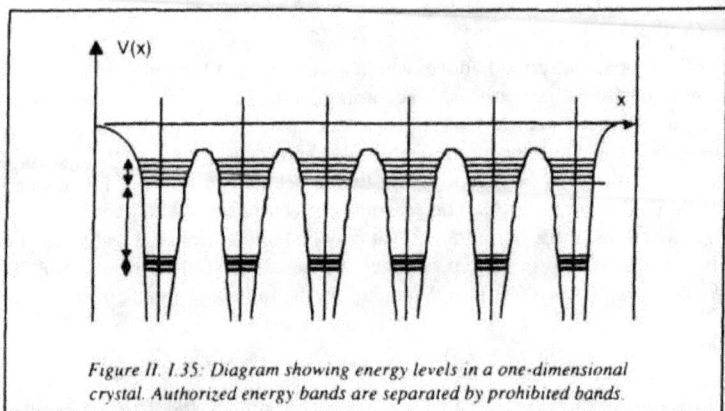

Figure *II. 1.35: Diagram showing energy levels in a one-dimensional crystal. Authorized energy bands are separated by prohibited bands.*

In order to describe the behavior of an electron in a crystal more accurately, we must transform the expression of E_n. Where b is the space between 2 atoms. Nb is therefore the total length of crystal L, and Nb/n takes all the possible values between 0 and L, when N tends towards infinity (and, of course, b tends towards zero). So $2\pi n/Nb$ represents all the possible wave numbers in a system of dimension L, and the energy expression becomes:

$$E_k = E_0 - 2A \cos (kb) \text{ with } k = \frac{2\pi n}{Nb}$$

If k is small enough, a limited development of cos kb and a readjustment of the energy scale so that $E_0 = 2A$, lead to

$$E_k = Ak^2b^2$$

We know (cf., Sec. II.1.7.3) that the speed of a wave packet equals the speed of group $v_g = d\omega/dk$. As $E = \hbar\omega$, we deduce that $v_g = 2Ab^2k/\hbar$ which gives the following results for the energy of an electron in the crystal:

$$E = \frac{1}{2} \left(\frac{\hbar^2}{2Ab^2} \right) v_g^2$$

We obtain the surprising result that an electron in a crystal moves inside a permitted energy band as a free particle, *but* with a different mass from that of an electron in a vacuum. In our highly simplified model this mass, known as "total mass", may be expressed as follows:

$$m_{eff} = \frac{\hbar^2}{2Ab^2}$$

In this way, we have been able to provide a qualitative description of electron behavior in molecules and crystals, using a very simple phenomenological model. It is quite clear that, although this approach may not give any correct quantitative data, it illustrates the capacity of quantum mechanics to account for the electron structure of molecules, in a quantitative way, by using more highly-developed models.

8. Appendix 3: How can atoms be represented?

8.1 Illustration of the probabilistic aspect of the electron density of a hydrogen atom

A description of the electron structure of atoms is introduced in the secondary curriculum for 15-year-old students. The probabilistic aspect of electron density is a difficult concept to understand at an elementary level. In order to facilitate this acquisition, two experiments, presented as games, were carried out at our request in a class of thirteen-year-olds. In fact, the aim is to prepare the students to achieve the following objectives concerning a correct understanding of the electron cloud of a hydrogen atom:

— The meaning of the probabilistic nature of an event.
— The concept of the most probable position.
— The concept of the most probable nucleus-electron distance.
— The fact that it is possible to draw a sphere surrounding a volume containing almost all the electron charge.

These two games are presented below, followed by details of how to use the results, in certain cases going beyond the scope of secondary school work.

First game: Throwing dice

Preparation: Each student in the class (approximately 30 students) has two dice of different colors and a transparency with a copy of the grid shown in Fig. II.1.36. The values shown by the dice at each throw give the coordinates of a point in a quadrant of the diagram. The points are positioned in each of the quadrants in turn: 1, 2, 3, 4, 1, 2 ....

Interpretation: When two grids are shown side-by-side on an overhead projector, the class realizes that the results obtained by the two students are not identical: Each node in the network may be marked, but not at will by each student. The teacher may also use the overhead projector to superpose the results of the entire class: Each node may be marked one or more times and no particular areas are more heavily marked than others. The surface density of the points tends to be evenly spread, demonstrating the notion of equiprobability of the various throws, when the number is large enough.

Second game: Throwing darts

Preparation: Each student throws several darts at a target 4 dm in diameter, with five rings, 0.8 dm apart, numbered from 1 to 5 from the center to the outside. The throwing distance is approximately two meters. Fig. II.1.37 shows the impacts obtained with 143 throws.

Interpretation: Counting the impacts in each ring gives Table II.1.4. This table shows, for each ring i, the surface area S_i, the number of impacts N_i, the number of impacts cumulated with those of the inner rings N_{ci}, as well as the quantity

N_i/S_i, for the number of impacts per unit area (surface density of the impacts) in each ring.

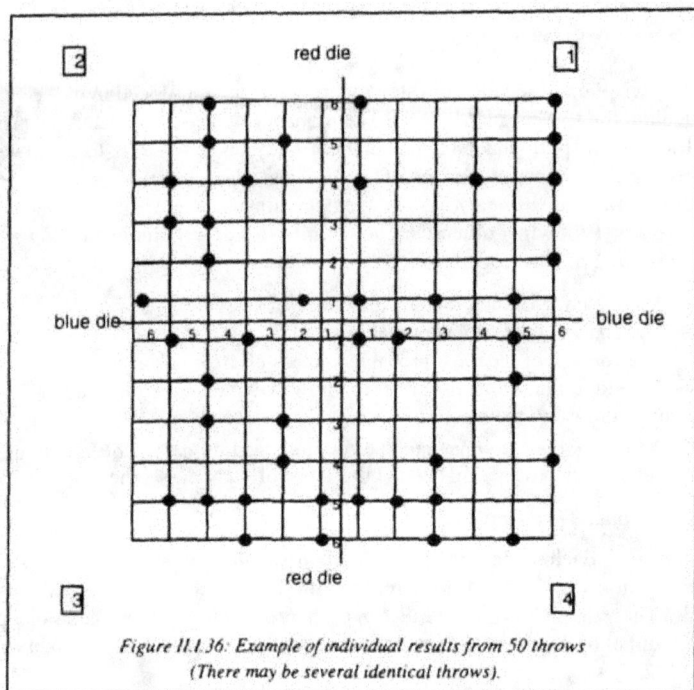

Figure II.1.36: Example of individual results from 50 throws
(There may be several identical throws).

TABLE II.1.4 — **Results obtained by throwing 143 darts.**

ring number i	1	2	3	4	5	off target
S_i (dm^2)	2	6	10	14	18	infinity
N_i	7	26	28	19	13	7
N_{ci}	7	33	61	80	93	100
N_i/S_i (dm^{-2})	5.0	6.1	4.0	1.9	1.0	0

Analysis of this table shows that, although the surface density N_i/S_i is highest in rings 1 and 2, the maximum number of impacts occurs in ring 3. This is due to the fact that, as the distance between rings is the same, the surface area increases as the rings are further from the center of the target. This increase more than

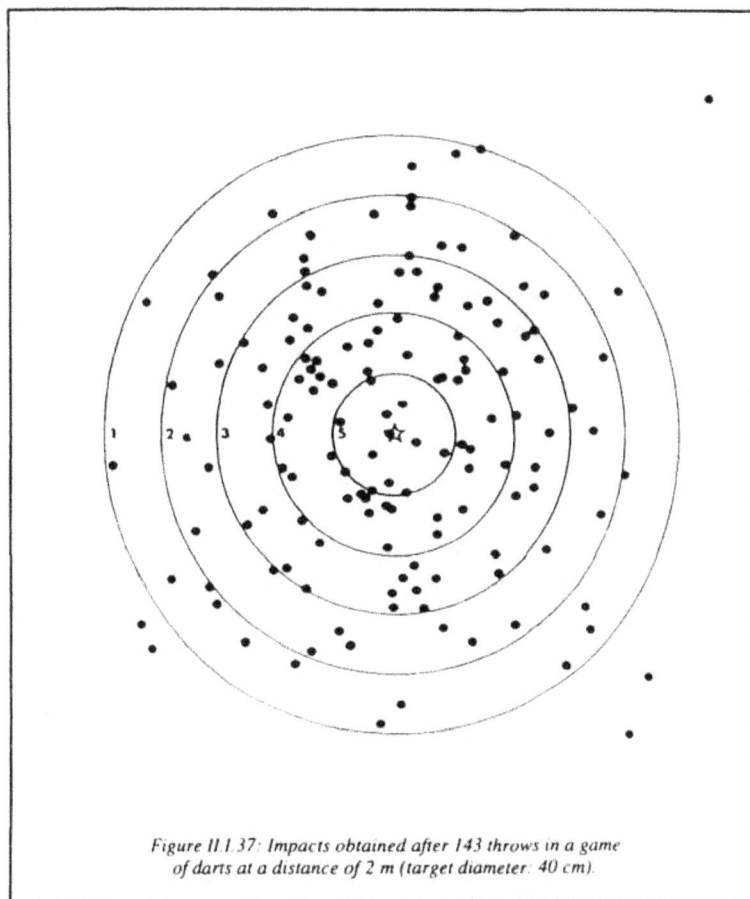

Figure II.1.37: Impacts obtained after 143 throws in a game
of darts at a distance of 2 m (target diameter: 40 cm).

compensates for the decrease in surface density between ring 2 and ring 3. On the other hand, the considerable decrease in surface density between ring 3 and rings 4 and 5, is not longer compensated by the increase in surface, so N_i decreases. In other words, the number of impacts N_i in the various rings is equal to the product of two terms: The surface density which, with the exception of the center circle, tends to decrease with the distance from the center, and the surface S_i which increases regularly.

The variation of N with i is shown in Fig. II.1.38.

It is now possible to try to estimate the average distance of the impacts from the center. For this purpose, we assume (to simplify matters) that all the impacts in ring i are the same distance from the center, $r_i = (0.4\,i - 0.2)$ dm.

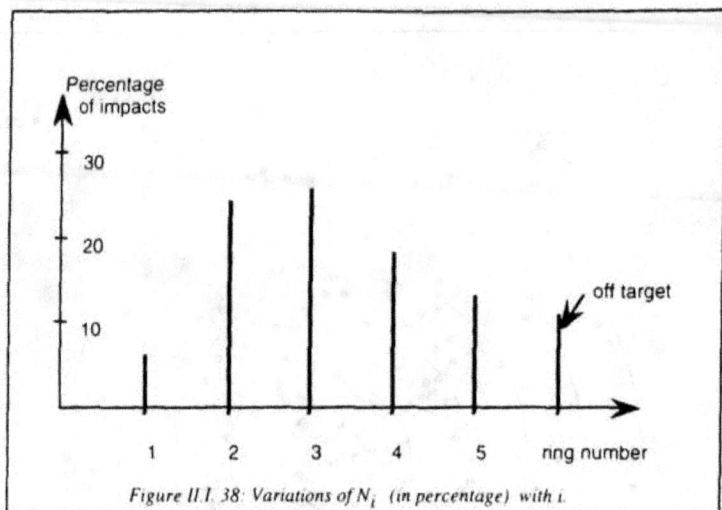

Figure II.1. 38: Variations of N_i (in percentage) with i.

If we ignore the off target impacts, the distance may be expressed as follows:

$$\bar{r} = \frac{\sum\limits_{i=1}^{5} N_i r_i}{\sum\limits_{i=1}^{5} N_i} = 1 \text{ dm} \qquad (II.I.157)$$

This radius is located, in this case, in ring 3 where there is a maximum number of impacts, but it should be noted that this is not necessarily the case. In a similar way, the standard deviation, Δr, is (using the same approximations)

$$\Delta r = \left[\frac{\sum\limits_{i=1}^{5} N_i (r_i - \bar{r}_i)^2}{\sum\limits_{i=1}^{5} N_i} \right]^{1/2} = 0.46 \text{ dm} \qquad (II.I.158)$$

a value close to the distance between two rings. The percentage of impacts inside the outside circle of ring i (with a radius of 0.4 i) is equal to $N_{ci}(100/143)$. Values on Table II.I.4 show that 61 % of the impacts are inside a circle with a radius of $0.4 \times 3 = 1.2$ dm and 80 % inside a circle with a radius of $0.4 \times 4 = 1.6$ dm.

These considerations are comparable to the quantities and functions used to represent atomic orbitals, although these calculations involve volume rather than surface area. The following is an example of how these quantities or functions are calculated, for an orbital 1s of a hydrogen atom $\left(1s = \exp(-r/a_0)/\sqrt{\pi a_0^3}\right)$

— The electron space density (probability of finding the electron per unit of volume), which, according to the postulates, is as follows:

$$| 1s |^2 = \exp(- 2r/a_o)/\pi a_o^3$$

i.e., a constantly decreasing function of r, with a maximum value of $1/\pi a_o^3$ at the nucleus (Fig. II.1.40).

— The maximum probability of finding the electron in a spherical space with a radius of r and a given thickness dr (i.e., volume $dv = 4\pi r^2 dr$) is

$$| 1s |^2 4\pi r^2 dr = \frac{4}{a_o^3} e^{-2r/a_o} r^2 dr$$

The function $4r^2 \exp(- 2r/a_o)/a_o^3$ is known as radial density. The maximum for this function occurs where $r = a_o = 0,053$ nm (Fig. II.1.39). This is the most probable electron-nucleus distance.

— The average electron-nucleus distance, which, according to the postulates, is given by the following expression (as the 1s orbital is spherically symmetrical):

$$\bar{r} = \int_o^\infty (1s \, r \, 1s) \, 4\pi r^2 dr = \frac{2}{a_o^3} \int_o^\infty e^{-2ar} r^3 dr = \frac{3}{2} a_o = 0.080 \text{ nm} \qquad (II.I.159)$$

— The average square electron-nucleus distance:

$$\left(\overline{r^2} \right)^{1/2} = \left[\int_o^\infty (1s \, r^2 1s) \, 4\pi \, r^2 dr \right]^{1/2} = \sqrt{3} \, a_o = 0.092 \text{ nm}$$

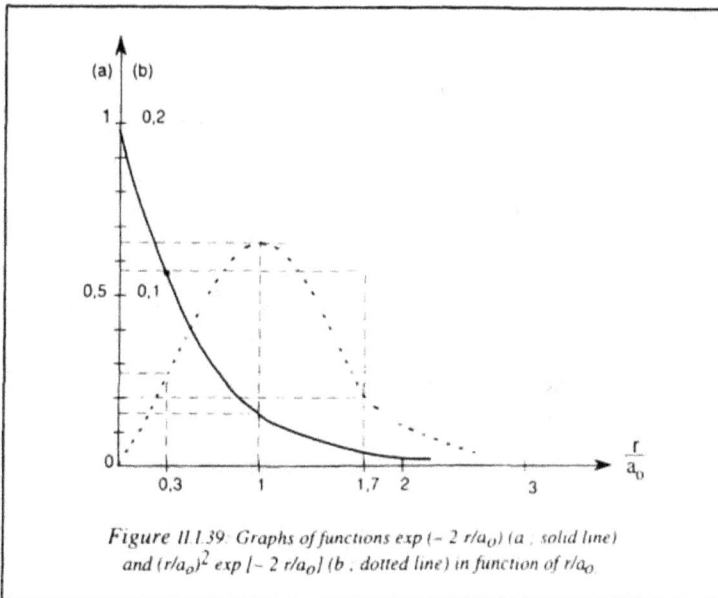

Figure II.1.39: Graphs of functions exp (– 2 r/a$_o$) (a , solid line) and (r/a$_o$)^2 exp [– 2 r/a$_o$] (b , dotted line) in function of r/a$_o$

— The standard deviation , Δr, which is easily deduced from the above results:

$$\Delta r = \left[\int_0^\infty \left[1s \, (r - \bar{r})^2 1s \right] 4\pi r^2 dr \right]^{1/2} = \frac{\sqrt{3}}{2} a_0 = 0.046 \text{ nm} \qquad (\text{II.I.160})$$

The value obtained (approximately half of $\bar{r}$ and on the order of the most probable electron-nucleus distance) shows that it is not possible, even as an approximation, to consider that the electron is at a fixed distance from the nucleus. This result was predictable, in view of the difference between the most probable distance and the average radius.

It should be noted that Eqs. II.I.159 and II.I.160 are the quantum equivalents, where r is a continuous variable, of discrete Eqs. II.I.157 and II.I.158.

— The radius $r_{0.9}$, with a probability of 0.9, i.e., the radius of the sphere where the probability of finding the electron equals 0.9, is given by the following integral equation:

$$\int_0^{r_{0.9}} |1s|^2 4\pi r^2 dr = \frac{4}{a_0^3} \int_0^{r_{0.9}} e^{-2ar} r^2 dr = 0.9$$

with the following solution: $r_{0.9} = 2.7 a_0 = 0.143$ nm.

These results are summarized in Table II.I.5.

TABLE II.I.5 — **Some characteristic values of an orbital 1s of a hydrogen atom.**

— maximum electron space density	$r = 0$ nm
— most probable electron-nucleus distance	$r = a_0 = 0.053$ nm
— average electron-nucleus distance	$\bar{r} = 3a_0/2 = 0.080$ nm
— average square distance	$\left(\bar{r^2}\right)^{1/2} = \sqrt{3} \, a_0 = 0.092$ nm
— standard deviation in measuring r	$\Delta r = \sqrt{3} a_0/2 = 0.046$ nm

8.2 Atomic orbitals (AOs): Radial and angular distribution

We saw, in Chapter II.I.4.4, that solving the equation for eigenvalues of Hamilton operator H, in the case of the hydrogen atom (or an hydrogenoid atom), leads to wave functions ψ of the following type, (in spherical polar coordinates):

$$\psi \, (r,\theta,\varphi) = R_n(r) \, Y_{l,m}(\theta,\varphi)$$

where $R_n(r)$ is the radial part and $Y_{l,m}(\theta,\varphi)$ the angular part .The wave function of a polyelectronic atom is much more complicated but an approximate and rather easy solution is a Slater determinant built with AOs ϕ characterized by three quantic numbers n, l, and m (see Eq. II.I.6)

$$\phi(r,\theta,\varphi) = R_n{}^*(r) \, Y_{l,m}(\theta,\varphi)$$

These AOs are only different from the hydrogenoid functions because of their radial part. An approximation added by Slater means that the following is often used (see Eq. II.I.8.3)

$$R_n{}^*(r) = \left|\frac{r}{a_o}\right|^{n^*-1} \exp(-Z^* \, r / n^* a_o)$$

A lot of atomic chemical properties may be explained or predicted according to the area and angular distributions of the AOs. Therefore, it is essential to have a clear image of these representations. The following examples correspond to hydrogenoid atoms, but it is not difficult to extend the results to Slater AOs.

Visualization of the electronic density as a cloud of points

To visualize the "electronic cloud" associated with the AO, we assume that the probability of finding the electron in an elementary volume dv around the point $M(r,\theta,\varphi)$ is

$$dP(r,\theta,\varphi) = |\psi(r,\theta,\varphi)|^2 \, dv$$

The value

$$\rho(r,\theta,\varphi) = \frac{dP(r,\theta)}{dv} = |\psi(r,\theta,\varphi)|^2$$

is often called volumic probability density. The representation of this function dependent on three coordinates is too complex, so we will only represent it in the plane corresponding to different values φ_k chosen for φ (Fig. II.I.40), by plotting a lot of points, in such a way that their surface density is proportional to $\rho(r,\theta,\varphi_k)$.

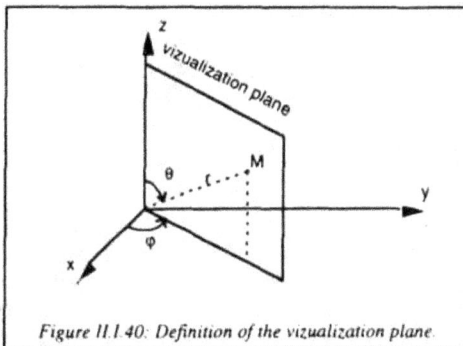

Figure II.I.40: Definition of the vizualization plane.

In order to do that, let us define this plane as a square, with sides of $2 r_{max}$ (this value is chosen *a priori* so that the probability of finding the electron outside the square is very low; cf., Fig. II.I.41) centered in O (the center of the atom).

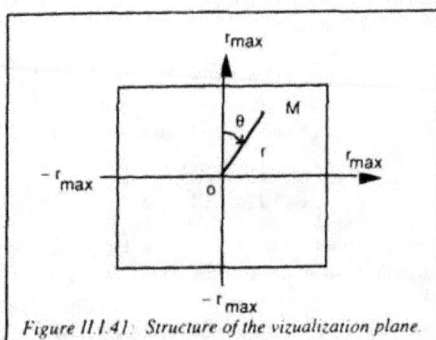

Figure II.I.41: Structure of the vizualization plane.

Let us search for the upper value ρ_{max} for the function $\rho(r,\theta,\phi_k)$ in this square [this value can be found approximately starting with a large number of randomly distributed points M, and comparing the corresponding values of $\rho(r,\theta,\phi_k)$].

The point M will be plotted (on a computer screen or sheet of paper) only if $\rho_o \leq \rho(M)$. As the probability of ρ_o being between 0 and $\rho(M)$ is $\rho(M)/\rho_{max}$, the point M will be plotted according to a probability proportional to the volumic density. Fig. II.I.42 illustrates this method.

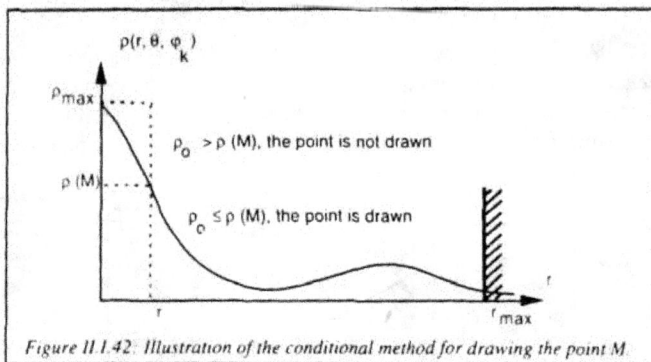

Figure II.I.42: Illustration of the conditional method for drawing the point M.

By repeating this procedure many times, a cloud of points will be obtained with a surface density proportional to the volumic probability density of the electron at this point. This electronic cloud may be visualized rapidly using a computer program (see flowchart below; N is the whole number of points admitted for the electronic cloud; r_{max} value will be chosen so that the size of the cloud is reasonable).

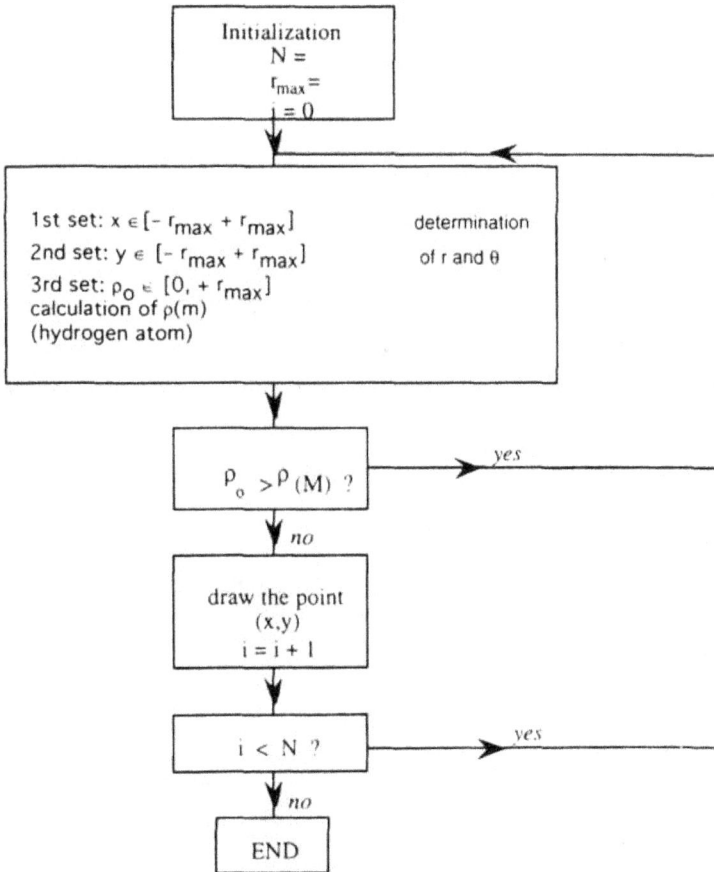

Initialization
N =
$r_{max} =$
i = 0

1st set: $x \in [-r_{max} + r_{max}]$ determination

2nd set: $y \in [-r_{max} + r_{max}]$ of r and θ

3rd set: $\rho_0 \in [0, + r_{max}]$
calculation of $\rho(m)$
(hydrogen atom)

$\rho_o > \rho(M)$?   yes

no

draw the point
(x,y)
i = i + 1

i < N ?   yes

no

END

Some results are shown in Figs. II.1.43 to II.1.48 .

Figure II.1.43: Visualization of 1s orbital; Z = 4.

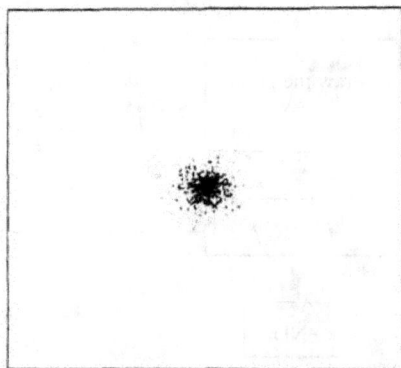

Figure II.1.44: Visualization of 1s orbital; Z = 10.

Orbital 2pz
rmax/ao : 1.00   M : 1000   z   10

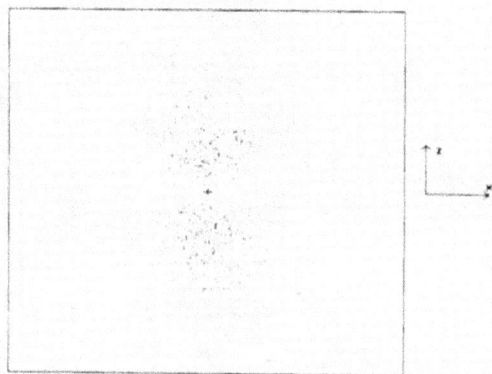

Figure II.1.45: Visualization of $2p_z$ orbital; Z = 10.

Orbital 3pz
rmax/ao : 2.00   M : 1000   z : 10

Figure II.1.46: Visualization of $3p_z$ orbital; Z = 10.

Figure II.1.47: Visualization of $3d_{z^2}$ orbital ; $Z = 10$.

Figure II.1.48: Visualization of $4d_{x^2-y^2}$ orbital; $Z = 10$.

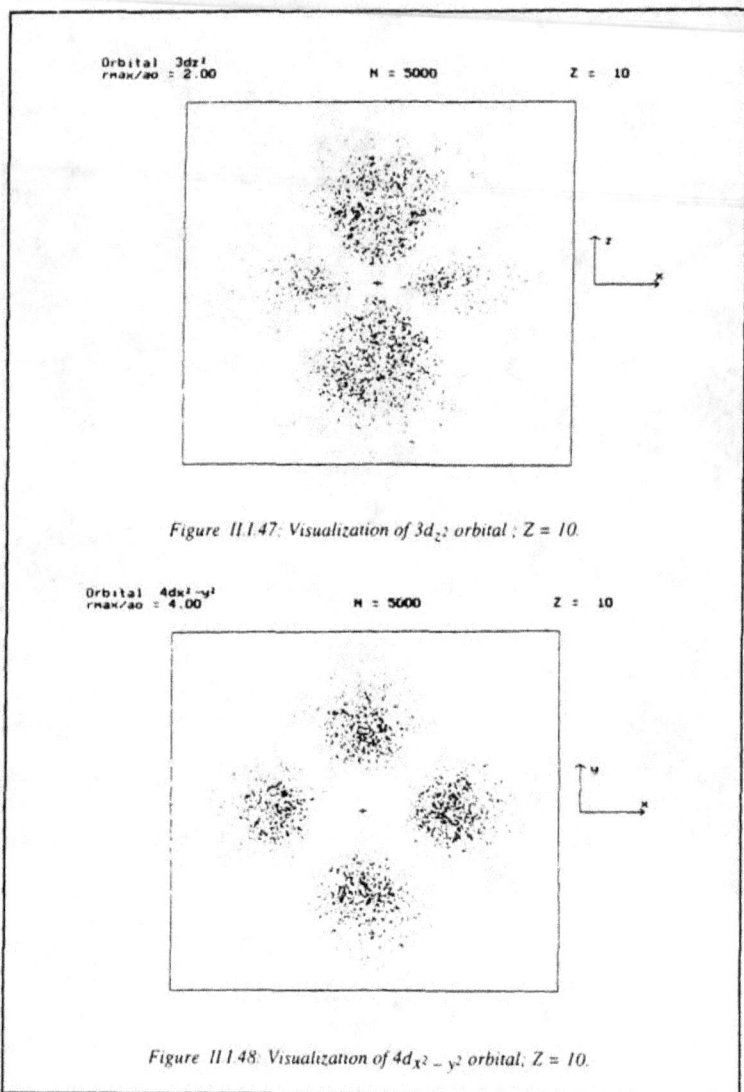

Representation of the radial density

The probability of finding the electron in the space between two spheres of radii r and $r + dr$ is calculated by integrating the product of the volumic density and the elementary volume expressed in spherical coordinates over the angle.

We find

$$r^2\,dr\,|R(r)|^2 \int_0^{2\pi} \int_0^\pi |\Theta\,(\theta)\,\phi\,(\phi)|^2 \sin\theta\;d\theta d\phi.$$

Taking the normalization of the function $\Theta\,(\theta)\,\phi\,(\phi)$ into consideration, the last equation reads $|R(r)|^2\,r^2 dr$, representing the probability of finding the electron between the spheres of radii r and r + dr.

$D(r) = |R(r)|^2\,r^2$ is called *radial probability density*

The following are examples of expressions of radial parts of AOs used later in this brilliant reference work.

TABLE II.1.6 — **Expressions of radial parts of hydrogenoid orbitals**

$$R_{n,\,l} = F_{n,\,l}\left(\frac{Z}{a_0}\right)^{3/2} \exp\left(-\,Zr/a_0\right)$$

Z is the atomic number; $a_0 = 0.053$ nm.

n	l	state	$F_{n,\,l}$
1	0	1s	$F_{1,\,0} = 2$
2	0	2s	$F_{2,0} = \dfrac{2\,a_0 - Zr}{2\sqrt{2}\,a_0}$
	1	2p	$F_{2,1} = \dfrac{Zr}{2\sqrt{6}\,a_0}$
3	0	3s	$F_{3,0} = \dfrac{54\,a_0^2 - 36\,Z\,a_0 r + 4\,Z^2 r^2}{81\sqrt{3}\,a_0^2}$
	1	3p	$F_{3,1} = \dfrac{4\,Zr\,(6\,a_0 - Zr)}{81\sqrt{6}\,a_0^2}$
	2	3d	$F_{3,2} = \dfrac{4\,Z^2 r^2}{81\sqrt{30}\,a_0^2}$

Figures II.1.49 and II.1.50 represent the radial density of the 1s and 2p orbitals of the hydrogen atom plotted using the expressions given in Table II.1.6. In the case of the 1s orbital, the maximum radial density occurs when $r = a_0 = 52.9$ pm.

Concerning the 2p orbital, the maximum occurs when r = $4a_0$. However, this representation gives no information about the electron angular distribution.

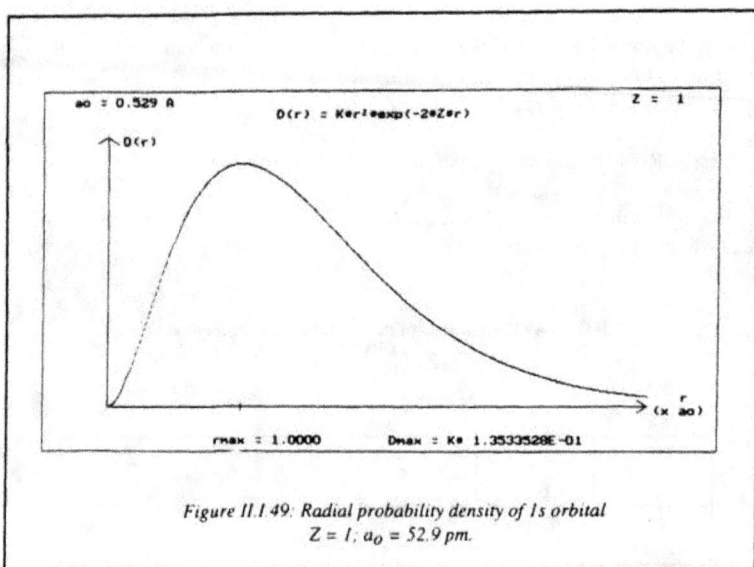

Figure II.1.49: *Radial probability density of 1s orbital*
$Z = 1$; $a_0 = 52.9$ pm.

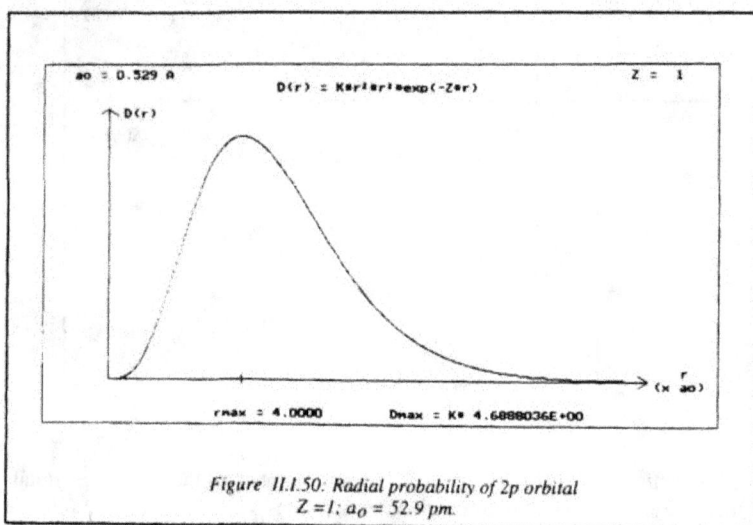

Figure II.1.50: *Radial probability of 2p orbital*
$Z = 1$; $a_0 = 52.9$ pm.

Representation of the angular dependence

In order to deal with this problem, let us set r and plot an $\overrightarrow{OM}$ vector in the direction defined by θ and φ so that $|\overrightarrow{OM}| = |Y_{lm}(\theta,\varphi)|^2$.

The following table gives the expressions of the angular parts of some AOs.

TABLE II.1.7 — **Expressions of the angular parts of some AOs.**

type of orbital	angular function
s	$1/2\ \sqrt{\pi}$
p_z	$(\sqrt{3}/2\ \sqrt{\pi}) \cos \theta$
p_x	$(\sqrt{3}/2\ \sqrt{\pi}) \sin \theta \cos \varphi$
p_y	$(\sqrt{3}/2\ \sqrt{\pi}) \sin \theta \sin \varphi$
d_{z^2}	$(\sqrt{5}/4\ \sqrt{\pi}) (3 \cos^2 \theta - 1)$
$d_{x^2-y^2}$	$(\sqrt{15}/4\ \sqrt{\pi}) \sin^2\theta \cos 2\varphi$
d_{xy}	$(\sqrt{15}/4\ \sqrt{\pi}) \sin^2\theta \sin 2\varphi$
d_{yz}	$(\sqrt{15}/2\ \sqrt{\pi}) \sin \theta \cos \theta \sin \varphi$
d_{zy}	$(\sqrt{15}/2\ \sqrt{\pi}) \sin \theta \cos \theta \cos \varphi$

The data in this table may be used to plot plane sections representing the angular distributions of $d_z{}^2$ and $d_{x^2-y^2}$ orbitals, e.g.,

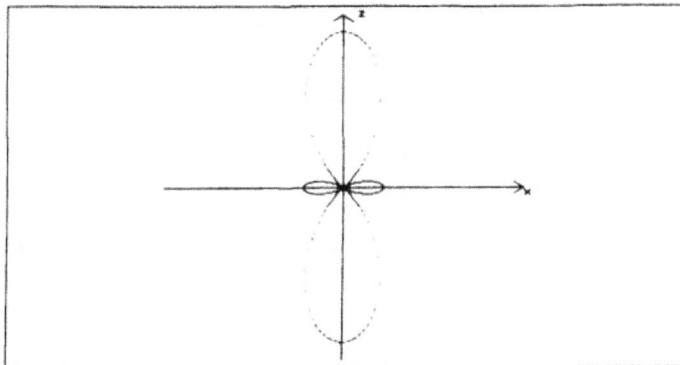

Figure II.1.51: *Angular dependence of $d_z{}^2$ orbital (representation in the xOz plane; $\varphi = 0$).*

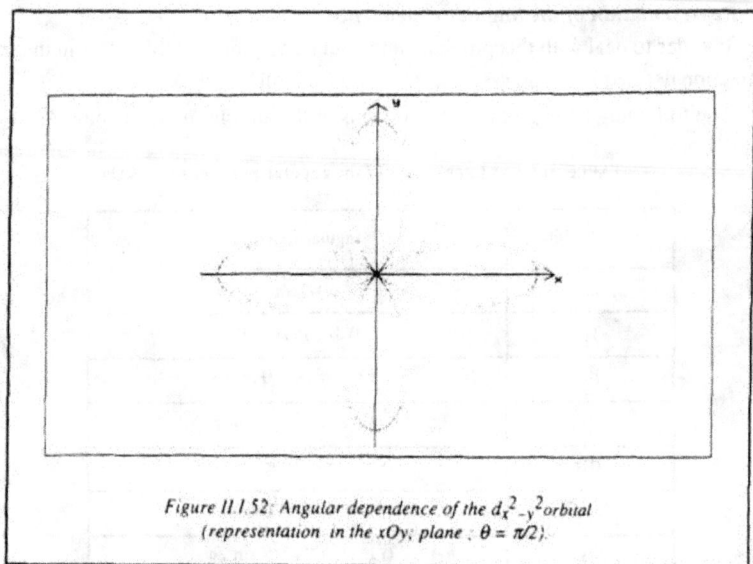

Figure II.1.52: Angular dependence of the $d_{x^2-y^2}$ orbital (representation in the xOy, plane ; $\theta = \pi/2$).

Indeed, we already mentioned in Sec.II.1.4.4 that those angular distributions give no information about the size of the electronic cloud.

8.3 How can atoms be modeled?

Prior to the formulation of quantum mechanics, this question would hardly have been justified. However, starting in the early 20th century, the chemists' indivisible, spherical atom was transformed, in the hands of physicists, into a "probable" cloud.

Today, the quantum concepts and their interpretation in terms of physics are still difficult to deal with and, even more so, to teach. This is why we shall start by presenting a brief *bibliographical approach*, illustrating the historical development of current ideas on atoms. The second part of this section gives details of how it is possible to *model* an atom by a sphere, both theoretically and experimentally. The limitations of these models will be clearly presented. Finally, we considered that it was very important to find out about the mental images the students built up from our teaching. The third part gives the somewhat surprising results of a *survey* of 12- to 15-year-old students.

Historical development of current ideas on atoms: Bibliographical approach

Most of these publications are in French but should not be difficult to find. Fig. II.1.53 below illustrates their distribution in time. It also shows the two categories into which the material may be divided:

— Publications by physicists, mainly around the 1930's.

— An explosion of articles, biographies, historical and philosophical discussions, as well as a number of publications aimed at the general public, as the late 1970's.

In our historical approach, we shall see to what extent these two groups correspond to two periods in quantum mechanics. It is also quite remarkable that publications and articles by physicists do not include any diagrams of atoms: only those publications aimed at the general public and school textbooks include this sort of illustration. Of course, this raises the question of the validity and pertinence of these images. Publications are mentioned in one of the following two ways:

— Extracts printed in the text, headed: (*Text n*).

— Bibliographical footnotes, giving the page number and a letter corresponding to the classification in the list of publications (e.g., B p.246).

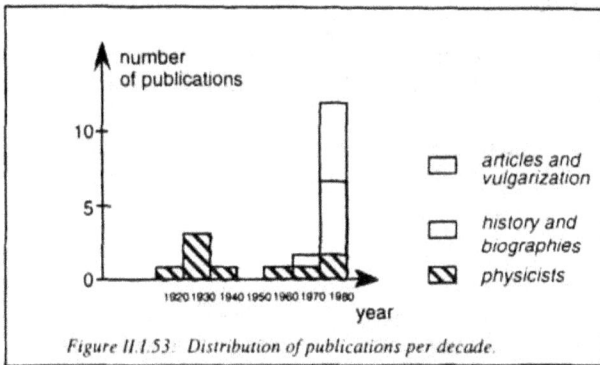

Figure II.1.53: Distribution of publications per decade.

The appearance in Greece of the idea of the atom is linked to a materialistic philosophy and does not seem to be based on a concern to describe nature (Text 1). This idea died out with the rise of spiritualistic thought.

Text 1: L p 15

"Atomic theory — For Leucippus, born about 480 BC, defining matter, and existence, i.e., Being, also implies defining nonbeing, nonmatter or a total vacuum. Matter and vacuum should therefore both be involved on the very level of the composition of the world. It follows that matter is not uniformly solid. It is neither homogeneous, nor continuous, but may be divided into solids and voids. The Greeks called these solid parts "atoms": That which may not be divided. Each atom is thus a complete, closed, universe, resembling Parmenides' global universe.

These ideas were developed in particular by Democritus (460-360 BC), a *student of Leucippus. Democritus thought that there was no reason to attribute a* specific shape to atoms. The immense diversity of geometric shapes that this implied, combined with the way the irregularity of atoms made it possible to assemble them, explained the formation of everything that existed".

In the 16th century, the atom reappeared, put forward by chemists to justify their interpretation of the behavior of matter (Text 2).

Text 2: L p 38

"In the 16th century, some interest began to be taken in atomism. Francis Bacon (1561-1626) seemed at one time to be favorable to atomic ideas dating back to Antiquity. Daniel Sennert (1572-1637) had more definite ideas, that were *clearly applied to chemistry. Sennert tried to explain the phenomena of chemical reactions by synthesizing ideas about corpuscular matter and the theory of elements*. In fact, he considered that the elements are made up of atoms and, in an advance on the ideas of Lucretius, these atoms are not restricted to a mechanical identity. They have a veritable chemical identity, that persists through transformations of matter. Water vapor is not water changed or transmuted into air, but particles of water escaped from the liquid. Ultimate particles may form second order combinations that Sennert calls *prima mixta*, prefiguring our molecules"

Chemical analysis even led to the combination of the idea of elements and atoms (Text 3), a foreshadowing of the Lavoisier system (L p.52).

Text 3: L p 39

"Junge thought that any substance that could not be decomposed should be considered as an element. This empirical concept would have clarified a good many ideas in chemistry if Junge's ideas had been followed. In the same way, he supposed that chemical bodies had a sort of power or appetite. Different bodies had different levels of "appetite" for one another, pushing them to form combinations"

The spiritualistic conception of the world was, however, still preponderant and gave rise to highly varied, and sometimes bizarre (L p.39), or even prophetic (I p.19) models.

The atom theory came back in force in the 19th century with the appearance of quantitative chemistry, proposed initial by Dalton (Text 4). It was later reinforced by Avogadro (H p.127), Cannizzaro (L p.66), and Mendeleev (H p.133).

Text 4: H p 123

"All bodies of an appreciable size, liquids as well as solids, are made up of a multitude of infinitesimal particles, or atoms of matter bound together by a more or less strong force of attraction, according to the circumstances.

It was, however, hardly possible to imagine how aggregations of dissimilar particles could be so uniformly identical. If certain water particles were heavier than others, and if, occasionally, a quantity of liquid consisted mainly of these heavier particles, it is to be supposed that this would affect the specific gravity of the mass, which would, therefore, be an unknown circumstance. [...] We must conclude that the ultimate particles of all homogeneous bodies are perfectly identical in terms of weight, pattern, etc. In other words, any water particle is the same as any other water particle ; any hydrogen particle is identical to all other hydrogen particles, etc."

This explanation met with strong opposition from " experimentalists " (Text 5 and H p.130, 0 p.199 and p.204) who criticized its speculative aspects and "energetists" who denied the existence of matter (Text 6 and P p.22).

Several, more marginal, opposing models were also put forward (R p.240 and M p.223). The "equivalentists" (R p.235) maintained their position until the end of the century.

Text 5: R p 229

"In the early 19th century, Dalton's theory received a rather cold welcome from the most influential scientists of the period: Laplace and Berthollet. In 1837, the chemist Jean-Baptiste Dumas declared: "If I were in charge, I would erase the word *atom* from science, as I am convinced it has wider implications than experiments allow" Wurtz, Dumas' successor at the Medical School, was a determined atomist. However, even after 1860, he had to fight to obtain the recognition of atoms and molecules. Among his opponents was Sainte-Claire Deville: "I do not accept Avogadro's law, nor atoms, nor molecules, as I refuse to believe in things I can neither see nor imagine" For many years, Marcelin Berthelot also maintained his obstinate denial"

Text 6: M p 222

"Matter is an invention, and a rather imperfect one, that we have forged to represent that which is permanent through all types of mutations. Reality, i.e., that which truly has an effect on us, is energy"

However, Dalton's initial theory bore fruit and was used by other authors (Text 7 and M p.204) long before it was generally accepted.

Text 7: R p 238

" People like Humphry Davy and Thomas Graham shared the same "unitarian" views. Indeed, Graham held that all matter was made of a single constituent; all the differences we observe between elements were due to differences in the conditions of the *movement* of this ultimate particle. He proposed a dynamic model, which also seems quite modern today: "The faster the movement, the larger the space occupied by the atom, in a similar way that the orbit of a planet depends on its speed"..."

At the turn of the century, Rutherford's experiment, in conjunction with the discovery of the electron a few years earlier, made it possible to put forward the first planetary model of the atom (Text 8).

Text 8: T p 23

"In order to understand the audacity and impact of Bohr's thought, as expressed in his 1913 article, we have to go back to the famous model of the atom proposed two years earlier by the Australian physicist Ernest Rutherford, winner of the Nobel prize for chemistry in 1908, teaching, at that time, at Manchester university. We know that Rutherford suggested considering atoms like solar systems, with electrons spinning around an atomic nucleus in exactly the same way as the planets orbit the sun. The force of attraction that took the place of gravity in this model was electrical attraction between the negatively charged electrons and the positively charged nucleus.

Starting from this model, a certain number of predictions could be made about the behavior of electrons. It is interesting to note that, in many instances, they turned out to be true. For example, the time taken by electrons to make one revolution, which could be deduced from the frequency of the light emitted by atoms, corresponded, more-or-less, to the value that was expected in view of the dimension of the orbits, deduced from the dimensions of the atoms themselves"

Perfecting the quantum model

New work in physics led to the current vision of the atom. For a simplified first approach, we may consider that the discovery of the basic principles was spread over the period from 1900 to 1927, date of the famous Solvay conference, where quantum mechanics was adopted and defined.

The most important dates are:

— 1900:   Planck introduced the quantification of energy.
— 1905:   Einstein introduced the quantification of light.
— 1910:   Rutherford presented his planetary model of the atom.
— 1913:   Bohr introduced quantification to the atom (Text 9 and T p.26, A p.296, 304 and 310, B p.215, 219 and 221).

Text 9: E p 183

" Any electrically neutral atom is made up of a certain number of protons, firmly attached to each other, and an equal number of electrons. Some of these electrons are strongly bonded to the protons to form the atomic nucleus, while the other electrons revolve around the atomic nucleus.

In this way, the smallest atom, hydrogen, is made up of a single proton and a single electron revolving around it, while the largest atom, uranium, consists of

238 protons and as many electrons. However, only 92 of the electrons move around the nucleus, as the others are in the nucleus itself. All the other elements, presenting all the possible combinations of protons and electrons, come somewhere between these two. The chemical nature of an element is not determined by its total number of protons and electrons, but only by the number of mobile electrons it has. This is known as its atomic number "

— 1923: De Broglie proposed wave-particle duality on the atomic scale (Text 10 and E p.190 F p.84, K p.29).

Text 10: T p 28

"What an unexpected, exciting situation: specific atomic states could be described as harmonic vibrations of electron waves, under the influence of the electric force of the nucleus. The specific properties of the various elements could thus be due to a natural property of these vibrations"

— 1925: Schrödinger and Heisenberg independently developed the first quantum mechanics (F p.143 and 157).
— 1927: Solvay Conference "Electrons and Photons" (Text 11).

Text 11: P p 129

" At the Como and Solvay meetings in the fall of 1927. Bohr presented the second half of the Copenhagen interpretation as a definitive statement.
It is based on three main postulates:

— A physics theory describes the results of experiments and not the nature of things.
— Reports must describe experimental protocols and results in the terms of classical physics.
— Finally, the double language of wave and particle theories describes all the possible interactions between experimenters and microscopic entities "

Quantum mechanics provides and accounts for the various microscopic phenomena not explained by classical physics. At the same time, it became harder and harder to understand, due to the fact that it abandoned the normal concepts of classical physics. The impossibility of producing a concrete representation raised misgivings for some of the founder members — Planck, Einstein, de Broglie and Schrödinger (Text 12 and C p.257, K p.37) — and continued to generate many attempts to produce representative models (Texts 13 and 14 and I p.93, L p.110, F p.37).

Text 12: K p 52

" Our scientific hopes have led us to opposing positions. You believe in the God who plays dice, and I believe only in the value of laws in a universe where something exists objectively, that I try to apprehend in a wildly speculative way [...]. The great success of quantum theory from the very beginning cannot make me believe in this fundamental game of dice, although I know that my younger colleagues see this as a sign of fossilization. One day, we shall find out which of these two instinctive attitudes was right " (letter from A. Einstein to M. Born, 7.9.1944) "

Text 13: F p 63

"Imagine the electron orbits to be trolley tracks and yourself the contractor who must build them. Deciding to start with the tracks for an electron of some particular speed, you take from your pocket a well-thumbed copy of *Everybody's Manual of Electron Trolley Track Construction* and look up the size of the circle needed for that speed. Next you telephone the factory and tell them to send you some track capable of withstanding the speed you are interested in. They send it to you in segments, and at once you realize that you have on your hands a harder problem than you thought. For, because of unusual manufacturing difficulties, the factory is able to make the track only in segments of a particular length depending on the speed to be withstood. The factory thus has a manual of its own, and it does not agree any too well with yours. For example, your track must have a length of seventeen units, but for this speed the factory can make track only in segments of three units length. Though flexible, the track is so tough as to be quite unbreakable ; and three doesn't go into seventeen. What will you do ? You cannot build the orbital track you want because there would be an overlap. You will simply have to declare that particular orbit impossible to construct. That is no way to make money, though. You will have to try your luck again with a different speed.

This one turns out to need a circle of length twenty-five units, but the factory happens to construct track for it only in lengths of four units. Again the segments don't fit the circle, and you begin to despair of ever being able to construct an orbit. However, a systematic search of both manuals reveals several possibilities . There is one in which the segment happens to be exactly the length of the circumference. There is a larger one where the corresponding segment is just half the circumference. Another, yet larger, where the segments step around the circumference just three times. Another four times, and so on without end. These are the permitted orbits, all others being forbidden.

... no two electrons may have the same set of four quantum numbers. It is as if the Bohr atom were a large city where electrons live in separate apartments. Each apartment has a different address, one quantum number indicating the street, another the house, a third the floor, and the fourth the apartment. These four quantum numbers are, then, the complete address of each apartment, and Pauli's principle is a regulation against overcrowding. Indeed, it is technically referred to as the exclusion principle. Because of it only one electron at a time may inhabit an apartment, another electron being forbidden entry until the first moves out."

Text 14: J p 83

" Nowadays, we tend to think that this famous "wave/particle duality" was perhaps also a transitory way of seeing things. We have realized that, from a certain point of view, electrons are like standard particles, or little pebbles, because we can count them, for example, as we can count pebbles. On the other hand, these same electrons are amazingly like waves — fuzzy, dispersed, and ubiquitous. In a way, electrons are everywhere, in the same way as you cannot locate one wave system in the sea— the sea is full of waves !

So, in some ways, electrons are like particles, while, in other ways, they seem like waves. We have therefore become resigned to thinking in terms of duality. Of course, this idea had its usefulness at the time, but it is as much of an oversimplification as describing a cylinder by what we could call a "rectangle/circle duality" Take a cylinder such as a can. If you look along it, you see a circle, but if you look perpendicularly across it, you see a rectangle. So, is it useful to describe a can by means of a circle/rectangle duality ? "

Among the major readjustments, we should mention the issues involved in renormalization, which was only really settled in the late 1940's (F p.263 and K p.106), and experimental verification of the existence of particles predicted in theory. At present, the main objectives concern the issues of unification, nucleon structure and microscopic/macroscopic boundaries (S p.78 and 115 and N p.118). Although the users of quantum mechanics now agree that it is pertinent, the issues of its interpretation or completeness are still open to debate (Text 15 16 and 17 and E p.182, F p.175, T p.30 and 56).

Text 15: F p 176

"That some prefer to swallow their quantum mechanics plain while others gag unless it be strongly seasoned with imagery and metaphysics is a matter of individual taste behind which lie certain fundamental facts which may not be disputed; hard, uncompromising, and at present inescapable facts of experiment and bitter experience, agreed upon by all and directly opposed to the classical way of thinking:

There is simply no satisfactory way at all of picturing the fundamental atomic processes of nature in terms of space and time and causality. The result of an experiment on an individual atomic particle generally cannot be predicted. Only a list of various possible results may be known beforehand. Nevertheless, the statistical result of performing the same individual experiment over and over again an enormous number of times may be predicted with virtual certainty."

Text 16: G p 22

"Another most interesting change in the ideas and philosophy of science brought about by quantum mechanics is this: it is not possible to predict *exactly* what will happen in any circumstance. For example, it is possible to arrange an atom which is ready to emit light, and we can measure when it has emitted light by picking up a photon particle, which we shall describe shortly. We cannot, however, predict *when* it is going to emit the light or, with several atoms, *which one* is going to.

Returning again to quantum mechanics and fundamental physics, we cannot go into details of the quantum-mechanical principles at this time, of course, because these are rather difficult to understand. We shall assume that they are there, and go on to describe what some of the consequences are. One of the consequences is that things which we used to consider as waves also behave like particles, and particles behave like waves; in fact everything behaves the same way. There is no distinction between a wave and a particle. So quantum mechanics *unifies* the idea of the field and its waves, and the particles, all into one. Now it is true that when the frequency is low, the field aspect of the phenomenon is more evident, or more useful as an approximate description in terms of everyday experiences. However, as the frequency increases, the particle aspects of the phenomenon become more evident with the equipment with which we usually make the measurements. In fact, although we mentioned many frequencies, no phenomenon directly involving a frequency has yet been detected above approximately 10^{12} cycles per second. We only *deduce* the higher frequencies from the energy of the particles, by a rule which assumes that the particle-wave idea of quantum mechanics is valid."

Text 17: J p 79

" It is not possible to attribute *one* position to an electron, in the same way as, to use a rather accurate metaphor, it is not possible to attribute *one* color to sunlight. The light that reaches us from the sun is not a pure, well-defined color, but a set of colors. If it passes through a prism, we see that it is a mixture of red, yellow, etc., through to blue and violet. It has a *spectrum* of colors. In quantum theory, it is the same for most physical quantities. Nowadays, we think that quantum objects are identified, not by specific numerical values for their physical quantities, but by *spectra*. An electron, to put it rather crudely, does not have *one* position but several positions at once "

As far as quantum atoms are concerned, we can summarize the following points:

— The most pertinent image of microscopic matter is a mathematical representation, divorced from sensory perceptions, called a "wave function" (Text 18 and C p.307, G vol. 1 p.451, Q p.126).

Text 18: D p 228

" In fact, our notions of space and time drawn from daily experience are only valid for large-scale phenomena. We need to substitute other basic concepts that are valid in microphysics, and which will lead us asymptotically back into the normal concepts of space and time, when we move from elementary phenomena to phenomena that are observable on our macroscopic scale. Do we need to say that this is an extremely difficult task ? We can even wonder if it could ever be possible to eliminate the very framework of our everyday lives, but the history of science demonstrates the extraordinary fertility of the human mind, so we should not give up trying. However, until we manage to widen our concepts in that direction, we will continue to try, more or less clumsily, to make microscopic phenomena fit into the space – time framework, and we will have the uncomfortable feeling that we are trying to put a precious stone in a setting that was never made for it ".

— A measurement can cause a considerable disturbance to a quantum object and implies a probabilistic knowledge of this object.

— The uncertainty principles exclude the possibility of knowing precisely and at the same time the position and speed of a particle, or of knowing its energy with any level of accuracy, unless it remains in the same stationary state for an infinite length of time. The very existence of atoms is bound up with these uncertainty principles (TEXT 19 and J p.77, K p.42). This is possible thanks to quantum interactions (TEXT 20 and N p.119).

Text 19: D p 260

"The Pauli exclusion principle expresses a very unusual property of electrons and other particles that are subject to this condition. It is, at present, practically impossible to understand how two identical particles may prevent each other from being in the same state. This type of interaction is totally different from those in classical physics and its physical nature is still unknown to us. It would seem that one very important, and certainly very difficult task, for theoretical physics in the future is to give us a clear idea of the physical origins of the exclusion principle.

To show how far we are in this field from the old concepts, let us consider the case of a gas made of particles of the same type, subject to the Pauli principle, e.g. an electron gas. According to the exclusion principle, it is impossible for two electrons in the gas to be in the same state of uniform movement in a straight line, as the states quantified here are uniform and straight line movement. According to classical concepts, this would mean that a particle at one point in the container of gas could prevent any other particle in the gas from being in the same state. This is totally paradoxical, as container full of gas may be assumed to be any size and, consequently, the molecules may be any distance apart. However, this paradox is intimately related to the Heisenberg uncertainty equations and disappears if we take them into account. In fact, the states of uniform and straight line movement of the particles correspond to specific energies of these particles. According to the uncertainty equations, we cannot talk about the states of movement of two particles and their positions at the same time. The fact that we consider the energy states of the particles to be defined means that we can no longer talk of their being a certain distance apart, as they can no longer be localized. This example shows that, in order to make a physical interpretation of the exclusion principle, we must leave classical images far behind ".

Text 20: G p 21

"Quantum mechanics has many aspects. In the first place, the idea that a particle has a definite location and a definite speed is no longer allowed; that is wrong. To give an example of how wrong classical physics is, there is a rule in quantum mechanics that says that one cannot know both where something is and how fast it is moving. The uncertainty of the momentum and the uncertainty of the position are complementary, and the product of the two is constant. We can write the law like this: $\Delta x\, \Delta p \geq h/2\pi$, but we shall explain it in more detail later. This rule is the explanation of a very mysterious paradox: if the atoms are made out of plus and minus charges, why don't the minus charges simply sit on top of the plus charges (they attract each other) and get so close as to completely cancel them out? *Why are atoms so big?* Why is the nucleus at the center with the electrons around it? It was first thought that this was because the nucleus was so big; but no, the nucleus is very *small*. An atom has a diameter of about 10^{-8} cm. The nucleus has a diameter of about 10^{-13} cm. If we had an atom and wished to see the nucleus, we would have to magnify it until the whole atom was the size of a large room, and then the nucleus would be a bare speck which you could just about make out with the eye, but very nearly *all the weight* of the atom is in that infinitesimal *nucleus*. What keeps the electrons from simply falling in? This principle: If they were in the nucleus, we would know their position precisely, and the uncertainty principle would then require that they have a very *large* (but uncertain) momentum, i.e.,, a very large *kinetic energy*. With this energy they would breack away from the nucleus. They make a compromise: they leave themselves a little room for this uncertainty and then jiggle with a certain amount of minimum motion in accordance with this rule. (Remember that when a crystal is cooled to absolute zero, we said that the atoms do not stop moving, they still jiggle. Why? If they stopped moving, we would know where they were and that they had zero motion, and that is against the uncertainty principle. We cannot know where they are and how fast they are moving, so they must be continually wiggling in there!)."

Text 21: N p 136

"The simplest atom, called hydrogen, is a proton and an electron. By exchanging photons, the proton keeps the electron nearby, dancing around it (see Fig. 65).

Now, I'd like to show you a diagram of an electron in a hydrogen atom scattering light (see Fig. 66). As the electron and the nucleus are exchanging photons, a photon comes from outside the atom, hits the electron and is absorbed;

then a new photon is emitted. (As usual, there are other possibilities to be considered, such as the new photon is emitted before the old photon is absorbed.)"

— In certain cases, atoms are only perceptible through exchanges of discrete quantities of energy (atomic spectra of bound states).

— If we make an approximate representation of a polyelectronic system , such as an atom, by superposing single-electron quantum states, Pauli's " exclusion principle " must be satisfied, prohibiting us from attributing the same state to two electrons (Text 21).

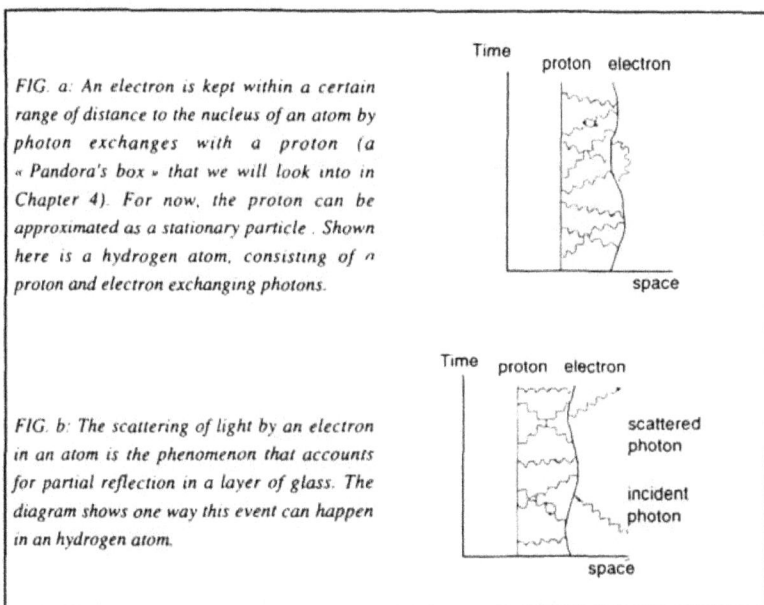

FIG. a: An electron is kept within a certain range of distance to the nucleus of an atom by photon exchanges with a proton (a « Pandora's box » that we will look into in Chapter 4). For now, the proton can be approximated as a stationary particle . Shown here is a hydrogen atom, consisting of a proton and electron exchanging photons.

FIG. b: The scattering of light by an electron in an atom is the phenomenon that accounts for partial reflection in a layer of glass. The diagram shows one way this event can happen in an hydrogen atom.

List of publications mentioned in the text

A   PERRIN J. — *Les Atomes.* Paris, Félix Alcan (1924).

B   BOUTARIC A. — *La vie des atomes.* Paris, Flammarion (1933).

C   EINSTEIN A., INFELD L. — *L'évolution des idées en physique.* (1936) et Paris, Flammarion (1983).

D :   DE BROGLIE L. — *La physique nouvelle et les quanta.* Paris, Flammarion (1937).

E :   PLANCK M. — *Initiations à la physique.* Paris, Flammarion (1941).

F   HOFFMANN B., PATY M. — *L'étrange histoire des quanta*. Paris, Seuil (1967).

G   FEYNMAN, LEIGHTON, SANDS — *Le cours de physique de Feynman, vol.1
         Mécanique*. Paris, Interéditions (1979).

H   CANGUILHEM G. — *Introduction à l'histoire des sciences, vol.1 éléments et
         instruments*. Paris, Hachette (1970).

I   THUILLIER P. et DELPECH J.F. — *La matière aujourd'hui*. Paris, Seuil (1981).

J   LEVY-LEBLOND J.M. — *L'espace et le temps aujourd'hui*. Paris, Seuil (1983).

K   DELIGEORGES S. et al. — *Le monde quantique*. Paris, Seuil/Sciences et
         Avenir (1984).

L   VIDAL B. — *Que sais-je ? Histoire de la chimie*. Paris, Presses Universitaires de France
         (1985).

M   ROSMORDUC J. — *Une histoire de la physique et de la chimie*. Paris, Seuil (1985).

N   FEYNMAN R. — *Lumière et matière*. Paris, Interéditions (1987).

O   JACQUES J. — *Berthelot*. Paris, Belin (1987).

P   HEILBRON J. L. — *Planck*. Paris, Belin (1988).

Q   CARATINI R. — *L'année de la science*. Paris, Seghers (1987 et 1988).

R   THUILLIER P. — *D'Archimède à Einstein*. Paris, Fayard (1988).

S   COHEN-TANNOUDJI G. — *La symétrie aujourd'hui*. Paris, Seuil (1989).

T   WEISSKOPF V. — *La révolution des quanta*. Paris, Hachette (1989).

A spherical model of an atom : Choosing the radius
Reducing the wave function to a single-electron function

As we have said, the most comprehensive representation of a quantum object such as an atom is the purely mathematical representation provided by its wave function ψ. (We assume in the following explanation that the atom is at its lowest energy— or basic— state and that it is possible to obtain accurate wave functions, which is certainly the case for light atoms.) It is difficult to make a more-or-less concrete representation of a wave function that can be perceived by our senses or at least one that would be easy to understand. In fact, it is a mathematical entity, that becomes ever more terrifyingly complex as the number " n " of electrons increases, even if we admit simplifying hypotheses (e.g. fixed, localized nucleus, nonrelativistic mechanics, etc.). The square of the module of this function gives the probability of finding at the same time a first electron at point M_1 with a spin α or β, a second electron at a point M_2 with a spin α or β, etc. It is therefore necessary to replace this exhaustive representation with a more limited one, at the risk of losing a lot of information, which is not essential for an initial approach.

One efficient method consists of determining the probability density of one electron, e.g. electron 1, whatever its spin, at a point in space, while the other electrons may be anywhere with any spin. From a mathematical point of view, this is equivalent to integrating $|\Psi|^2$ in the entire space over the volume elements of the other electrons, and finding the sum of the spin variables σ of all the electrons. This gives the following equation:

$$\rho_1(\mathbf{r}_1) = \sum_{\sigma_1} \cdots \sum_{\sigma_n} \int \cdots \int |\psi|^2 dv_2 \cdots dv_n$$

The function for just the space coordinates of electron 1 obtained here is the same (other than the name of the variables) as that we would have obtained for the other electrons. This property, resulting from the anti-symmetricality of the wave function, is linked to the indistinguishability of electrons. It thus appears that the function

$$\rho(\mathbf{r}) = n\rho_1(\mathbf{r})$$

which will be used in Chapter III (cf., Secs. III.I.2 and III.II.4), represents the probability space density of finding any one electron at a point M, defined by the radius vector $\mathbf{r}$. The product of this function multiplied by the electron charge q, that is often wrongly assimilated with the electric charge density, is, in fact, the average charge density at point M that would be obtained by making a very large number of measurements.

Although it is much simpler to represent the $\rho(\mathbf{r})$ function than that the ψ function, it is still no trivial matter, as it is necessary to represent the values taken by ρ in function of the three variables r, θ, φ (in polar coordinates). However, as the nucleus potential has a spherical symmetry, there is *a priori* no reason to choose a function that does not have spherical symmetry. This is, in effect, what happens in the case of atoms with " full layers " or " full layers + s electrons ". For other atoms, which, following symmetry, are in a degenerate state, it is always possible to choose an adequate linear combination of the wave functions defining this state to ensure that this is so. If $\mathbf{r}$ is expressed in polar coordinates r, θ, φ, the probability space density $\rho(\mathbf{r}) = \rho(r)$ is no longer dependent solely on the r variable. We can also introduce another function of r, the probability radial density, $\rho_r(r)$, defined in such a way that $\rho_r(r)\, dr$ is the probability of finding one electron at a distance between r and r + dr, whatever the spherical polar variables θ and φ. This function is linked to the previous one, as $\rho(r)$ is assumed to be independent of θ and φ and, therefore, only depends on r, according to the following equation:

$$\rho_r(r)dr = \int_0^{2\pi} \int_0^{\pi} \rho(r) r^2 \sin\theta \, dr \, d\theta \, d\varphi$$

or
$$\rho_r(r) = 4\pi r^2 \rho(r)$$

All that remains to be done at this point is to plot the functions $\rho(r)$ and $\rho_r(r)$ to obtain additional information. The characteristics of the statistical distributions corresponding to these functions for the hydrogen atom, are given in Appendix 3 of Sec. II.I.8 where $\rho(r) = |1s|^2$ They highlight the differences between values corresponding to the most probable electron-nucleus distance [maximum value of $\rho_r(r)$], average radius $\bar{r}$ and average square radius $\sqrt{r^2}$. The high standard deviation, as compared to the values for $\bar{r}$ and $\sqrt{r^2}$, also shows that the electron cloud is much too diffuse to make it possible to reduce the representation to a simple sphere corresponding to the average radius or the most probable electron-nucleus distance. A study of the electron densities of the other atoms leads to similar conclusions.

The preceding conclusions, coupled with the fact that the $\rho(r)$ and $\rho_r(r)$ functions only cancel each other out at infinity [and at $r = 0$ in the case of $\rho_r(r)$], mean that it is difficult to attribute accurate dimensions to an atom. There is simply an area in space where these probabilities have a considerable value. It is quite evocative to plot the spheres inside which the probability p of finding the electron(s) equals $p = 0.9 ; 0.95 ; 0.99....$

The radii, r_p, of these spheres are calculated from the integral equation

$$\int_0^{r_p} \rho(r)4\pi r^2 dr = p$$

Of course, the greater the number of electrons the closer the chosen value of p must be to 1, so that the outer electrons, which play a fundamental role in many properties, are included in the sphere.

Estimating atom radii using an approximate quantum approach: Slater radii

Slater suggested modeling the radial part of polyelectron atom orbitals by a function that is expressed as follows, with the exception of the standardization factor:

$$R_n(r) = \left(\frac{r}{a_0}\right)^{n-1} \exp\left(- Z^*r/na_0\right)$$

where n is the main quantum number (above $n = 3$, it must be replaced by lower effective values of n^*), a_0 is equal to 1 Bohr and Z^* is the " total charge " of the nucleus (the real charge less a dummy charge representing the average effect of the other electrons, expressed by taking |q| as a unit and depending on the orbital considered). The reader will find the rules for calculating n^* and de Z^* in specialized publications. The Slater radius is the name given to a radius $R_S(n)$ of an orbital with a quantum number n, with the value of r that maximizes the radial density $4\pi r^2|R_n(r)|^2$, as shown in Appendix 3.8.1. We find the following:

$$R_S(n) = \frac{n^2}{Z^*} a_0$$

For example, for a sodium atom ($Z_{1s}^* = 10,6$ et $Z_{3s}^* = 2.2$) , the corresponding Slater radii are, respectively,

$$R_S(1) = 5 \text{ pm} \qquad \text{and} \qquad R_S(3) = 216 \text{ pm}$$

I his approximation is a very simple method for calculating the radial density of valence orbitals (still known as outside orbitals), from which we deduce the dimensions of atoms, and their order of magnitude.

Experimental modeling

Another way of perceiving the dimensions of an atom consists of studying the approach distance of another atom not chemically bound to the first. This type of study may naturally be carried out theoretically, but certain experiments (diffraction of X-rays by crystals, compressibility of gases, atom collisions, etc.) also provide useful information, and, historically speaking, the first data on this subject was experimental. The results show that two atoms that are not capable of creating a chemical bond are weakly attracted to one another when they are relatively far apart, by forces, known as Van der Waals forces, which vary in inverse proportion to the seventh power of their distance R. This corresponds to a potential attraction energy, expressed as: $V_{att} = - A/R^6$. On the other hand, when they are very close, they are subject to repulsive forces derived from a potential energy that is often expressed as $V_{rep} = Be^{-\beta R}$. The total potential energy $V(R) = Be^{-\beta R} - A/R^6$ is represented in function of R in Fig. II.I.54.

The graph (a) is similar to the one corresponding to a chemical bond, but with a much weaker bonding energy ΔE and a much larger equilibrium distance R_{eq}. Of course, this distance depends on the atom that is approaching the atom under investigation, but the repulsive part of the potential energy varies so rapidly with the distance R that, on first approximation, this part of the curve seems vertical [Fig. II.I.54 (b)]. In other words, the atoms behave like two balls that are not capable of interpenetration: This is known as the *"hard spheres* approximation " Experimental data is then interpreted by assuming that atoms may be considered as balls with a constant radius, known as the *Van der Waals radius.* Some examples are given in Table II.I.8.

Two atoms that are not bound to each other, but may be bound to other atoms, in the same molecule or not, behave in a similar way. On the other hand, when two atoms are chemically bound, the interatomic distance is always less than the sum of their Van der Waals radii. These interatomic distances, depending on multiple bonds, make it possible to determine the *covalent radii* (see Table II.I.8).

These rather rough approximations are used as the base for concrete models to represent molecules, and are used to show that certain combinations are impossible due to steric overload.

Modeling and molecular mechanics software, using the above types of potential energy functions, corrected by additional interaction terms, lead to more realistic and increasingly successful molecular geometries and energies corresponding to the various combinations. They also make it possible to display the patterns under study.

TABLE II.1.8 — Van der Waals radii and covalent " radii " for several atoms (according
to L. Pauling, The nature of chemical bond,
Presses Universitaires de France, Paris 1949).

Atoms	Van der Waals atomic radii (nm)	Covalent atomic " radii " (nm)		
		Single bonds	Double bonds	Triple bonds
H	0.12	0.030		
C		0.077	0.0665	0.0602
N	0.15	0.070	0.060	0.0547
O	0.14	0.066	0.055	
F	0.135	0.064		
Cl	0.180	0.099		

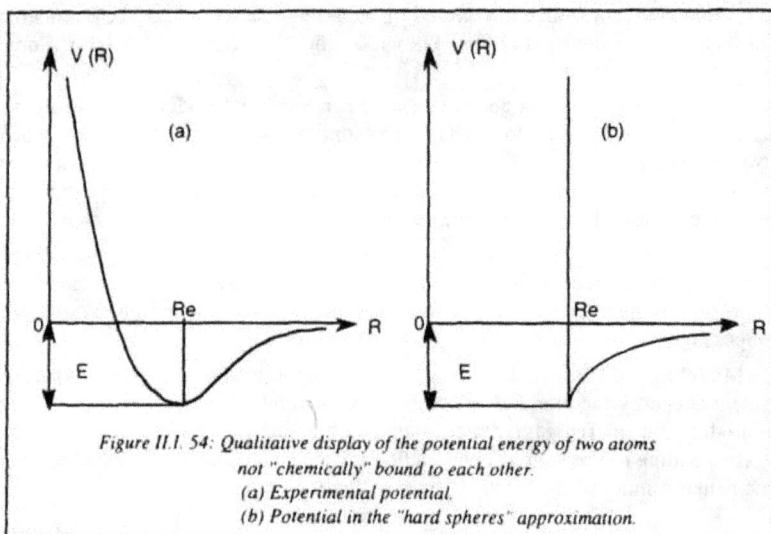

Figure II.1. 54: Qualitative display of the potential energy of two atoms
not "chemically" bound to each other.
(a) Experimental potential.
(b) Potential in the "hard spheres" approximation.

How do students from 12- to 15-years old see atoms ?
(survey of 12- to 15-year old students)

In 1989, we carried out a survey of 200 students to find out how much they had learned about atomic structure. The question was: *" How do you imagine the structure of an atom ? "*

The survey was answered by 55 students aged 12, 69 students aged 13 and 78 students aged 14. The 12 - 14-year-old students were taught according to the 1988 program, while the older group followed the previous syllabus.

The answers were divided into three main categories:

— *" Internal "* representation: Students imagined the structure of the atom in a more or less pertinent way.

— *" External "* representation: Students had no definite mental image of the structure of the atom, only a generalized picture.

— *" Complex or ambiguous"* representation: Students used knowledge from outside the subject under discussion to imagine the atom.

The way the students chose to answer (diagram, text, or both) is shown by the graph in Fig. II.I.55.

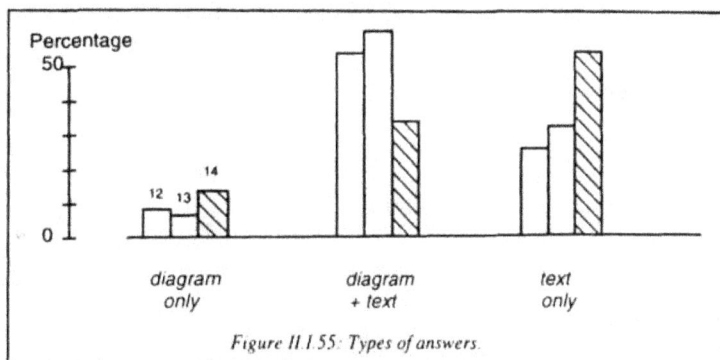

Figure II.I.55: Types of answers.

The following are some examples of the results:

12-year-olds

— *" Internal "* representation (Fig. II.I.56).

As they had not been given any information on atomic structure in class, the students' images reflect their general knowledge and (or) their imagination.

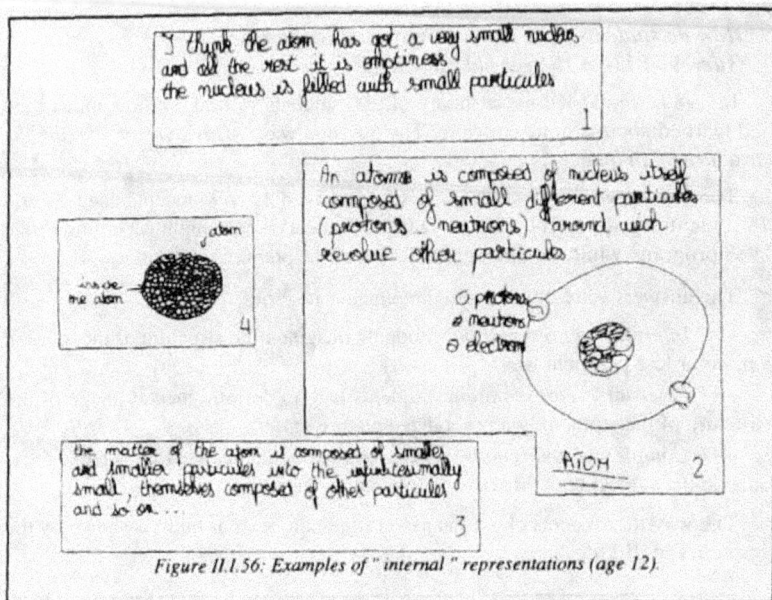

Figure II.1.56: Examples of " internal " representations (age 12).

Note:
— The existence of specific structures, sometimes given names (nucleus, electrons) (1).
— The use of a planetary model in some cases (2).
— The very frequent occurrence of recurrent reasoning, based on interlocking or clustered structures of particles inside the atom (3 and 4).

— *" External "* *representation* (Fig. II.1.57).

In accordance with their acquired knowledge, some of the students based their representation on new information. In this case, their answers to the question were biased, the students prudently restricted themselves to picturing an elementary ball with a clear, rigid outline, that may be solid or hollow, colored or transparent (5 and 6).

We recognized the characteristics of the plastics models used in class. Groups of more or less assembled atoms were also presented, images of compact molecular models (7).

The implicit progression suggested by the syllabus, meant that some students apparently assimilated atom to gas or even dust particles, in suspension in the air, flying around in Brownian motion (8).

This confusion was bound to cause problems the following year, when the curriculum included the atomic structure of metals.

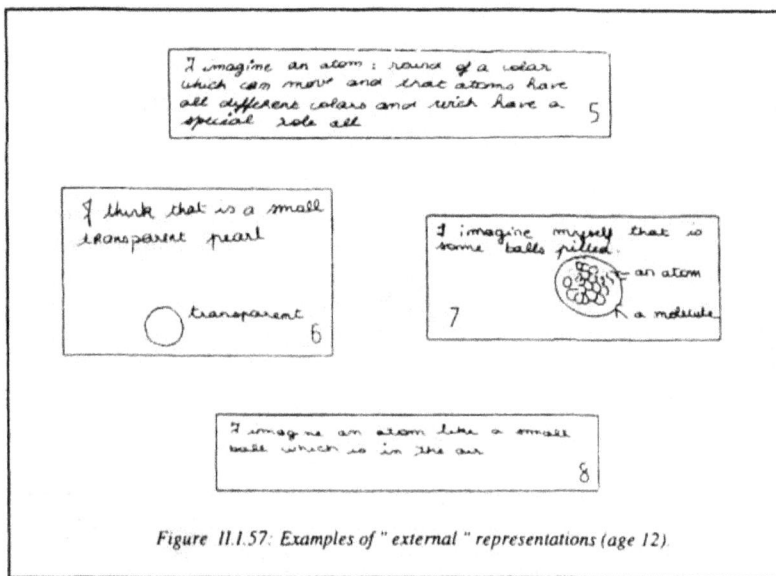

Figure II.1.57: Examples of " external " representations (age 12).

— " Complex or ambiguous' representation (Fig. II.1.58).

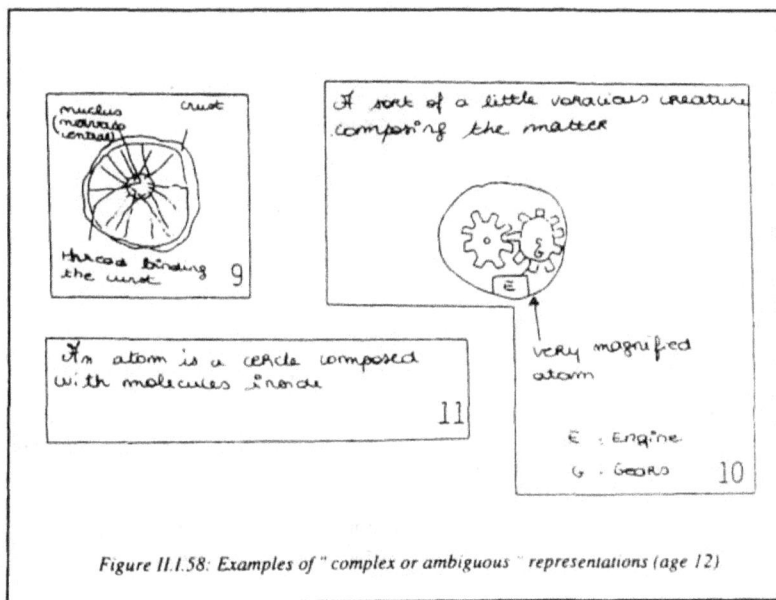

Figure II.1.58: Examples of " complex or ambiguous " representations (age 12)

The distortions were of two types:

— The " complex " images often tried to build up a rather biological, cell-like picture of the atom. These included typical characteristics of living matter: an inner world, organoids, fruit pits or cell nuclei, fuzzy or irregular outlines, deformability, the presence of superficial increaseths or tentacles, chemical mixtures, etc. (9) There was even a biomechanical model, curiously reminiscent of Maxwell's vision ! (10).

— The " ambiguous " images were fragmentary and connected with attempts to explain how atoms function. There was also considerable confusion in terms of vocabulary, mixing up " particles ", " atoms " and " molecules ".

This type of confusion was also found in some of the older students' answers. This may well be due to difficulties in assimilating new ideas and vocabulary.

14-year-olds

— *" Internal " representation* (Fig. II.1.59).

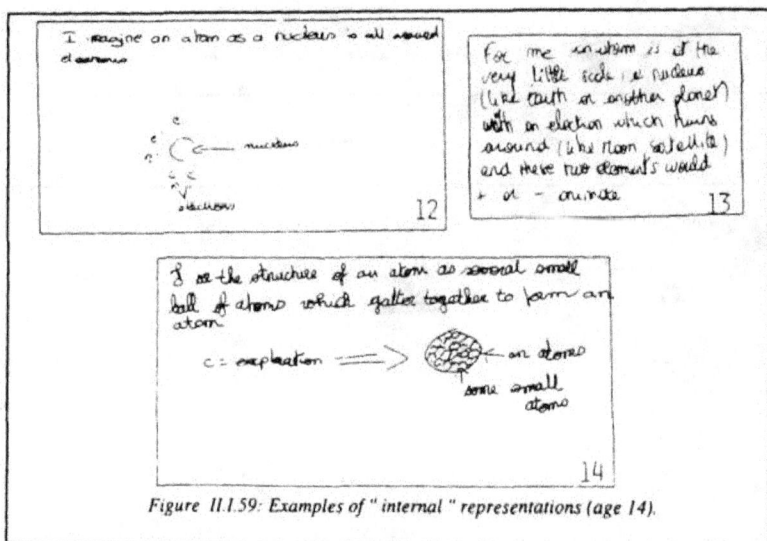

Figure II.1.59: Examples of " internal " representations (age 14).

This group had a more comprehensive idea than the 12-year-old students, usually involving concepts of " nuclei " and " electrons " (12), although these words were not always used explicitly.

Planetary models were sometimes mentioned (13), but relatively infrequently, in view of the many pictures of this type in the textbooks for the old syllabus (that this class was the last to use). Models showing " clusters " of particles were much more common, used to justify a view of the repetitive structure of matter (14).

— *" External " representation* (Fig. II.1.60).

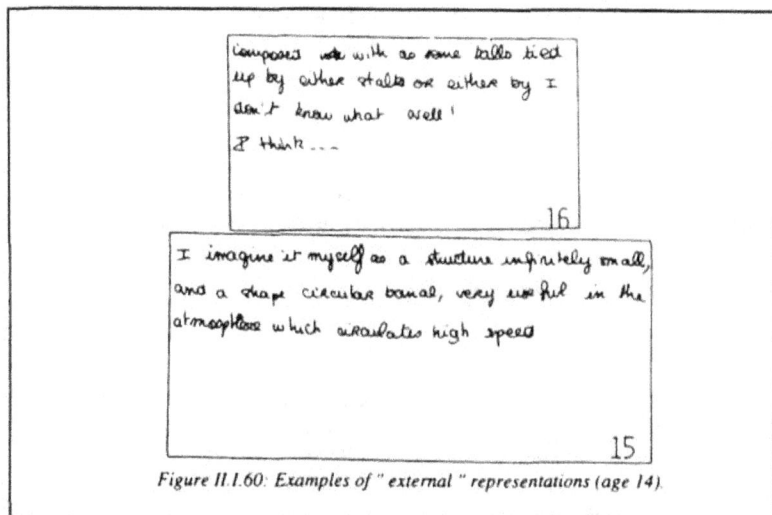

Figure II.1.60: Examples of " external " representations (age 14).

Although this image may have been pertinent for younger students, it did not seem adequate for this age group, and doubtless indicated that students had difficulty assimilating what they were taught, or that they were simply not interested. They still thought that atoms were elementary balls — even, in some cases, indivisible! — and also confused them with gas particles (15). They sometimes added linking rods (16) or pictured them in clumps. These last two characteristics were due to poorly understood handling of exploded and compact molecular models.

— *" Complex or ambiguous" representation* (Fig. II.1.61).

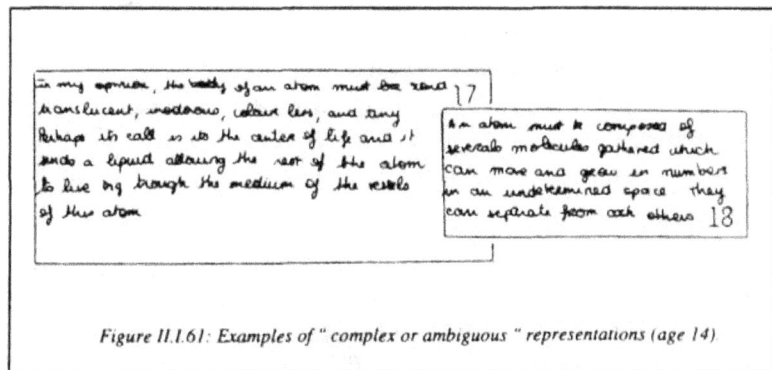

Figure II.1.61: Examples of " complex or ambiguous " representations (age 14).

This view was also based on the concepts previously mentioned for the 12-year-olds: biological (17), functional, and fragmentary aspects, but with more frequent confusions between " atoms " and " molecules " (18) probably due to rather careless use of vocabulary in the classroom.

13-year-olds

The program for this class, with its new ideas had made a strong impression on the students. This gave rise to more detailed, pertinent images, but also an increase in ambiguous impressions, due to the multiplicity of concepts.

— " *Internal* " *representation* (Fig. II.I.62).

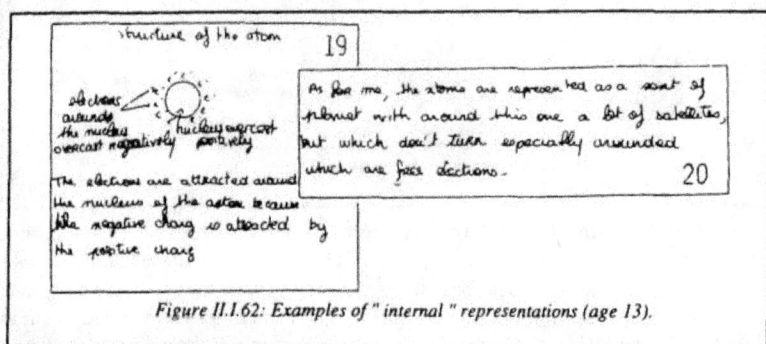

Figure II.I.62: Examples of " internal " representations (age 13).

There were still descriptions of interlocking particles or " clusters " of particles, enriched with the students' newly acquired knowledge of:
— Nuclei and electrons.
— Complementary electrical charges (19).
— A model amalgamating planetary and probabilistic views (20 and 21).
— And a reasoned criticism of outdated planetary models (22).

— " *External* " *representation* (Fig. II.I.63).

This was clearly a minority view, restricted to its essentials: Groups of balls or single balls.

— " *Complex or ambiguous*" *representation* (Fig. II.I.64).

Although there were still some biological models, they occurred much less frequently.

On the other hand, concepts were much more frequently confused. The confusions between atom/molecule/nucleus/electron (23) were still apparent, but the most frequent mistake consisted of thinking that ions were parts of atoms (24 and 25). Extra precautions should be taken to avoid this type of confusion.

An atom include one or several electrons which are anywhere (we can't know their trajectories) which are around the nucleus.

21

22

Hn atom. There are a positive nucleus and negative electron which circulate around

The next draw is false. We can't say the sun of an electron

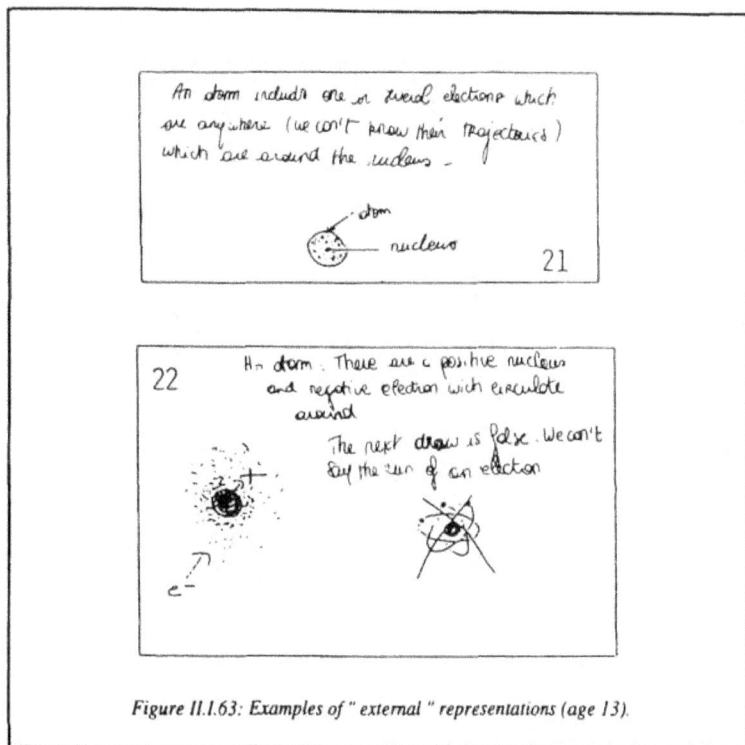

Figure II.1.63: Examples of " external " representations (age 13).

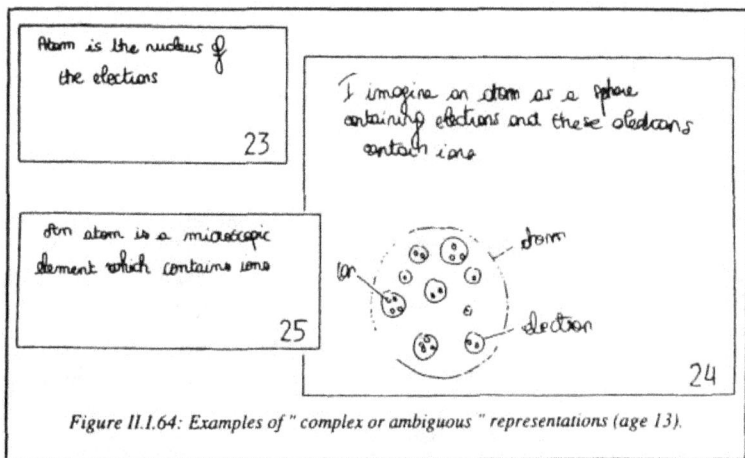

Atom is the nucleus of the electrons

23

I imagine an atom as a sphere containing electrons and these electrons contain ions

An atom is a microscopic element which contains ions

25

24

Figure II.1.64: Examples of " complex or ambiguous " representations (age 13).

II. Molecular symmetry, its description and consequences

1. On the role of symmetry in the description of electronic structure and chemical bonding

The geometrical arrangements of atoms in many molecules (hydrogen, water, ammonia, benzene, sulfur hexafluoride, etc.) are symmetrical. It seems natural that some properties, like electronic density, should also be symmetrical, as is indeed the case. However, symmetry is also responsible for other, less obvious, properties, such as *state degeneracy, allowed or forbidden optical transitions, polarization of emitted or absorbed optical radiations, reaction paths, certain characteristics of vector properties*, etc. Therefore, the study of symmetry is a valuable tool for *predicting* these characteristics and *explaining* them. This makes it possible to distinguish between effects arising from symmetry or from other causes such as energy. Furthermore, symmetry considerations often simplify the calculations necessary to determine wave functions or molecular properties.

The purpose of this text is to explain the base for the use of symmetry in theoretical chemistry and describe some derived properties.

2. Description of molecular symmetry

2.1 Symmetry elements and operations; transformation of a function by a symmetry operation

First, we must distinguish between symmetry *elements*, such as centers, axes and planes, which are geometrical objects, and symmetry *operations* that these objects enable us to define. For instance, let us consider the following set of three symmetry elements (Fig. II.II.1):

— A two-fold axis, called C_2, directed along the vertical z-axis,

— Two perpendicular vertical symmetry planes, including C_2, called σ_v and σ_v'.

These elements generate symmetry operations obtained by observing the movement of a point outside the symmetry elements.

In the first step, we consider the rotation operation via an angle π (twice $\pi/2$) about axis C_2. We shall use the symbol C_2 for this operation (the italicized letter specifying that it is an operation and not a symmetry element). Operations corresponding to any number n of successive rotations are expressed as

$$C_2{}^n = C_2 C_2 C_2 ... C_2$$

For every plane σ (σ_v or σ_v'), we shall similarly keep the reflection in this plane, represented by the symbol σ (italicized). Operations corresponding to any number n of reflections are expressed as

$$\sigma^n = \sigma\sigma\sigma...\sigma$$

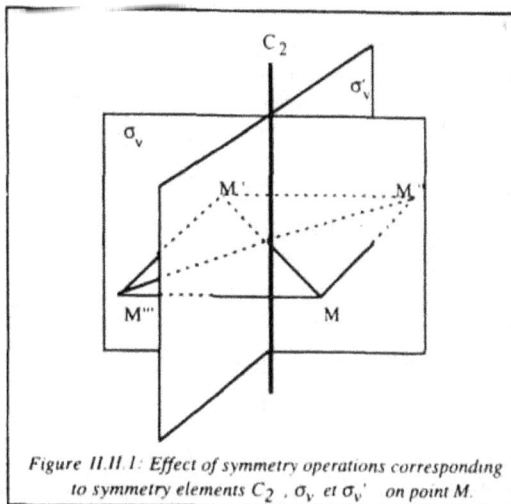

Figure II.II.1: Effect of symmetry operations corresponding to symmetry elements C_2, σ_v et σ_v' on point M.

It is then easy to build up Table II.II.1, indicating the points obtained following the movement of point M. In fact, operations C_2, C_2^3, C_2^5,...., which lead to the same point M', are considered equivalent and denoted by the simplest symbol: C_2. Similarly, operations σ_v, σ_v^3, σ_v^5,...., which lead to the same point M" are denoted σ_v, and operations σ_v', $\sigma_v'^3$, $\sigma_v'^5$,...., which lead to the same point M"' are denoted σ_v'. Operations such as C_2^2, σ_v^2, $\sigma_v'^2$,...., which leave point M invariant, are considered equivalent to the *identity operation*, which generates no movement. This is denoted by the symbol E.

TABLE II.II.1 — Action on point M of symmetry operations associated with symmetry elements C_2, σ_v et σ_v'

Symmetry elements	Symmetry operations	Results
C_2	C_2 $C_2^2 = C_2 C_2$ $C_2^3 = C_2 C_2 C_2$ etc.	$C_2 M = M'$ $C_2^2 M = M$ $C_2^3 M = M'$ etc.
σ_v	σ_v $\sigma_v^2 = \sigma_v \sigma_v$ $\sigma_v^3 = \sigma_v \sigma_v \sigma_v$ etc.	$\sigma_v M = M''$ $\sigma_v^2 M = M$ $\sigma_v^3 M = M''$ etc.
σ_v'	σ_v' $\sigma_v'^2 = \sigma_v' \sigma_v'$ $\sigma_v'^3 = \sigma_v' \sigma_v' \sigma_v'$ etc.	$\sigma_v' M = M'''$ $\sigma_v'^2 M = M$ $\sigma_v'^3 M = M'''$ etc.

Table II.II.2 shows the different symmetry elements in molecules (planes, centers, proper n-fold axes, improper n-fold axes) and associated operations. A proper n-fold axis, C_n, generates operations C_n^m, rotations through the angle m times $2\pi/n$ about this axis. An improper n-fold axis, S_n, generates operations S_n^m corresponding to m times the sequence σC_n (operation C_n followed by reflection σ in a perpendicular plane). The effect of operation S_3 on point M is shown in Fig. II.II.2.

TABLE II.II.2 — **Summary of operations generated by the different symmetry elements.**

Elements	Operations
none i (inversion center) σ (plane) C_n (axis) S_n (improper axis)	E (identity) i (inversion), E σ (reflection), E $C_n, C_n^2, ..., C_n^n = E$ $S_n = \sigma C_n, S_n^2 = C_n^2, ...,$ (n operations if n is even $2n$ operations if n is odd)

Having defined the effect of symmetry operations on the various points in space, we have to define their effect on functions such as molecular orbitals or wave functions.

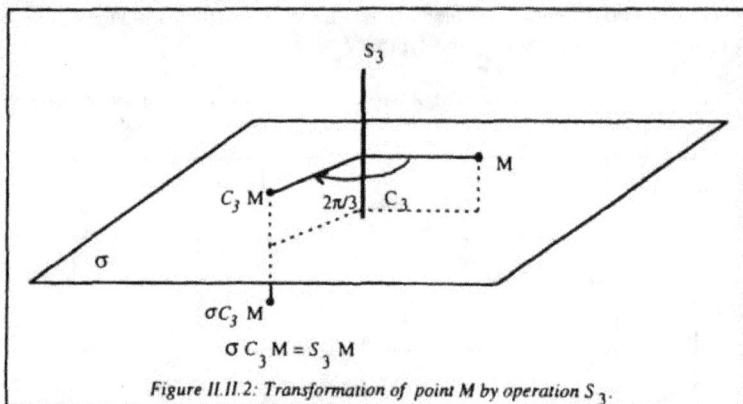

$$\sigma C_3 M = S_3 M$$

Figure II.II.2: Transformation of point M by operation S_3.

Therefore, consider a function f with a well defined value at each point M in space, as is the case of an atomic or molecular orbital. We define function φ, as the result of operation R on function f (symbolically expressed as: $\varphi = Rf$), the function which, at point RM, has the same value as function f at point M

$$\varphi(R\text{M}) = Rf(R\text{M}) = f(\text{M})$$

It may be demonstrated that the diagram used by chemists to represent function φ (polar diagram, isodensity contour, etc.) is deduced from the corresponding diagram of function f on application of operation R (Figs. II.II.3a and II.II.3b).

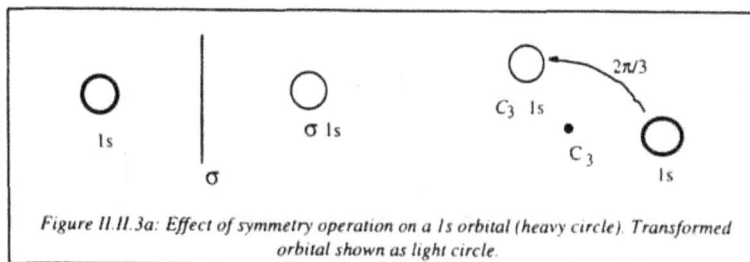

Figure II.II.3a: *Effect of symmetry operation on a 1s orbital (heavy circle). Transformed orbital shown as light circle.*

In order to obtain an analytical form of the function, it is convenient to replace point M by point $R^{-1}M$ in the previous relation. As $RR^{-1}M = M$ we obtain

$$\varphi[R(R^{-1}M)] = \varphi(M) = f(R^{-1}M).$$

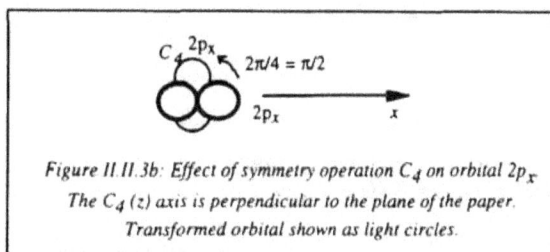

Figure II.II.3b: *Effect of symmetry operation C_4 on orbital $2p_x$.*
The C_4 (z) axis is perpendicular to the plane of the paper.
Transformed orbital shown as light circles.

For instance, if (x, y, z) and (x', y', z') are the cartesian coordinates of points M et $R^{-1}M$, respectively,

$$\varphi(x, y, z) = f[x'(x, y, z), y'(x, y, z), z'(x, y, z)].$$

If $R = C_2$ (two-fold axis directed along Oz), $x' = -x, y' = -y, z' = z$. (see Fig. II.II.4). We deduce from this $\varphi(x, y, z) = f(-x, -y, z)$.

2.2 Symmetry groups

The set of operations defined previously, $E, C_2, \sigma_v, \sigma_v'$ is sufficient to represent the result of any sequence of operations generated by the symmetry elements C_2, σ_v, σ_v'. The sequential actions of two of these elementary operations are shown in Table II.II.3. The intersection of each row and column represents the result of the operation in the column heading, followed by the operation for the row. For example, element σ_v' at the intersection of row C_2 and column σ_v means that $C_2\sigma_vM = C_2M'' = M'''$ gives the same result as $\sigma_v'M$.

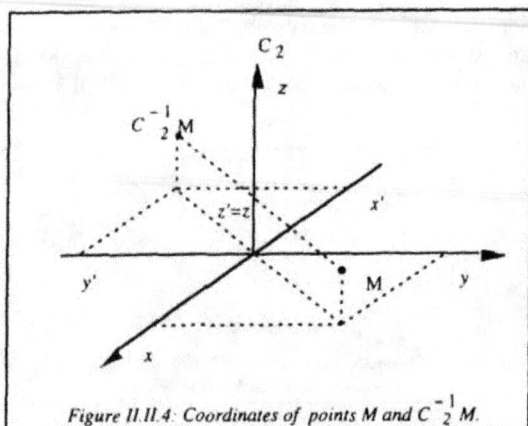

Figure II.II.4: Coordinates of points M and $C_2^{-1} M$.

TABLE II.II.3 — **Composition table of operations E, C_2, σ_v, σ_v'.**

	E	C_2	σ_v	σ_v'
E	E	C_2	σ_v	σ_v'
C_2	C_2	E	σ_v'	σ_v
σ_v	σ_v	σ_v'	E	C_2
σ_v'	σ_v'	σ_v	C_2	E

This table, known as the *composition table* of the set formed by the four preceding operations, demonstrates that:

— The internal composition law is associative [(AB)C = A(BC)].
— E is the identity element ($EA = AE = A$, whatever A).
— Each element A has an inverse A^{-1}, such as $A^{-1}A = E$

$$(E^{-1} = E, \quad C_2^{-1} = C_2, \quad \sigma_v^{-1} = \sigma_v, \quad \sigma_v'^{-1} = \sigma_v').$$

This set forms what is called a *group* in mathematics. It is a symmetry group called C_{2v} (*Schoenflies* symbol), to signify the existence of an axis C_2 and a plane σ_v. The symmetry groups of atoms and molecules contain at least one invariant point and are called *point groups*.

A molecule is said to belong to a given symmetry group if the representation of the atomic nuclei (chemical formula) is globally invariant with respect to all the operations in a group, i.e., if these operations transform this representation into an identical representation (where the position of each nucleus remains unchanged) or into an equivalent representation (where each nucleus that may have shifted is in the initial position of a nucleus of the same type). It is, therefore, easy to verify that the water molecule belongs to group C_{2v} (cf., Fig. II.II.5). Operations C_2 and σ_v' leave the position of the oxygen atom nucleus unchanged and shift nucleus H_1 into the position of nucleus H_2 and vice-versa. Operations E and σ_v leave the three atoms in their original positions.

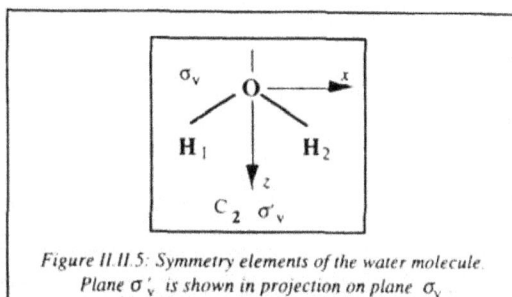

Figure II.II.5: *Symmetry elements of the water molecule. Plane σ'_v is shown in projection on plane σ_v*

Every molecule with assumed fixed nuclei belongs to a symmetry group. If the molecule has no symmetry element, it belongs to group C_1, formed by the identity operator E only. In order to determine the symmetry group of a molecule, it is convenient to use a flow chart, as shown in Fig. II.II.6.

Table II.II.4 shows symmetry elements for several molecules, found by systematic application of the flow chart in Fig. II.II.6. Figure II.II.7 gives the graphic representation of these molecules and the symmetry elements.

TABLE II.II.4 — **Symmetry elements found in the search of the symmetry group of some molecules.**

Molecule	Symmetry elements	Group
HCl	(linear)	$C_{\infty v}$
F_2	(linear), i	$D_{\infty h}$
CHFClBr	none	C_1
NOCl	σ	C_s
H_2O	C_2, σ_v	C_{2v}
trans-$C_2H_2F_2$	C_2, σ_h	C_{2h}
BF_3	C_3, C'_2, σ_h	D_{3h}
CH_4	C_3, C'_3, σ_d	T_d

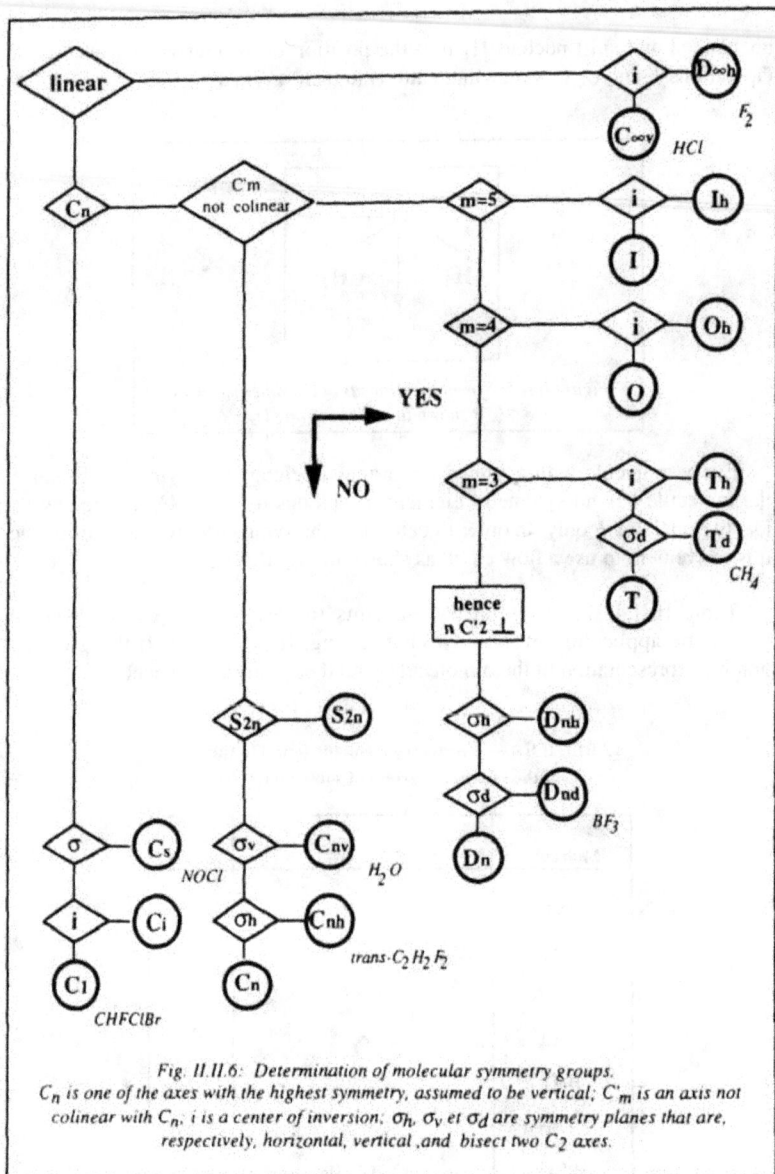

Fig. II.II.6: Determination of molecular symmetry groups.
C_n is one of the axes with the highest symmetry, assumed to be vertical; C'_m is an axis not colinear with C_n; i is a center of inversion; σ_h, σ_v et σ_d are symmetry planes that are, respectively, horizontal, vertical, and bisect two C_2 axes.

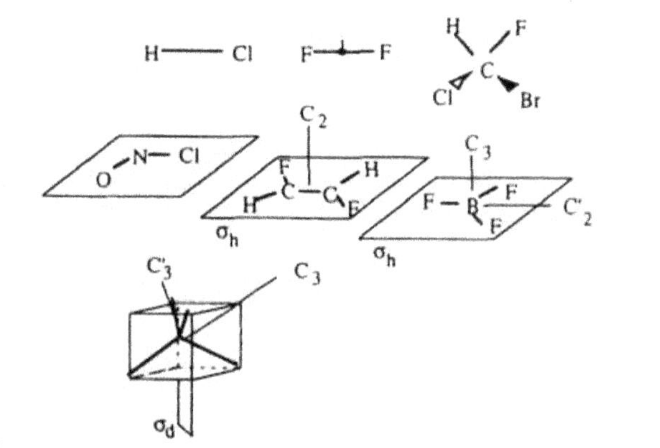

Figure II.II.7: *Spatial representation of molecules in Table II.II.4.*

2.3 Representation of a symmetry operation by matrices; choice of base; invariance of the trace; characters

Let us study in more detail the effect of symmetry group operations of a molecule on the atomic orbitals of its atoms. For example, the action of operation C_2 on the atomic orbitals $1s_1$ and $1s_2$ of the two hydrogen atoms of the water molecule is expressed by two linear combinations of the two functions $1s_1$ and $1s_2$

$$C_2 1s_1 = 1s_2 = 0 \times 1s_1 + 1 \times 1s_2 \quad \text{and} \quad C_2 1s_2 = 1s_1 = 1 \times 1s_1 + 0 \times 1s_2$$

The pair of functions $\{1s_1, 1s_2\}$ is said to form a base for a *two-dimensional matrix representation* of the operation C_2. The corresponding matrix is built by entering the coefficients of the base functions $1s_1$ and $1s_2$ of the expansion of $C_2 1s_1$ (0 and 1) in the first column and the coefficients of the expansion of $C_2 1s_2$ (1 and 0) in the second column.

$$(C_2) = \begin{pmatrix} 0 & 1 \\ 1 & 0 \end{pmatrix}$$

The *trace* of this matrix (sum of its diagonal elements) is called the *character* of the operation and is expressed as $\chi(C_2)$. In our example, $\chi(C_2) = 0$.

If the new base

$$\{f_+ = (1s_1 + 1s_2)/\sqrt{2} \quad , \quad f_- = (1s_1 - 1s_2)/\sqrt{2}\},$$

is used in place of the previous one, a similar method leads to

$$C_2 f_+ = (1s_1 + 1s_2)/\sqrt{2} = f_+ = 1 \times f_+ + 0 \times f_-$$
$$C_2 f_- = (1s_2 - 1s_1)/\sqrt{2} = -f_- = 0 \times f_+ - 1 \times f_-$$

$$(C_2)^{+/-} = \begin{pmatrix} 1 & 0 \\ 0 & -1 \end{pmatrix}$$

Thus, we obtain a new bidimensional matrix representation. Its character $\chi_{+/-}(C_2)$ is also equal to 0. This verifies an important property of matrix representation, namely the invariance of the trace after base changes. This new base is interesting because

$$C_2 f_+ = f_+ = 1 \times f_+$$

This means that function f_+ itself forms a base for matrix representation of the operation C_2, the corresponding 1×1 matrix being: $(C_2)^+ = (1)$; its character is $\chi_+(C_2) = 1$. Likewise, as

$$C_2 f_- = -f_- = -1 \times f_-,$$

matrix $(C_2)^- = (-1)$ is the one-dimensional representation obtained taking f_- as a base function. Its character is $\chi_-(C_2) = -1$.

When the base of a representation can be separated in this manner into several independent bases, the representation is said to be *reducible*. In our example, the two-dimensional representation (C_2) has been reduced to two monodimensional representations: $(C_2)^+$ et $(C_2)^-$. We note that the character of the initial representation is equal to the sum of the characters of the reduced representations

$$\chi(C_2) = \chi_+(C_2) + \chi_-(C_2)$$

As the last two representations are one-dimensional, they obviously cannot be reduced, and are known as *irreducible*. Functions used as a base for irreducible representations, such as f_+ and f_- are sometimes called *symmetry adapted functions* (or more simply *symmetry functions*). Function f_+, corresponding to a character equal to 1, is said to be *symmetrical* with respect to C_2. Function f_-, corresponding to a character equal to -1, is said to be *antisymmetrical* with respect to C_2.

The reader can verify, as an exercise, that, for all these bases, the *multiplication* table of the matrices associated with different operations is analogous to the *composition* table of operations given in Table II.II.3.

Similar calculations performed for the four operations in group C_{2v} produce Table II.II.5. The previous base, representing *all* group operations, is called *base for a group representation*. Representations Γ, $\Gamma_{+/-}$, Γ_+, Γ_- are formed by the totality of matrices corresponding to a given base set. Table II.II.5 shows that representation Γ derived from the two-dimensional base $\{1s_1, 1s_2\}$ is reducible into two one-dimensional irreducible representations of the group, conventionally expressed as

$$\Gamma = \Gamma_+ + \Gamma_-$$

We note that, for each operation R, the relation

$$\chi(R) = \chi_+(R) + \chi_-(R)$$

is fulfilled. For some groups, the dimension n of some irreducible representations may be greater than 1. As the matrix representing E is the unit matrix of dimension $n \times n$, the character associated with this operation is then

$$\chi_i(E) = n$$

The characters of the irreducible representation of several groups are listed in *character tables* [examples are given in Sec.II.II.5 (Appendix 4)]. The last two columns of these character tables show how the simple functions of coordinates x, y and z, centered at an invariant point, are changed. These properties are useful in applications of the theory, as atomic orbitals of an atom at such a point transform in the same way as these functions (for instance, orbitals p_x, p_y et p_z are transformed as functions x, y et z, respectively, and an s orbital transforms as function $x^2 + y^2 + z^2$). Irreducible representations are designated by *Mulliken symbols* [depending on their dimensions (usually A or B for dimension one, E for dimension two, T for dimension three, etc.)], with the possible addition of a subscript or superscript. A key is given in Table II.II.6 (Appendix 5). Upper case letters are used for atomic or molecular states, while irreducible representations of orbitals are expressed in lower case letters.

TABLE II.II.5 — **Representations of group C_{2v} in different base sets.**

base		E	C_2	σ_v	σ_v'	
$\{1s_1, 1s_2\}$	matrices	$\begin{pmatrix} 1 & 0 \\ 0 & 1 \end{pmatrix}$	$\begin{pmatrix} 0 & 1 \\ 1 & 0 \end{pmatrix}$	$\begin{pmatrix} 1 & 0 \\ 0 & 1 \end{pmatrix}$	$\begin{pmatrix} 0 & 1 \\ 1 & 0 \end{pmatrix}$	Γ
	characters	2	0	2	0	
$\{f_+, f_-\}$	matrices	$\begin{pmatrix} 1 & 0 \\ 0 & 1 \end{pmatrix}$	$\begin{pmatrix} 1 & 0 \\ 0 & -1 \end{pmatrix}$	$\begin{pmatrix} 1 & 0 \\ 0 & 1 \end{pmatrix}$	$\begin{pmatrix} 1 & 0 \\ 0 & -1 \end{pmatrix}$	$\Gamma_{+/-}$
	characters	2	0	2	0	
$\{f_+\}$	matrices	(1)	(1)	(1)	(1)	Γ_+
	characters	1	1	1	1	
$\{f_-\}$	matrices	(1)	(-1)	(1)	(-1)	Γ_-
	characters	1	-1	1	-1	

2.4 Fundamental properties in group theory and their practical use; group representations; reduction of representations; symmetry functions

The importance of character tables is that these characters have some mathematical properties. These properties may be used to find irreducible bases from an initial reducible base (*reduction of representation*). This greatly simplifies the search for symmetry functions. The following principles are used for simplification.

Base for a group representation

Let us take a set of N functions $F = (\Phi_1, ..., \Phi_p, ..., \Phi_N)$. These functions form a base set for a matrix representation of an operation R if, for any p, the function $R\Phi_p$ can be expressed as a linear combination of these functions. The p column of the corresponding $N \times N$ matrix is formed by the coefficients of these functions in the given order. The character $\chi(R)$, i.e., the sum of diagonal elements of the matrix, is associated with the representation. If set F is a base for all the operations of the group, it forms a base for the representation Γ of the group. This representation is labeled with all the characters of each representation.

Reduction of a representation

An important problem in group theory is the determination of functions of the functional space defined by a set such as F, that forms the base of an irreducible representation Γ_i, of a group. This is facilitated by knowing the number n_i of independent bases for the irreducible representations Γ_i generating different functional spaces. It has been proved that this number n_i is given by the following relation, due to certain properties of characters, that we will now discuss:

$$n_i = (1/h)\Sigma_R\chi_i(R)^*\chi(R) \tag{II.II.1}$$

where $\chi_i(R)$ is the character associated with the operation R in the irreducible representation Γ_i and h is the number of operations in the group (group order). These characters satisfy the relation

$$\chi(R) = \Sigma_i n_i \chi_i(R) \tag{II.II.2}$$

formally expressed as

$$\Gamma = \Sigma_i n_i \Gamma_i \tag{II.II.3}$$

and the irreducible representation Γ_i is said to occur n_i times in the reducible representation Γ.

Determining symmetry functions

Those functions ψ of functional space F that form the base for an irreducible representation of a given group

$$\psi = \Sigma_p c_p \Phi_p \tag{II.II.4}$$

are known as symmetry functions. If the irreducible representation is one-dimensional, for any R:

$$R\psi = \chi_i(R)\psi, \qquad (II.II.5)$$

In this case, the representation of R is a 1×1 matrix with a single element, $\chi_i(R)$. In addition, we have

$$R\psi = R\Sigma_p c_p \Phi_p = \Sigma_p c_p R\Phi_p. \qquad (II.II.6)$$

Identifying the right hand sides of Eqs. II.II.5 and II.II.6 it is easy to see that :

— If the operation R transforms a base function Φ_p into another base function Φ_q, (or $-\Phi_q$), i.e., $R\Phi_p = \pm \Phi_q$, this implies

$$c_p = \pm \chi_i(R)c_q \qquad (II.II.7)$$

— If the operation R leaves the base function Φ_q invariant, without regard to sign, i.e., $R\Phi_p = +\Phi_p$ (or $-\Phi_p$) and if $\chi_i(R) = -1$ (or $+1$), respectively, this implies that $c_p = -c_p$ and therefore, that

$$c_p = 0. \qquad (II.II.8)$$

Some exercises using these relations are given below in Secs. II.II.4.4 and IV.III.6.3.

In the case of irreducible representations with dimensions higher than 1, the search for symmetry functions is more complicated due to the indeterminacy of the irreducible bases. However, it is often possible to reduce the problem to a monodimensional representation, by using a subgroup of the molecule [for instance, the subgroup C_6 instead of group D_{6h} of the benzene molecule (cf., IV.III.6.3)].

3. Consequences of molecular symmetry concerning electronic properties

The preceding sections are particularly important due to certain essential properties. Unfortunately, it is not possible to detail them in this book. Consequently, the following are only a few of these major properties, and proofs are not given. Several related problems are, however, included in the next section.

— The *Hamiltonian operator H and the Hartree-Fock operator F* of an atom or a molecule are invariant with regard to all operations in their symmetry group.

— Their *eigenfunctions* representing stationary states and *atomic* or *molecular orbitals* form *irreducible base representations* of this group.

— Except for "fortuitous" degeneracy, i.e., not due to symmetry (rather unusual), the *degeneracy* of a mono or polyelectronic state is equal to the *dimension* of the *corresponding irreducible representation*.

— *Vector properties*, such as electric moment, are represented by vectors which remain *invariant with respect to all operations in the group* (for instance, for the water molecule, in group C_{2v}, these vectors are necessary directed along axis C_2). If no direction remains invariant, vectors characterizing these properties vanish.

— When a quantum system undergoes a transition from state ψ_n to state $\psi_{n'}$, with *emission* or *absorption* of light, a rapid change in the electronic distribution occurs. This change is characterized by the *transition moment* $\mu_{nn'}$. Its components in direction u'u (u = x, y or z) are given by the expression

$$(\mu_{nn'})_u = \int \psi_n{}^* \, \mu_u \, \psi_{n'} \, d\tau$$

where the electric moment operator has the same symmetry as the coordinate u. In order for the transition be *allowed* (in dipolar electric approximation), it is necessary that at least one of the components of the moment vector $\mu_{nn'}$ should be nonvanishing. If one of these components is nonvanishing, the emitted or absorbed radiation is polarized in this direction. If only two components are different from zero, the radiation emitted or absorbed is polarized in the plane defined by these two components. These components often vanish because of symmetry. It is then possible to predict the nonvanishing components from the symmetries of functions ψ_n and $\psi_{n'}$.

— When, in a molecular transformation, some symmetry elements corresponding to a given symmetry group are preserved, the system states must obey the symmetry conservation rule (cf., Sec. IV.III.6.4).

4. Selected problems

4.1 Atomic Orbitals (AOs) symmetry and atomic quantum numbers

For chemists, an atomic nucleus is depicted as a point. Its symmetry elements are those of a little sphere centered at this point, i.e., all rotation axes $C(\phi)$ (for any angle ϕ) passing through this point for any orientation, all corresponding improper axes $S(\phi)$ and all planes σ containing this point, also an inversion center i. The number of corresponding operations is infinite; these form a group called $O(3)$. It includes, as the sub group of rotations $C(\phi)$, $R(3)$, an infinity of irreducible representations with odd dimensions ($n = 2L + 1$, with L positive integer or zero). Chemists use the symbols S, P, D, and F for these representations (Table II.II.6) completed by indices, that may be ignored for our purposes.

TABLE II.II.6 — **Symbols for the irreducible representations of groups O(3) and R(3).**

L	0	1	2	3 ...
dimension: 2L + 1	1	3	5	7 ...
symbol	S	P	D	F ...

The following properties may be derived :

— The wave functions of stationary states of atoms are bases for these representations, indicated by capital letters (S, P, D, F, ...). The level degeneracy is usually equal to the dimension of the representation (for the hydrogen atom there are fortuitous degeneracies in the nonrelativistic approximation among the states $2S$ and $2P$, $3S$, $3P$ and $3D$, etc.).

— As a result of the symmetry of the electronic dipole moment operator, the transition moment from an initial state to a final state vanishes, if the difference of L values associated with these two states is different from +1 or −1. This is the selection rule: $\Delta L = \pm 1$.

— Atomic orbitals also satisfy these symmetry properties and may then be designated with symbols $s, p, d, f, ...$ (as already mentioned, lower case letters are used to indicate one-electron functions, such as orbitals). The azimuthal quantum number l, associated with the degeneracy of the electron level, then effectively characterizes their symmetries.

4.2 Symmetry of linear molecules

A linear molecule has an infinite number of symmetry operations, generated by the rotation axis $C(\phi)$ passing through the atomic nuclei and the symmetry planes σ_v containing this axis. If there is no other symmetry element, the symmetry group is called $C_{\infty v}$ (this is the case of heteronuclear diatomic molecules, such as FH, ClH, CO, NO, etc., or the monochloroacetylene molecule). If the molecule has an inversion center i, there is also a plane σ_h perpendicular to axis $C(\phi)$ and the symmetry group is called $D_{\infty h}$ (this is the case of homonuclear diatomic molecules such as H_2, F_2, Cl_2, O_2, etc., as well as carbon dioxide and acetylene). In both cases, there is an infinity of irreducible representations: Some one-dimensional, called Σ, and others multidimensional, called Π, Δ, Φ, etc. Indices may also be added to these symbols. In particular, in group $D_{\infty h}$, the subscripts g (gerade) or u (ungerade) specify parity (symmetry or antisymmetry) with regard to inversion.

Wave functions of these molecules, and the molecular orbitals used to construct them, should have the proper symmetry and degeneracy. Symbols σ, π, δ, etc., are used to identify molecular orbitals. The σ symmetry corresponds to an axial symmetry around the bond axis. The π symmetry corresponds to a pair of functions such as (p_x, p_y). (It should be noted that, for planar molecules, chemists often give a different meaning to these two symbols, using σ and π for orbitals that are symmetrical and antisymmetrical, respectively, with regard to the molecular plane.)

A molecular orbital of symmetry σ for a diatomic molecule A-B (assumed to be directed along z'z), may only be constructed from two atomic orbitals s or p centered on these two atoms, if the two atomic orbitals have symmetry σ (i.e., an axial symmetry about AB). We can then use obitals s_A and p_{zA} for atom A and s_B and p_{zB} for atom B. However, to build a molecular orbital of symmetry π, we have to take the atomic orbitals p_{xA} and p_{xB} or p_{yA} and p_{yB} (Fig. II.II.8).

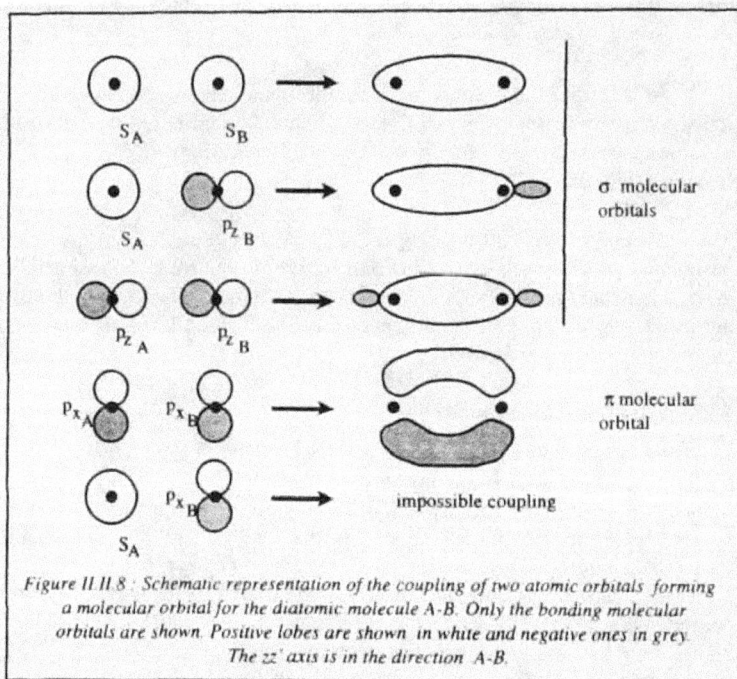

Figure II.II.8 : Schematic representation of the coupling of two atomic orbitals forming a molecular orbital for the diatomic molecule A-B. Only the bonding molecular orbitals are shown. Positive lobes are shown in white and negative ones in grey. The zz' axis is in the direction A-B.

4.3 Simplification of quantum calculations

The quality of a quantum calculation increases with the dimension of the base set, i.e., as the complexity and computation time increase. Symmetry properties are useful for the following reasons:

— They make it possible to predict symmetries of molecular orbitals and state functions, built from a given base set.

— They make it possible to replace this base set with an equivalent base of symmetry functions, i.e., the base of an irreducible representation.

— They make it possible to separate certain steps of the calculation into several problems with smaller dimensions.

— They simplify the interpretations.

For instance, if we wish to construct the molecular orbitals of the water molecule from the minimal base valence set $\{2s_O, 2p_{xO}, 2p_{yO}, 2p_{zO}, 1s_{H1}, 1s_{H2}\}$, a straightforward method, explained in the following section, enables us to predict that this 6-dimensional base set is reducible into three base sets, with dimensions of 3, 2 and 1, respectively, corresponding to the irreducible representations a_1, b_1 and b_2 (Table II.II.7). The irreducible representations of group C_{2v} are defined in the corresponding character table (see Table II.II.5 : Remember that lower case letters are used for orbitals).

It is thus possible to reduce some stages of the calculation into simpler problems. In particular, the diagonalization of 3×3, 2×2 and 1×1 matrices replaces the diagonalization of 6×6 matrices. As described in Sec. IV.III.6.2, this method leads to better understanding of the coupling between the oxygen atom and the two hydrogen atoms.

TABLE II.II.7 — **Symmetry functions for the water molecule (minimal base valence set).**

irreducible representations	symmetry functions
a_1	$2s_O, 2p_{zO}, (1s_{H1}+1s_{H2})/\sqrt{2}$
b_1	$2p_{xO}, (1s_{H1}-1s_{H2})/\sqrt{2}$
b_2	$2p_{yO}$

4.4 Study of H_2O Molecular Orbitals (MO's) using group theory

The purpose of this section is to apply the properties described in Sec. II.II.2.4 to the molecular orbitals of the water molecule depicted in Fig. II.II.5. We will restrict the base set to valence molecular orbitals: Orbitals $1s_1$ and $1s_2$ of the two hydrogen atoms and orbitals $2p_O$, $2p_{xO}$, $2p_{yO}$ et $2p_{zO}$ of the oxygen atom.

— *Let us show that these functions form a base for a representation of the group C_{2v} of the molecule.* (Note that, if this is not the case, the base set is ill-adapted to the study and should be changed). For this purpose, we apply operations E, C_2, σ_v, σ_v' of group to each base function:

$$E1s_1 = 1s_1 \qquad E1s_2 = 1s_2 \qquad E2s_O = 2s_O$$

$E2p_xO = 2p_xO \quad E2p_yO = 2p_yO \quad E2p_zO = 2p_zO$

$C_2 1s_1 = 1s_2 \qquad\qquad C_2 1s_2 = 1s_1 \qquad\qquad C_2 2s_O = 2s_O$
$C_2 2p_xO = -2p_xO \qquad C_2 2p_yO = -2p_yO \qquad C_2 2p_zO = 2p_zO$

$\sigma_v 1s_1 = 1s_1 \qquad\qquad \sigma_v 1s_2 = 1s_2 \qquad\qquad \sigma_v 2s_O = 2s_O$
$\sigma_v 2p_xO = 2p_xO \qquad\quad \sigma_v 2p_yO = -2p_yO \qquad \sigma_v 2p_zO = 2p_zO$

$\sigma_v' 1s_1 = 1s_2 \quad \sigma_v' 1s_2 = 1s_1 \quad \sigma_v' 2s_O = 2s_O \quad \sigma_v' 2p_xO = -2p_xO$
$\sigma_v' 2p_yO = 2p_yO \qquad \sigma_v' 2p_zO = 2p_zO$

The functions obtained in this way may all be expressed as linear combinations of base functions. The initial base set is a real base for the representation of a group.

— *Let us determine the matrices and characters of this representation.* The corresponding matrices are easily deduced. For example

$$C_2 1s_1 = 0\times1s_1 + 1\times1s_2 + 0\times2s_O + 0\times2p_xO + 0\times2p_yO + 0\times2p_zO$$

produces elements (0, 1, 0, 0, 0, 0) in the first column of the matrix (C_2). In the same way

$$C_2 1s_2 = 1\times1s_1 + 0\times1s_2 + 0\times2s_O + 0\times2p_xO + 0\times2p_yO + 0\times2p_zO$$

produces elements (1, 0, 0, 0, 0, 0) in the second column of the same matrix and so on. Finally, we obtain the following matrices:

$$(E) = \begin{pmatrix} 1 & 0 & 0 & 0 & 0 & 0 \\ 0 & 1 & 0 & 0 & 0 & 0 \\ 0 & 0 & 1 & 0 & 0 & 0 \\ 0 & 0 & 0 & 1 & 0 & 0 \\ 0 & 0 & 0 & 0 & 1 & 0 \\ 0 & 0 & 0 & 0 & 0 & 1 \end{pmatrix} \qquad (C_2) = \begin{pmatrix} 0 & 1 & 0 & 0 & 0 & 0 \\ 1 & 0 & 0 & 0 & 0 & 0 \\ 0 & 0 & 1 & 0 & 0 & 0 \\ 0 & 0 & 0 & -1 & 0 & 0 \\ 0 & 0 & 0 & 0 & -1 & 0 \\ 0 & 0 & 0 & 0 & 0 & 1 \end{pmatrix}$$

$$(\sigma_v) = \begin{pmatrix} 1 & 0 & 0 & 0 & 0 & 0 \\ 0 & 1 & 0 & 0 & 0 & 0 \\ 0 & 0 & 1 & 0 & 0 & 0 \\ 0 & 0 & 0 & 1 & 0 & 0 \\ 0 & 0 & 0 & 0 & -1 & 0 \\ 0 & 0 & 0 & 0 & 0 & 1 \end{pmatrix} \qquad (\sigma_v') = \begin{pmatrix} 0 & 1 & 0 & 0 & 0 & 0 \\ 1 & 0 & 0 & 0 & 0 & 0 \\ 0 & 0 & 1 & 0 & 0 & 0 \\ 0 & 0 & 0 & -1 & 0 & 0 \\ 0 & 0 & 0 & 0 & 1 & 0 \\ 0 & 0 & 0 & 0 & 0 & 1 \end{pmatrix}$$

and characters:

$$\chi(E) = 6 \qquad \chi(C_2) = 0 \qquad \chi(\sigma_v) = 4 \qquad \chi(\sigma_v') = 2$$

In this example, symmetry operations change each orbital into an orbital of the base set. The character associated with each operation is then easy to obtain. Each unchanged orbital contributes 1 to the character, orbitals which change sign contribute -1 and transformed orbitals do not contribute at all. When an orbital is changed into a linear combination of several base orbitals, the character calculation is more complex.

— *Reduction of the representation.* The previous characters do not constitute an irreducible representation, as shown in the character table (of irreducible representations) of group C_{2v} given in Appendix 4, cf., Sec.II.II.5. The representation Γ is therefore reducible. By applying Eq. II.II.1, we obtain

$$n_{a_1} = (1/4)(1 \times 6 + 1 \times 0 + 1 \times 4 + 1 \times 2) = 3$$
$$n_{a_2} = (1/4)(1 \times 6 + 1 \times 0 - 1 \times 4 - 1 \times 2) = 0$$
$$n_{b_1} = (1/4)(1 \times 6 - 1 \times 0 + 1 \times 4 - 1 \times 2) = 2$$
$$n_{b_2} = (1/4)(1 \times 6 - 1 \times 0 - 1 \times 4 + 1 \times 2) = 1,$$

From the initial base set, it is therefore possible to form three functions of symmetry a_1, two functions of symmetry b_1 and one of symmetry b_2, but none of symmetry a_2. This property is expressed as

$$\Gamma = 3 a_1 + 2 b_1 + b_2$$

— *Determine the symmetry functions.* For this purpose, we must determine the relations among coefficients c_p in the general expression

$$\psi = c_1 1s_1 + c_2 1s_2 + c_3 2s_O + c_4 2p_xO + c_5 2p_yO + c_6 2p_zO$$

As these representations are one-dimensional, we can apply Eqs. II.II.7 and II.II.8 to several atomic orbitals (only a few of these relations are required).

Functions of symmetry a_1

$\chi_{a_1}(\sigma_v) = 1$	and $\sigma_v 2p_yO = -2p_yO$	$\Rightarrow$	$c_5 = 0$
$\chi_{a_1}(\sigma_v') = 1$	and $\sigma_v' 2p_xO = -2p_xO$	$\Rightarrow$	$c_4 = 0$
	$\sigma_v' 1s_2 = 1s_1$	$\Rightarrow$	$c_2 = c_1$

As the other relations generate no additional information, the functions of symmetry a_1 have the following form:

$$\psi_{a_1} = c_1(1s_1 + 1s_2) + c_3 2s_O + c_6 2p_zO$$

Functions of symmetry b_1

$\chi_{b_1}(\sigma_v) = 1$	and $\sigma_v 2p_yO = -2p_yO$	$\Rightarrow$	$c_5 = 0$
$\chi_{b_1}(\sigma_v') = -1$	and $\sigma_v' 2s_O = 2s_O$	$\Rightarrow$	$c_3 = 0$
	$\sigma_v' 2p_zO = 2p_zO$	$\Rightarrow$	$c_6 = 0$
	$\sigma_v' 1s_2 = 1s_1$	$\Rightarrow$	$c_2 = -c_1$

whence:

$$\psi_{b_1} = c_1(1s_1 - 1s_2) + c_4 2p_xO$$

Functions of symmetry b_2

$\chi_{b_2}(\sigma_v) = -1$ and $\sigma_v 2s_O = 2s_O$ $\Rightarrow$ $c_3 = 0$

$\sigma_v 2p_xO = 2p_xO$ $\Rightarrow$ $c_4 = 0$

$\sigma_v 2p_zO = 2p_zO$ $\Rightarrow$ $c_6 = 0$

$\sigma_v 1s_1 = 1s_1$ $\Rightarrow$ $c_1 = 0$

$\sigma_v 1s_2 = 1s_2$ $\Rightarrow$ $c_2 = 0$

whence:

$$\psi_{b_2} = 2p_yO$$

The reader will verify that applying this method to the functions of symmetry a_2 gives $\psi_{a_2} = 0$, confirming that the representation a_2 does not occur in Γ.

Another way to find this result is to examine the symmetries of atomic orbitals of the oxygen atom, located at an invariant point. Indeed, in this case, these symmetries are given directly by the character table reproduced in Sec.II.II.5. The orbital $2s_O$ changes in the same way as $x^2 + y^2 + z^2$, i.e., has symmetry a_1. The same is true of $2p_zO$, which changes in the same way as z. However,, $2p_xO$ and $2p_yO$, which change like x and y, respectively, have symmetries b_1 and b_2. Now, we have find the possible symmetries for the base $\{1s_1, 1s_2\}$. We obtain a_1 [for $(1s_1 + 1s_2)$] and b_1 [for $(1s_1 - 1s_2)$]. It is then easy to describe the functions of allowed symmetries. The coupling between these symmetry-adapted orbitals will be examined more completely in Sec. IV.III.6.2.

In this simple case, a third presentation consists of noting that the previous matrices are block diagonal matrices They may be decomposed into blocks of dimension 2×2 for the first and of dimension 1×1 for the four others. These blocks show representations of symmetry operations using bases $\{1s_1, 1s_2\}$, $\{2s_O\}$, $\{2p_xO\}$, $\{2p_yO\}$, and $\{2p_zO\}$, respectively. The last four are one-dimensional, and irreducible. The character table shows that they belong to representations a_1, b_1, b_2 and a_1, respectively. The first is reducible and may therefore be reduced, as shown in Sec. II.II.2.3, taking as a new base set the functions $(1s_1 + 1s_2)/\sqrt{2}$ and $(1s_1 - 1s_2)/\sqrt{2}$, from irreducible representations a_1 et b_1, respectively. Of course, this gives the same decomposition and form of symmetry functions.

$$(E) = \begin{pmatrix} 1 & 0 & 0 & 0 & 0 & 0 \\ 0 & 1 & 0 & 0 & 0 & 0 \\ 0 & 0 & 1 & 0 & 0 & 0 \\ 0 & 0 & 0 & 1 & 0 & 0 \\ 0 & 0 & 0 & 0 & 1 & 0 \\ 0 & 0 & 0 & 0 & 0 & 1 \end{pmatrix} \qquad (C_2) = \begin{pmatrix} 0 & 1 & 0 & 0 & 0 & 0 \\ 1 & 0 & 0 & 0 & 0 & 0 \\ 0 & 0 & 1 & 0 & 0 & 0 \\ 0 & 0 & 0 & -1 & 0 & 0 \\ 0 & 0 & 0 & -1 & 0 \\ 0 & 0 & 0 & 0 & 0 & 1 \end{pmatrix}$$

$$(\sigma_v) = \begin{pmatrix} 1 & 0 & 0 & 0 & 0 & 0 \\ 0 & 1 & 0 & 0 & 0 & 0 \\ 0 & 0 & 1 & 0 & 0 & 0 \\ 0 & 0 & 0 & 1 & 0 & 0 \\ 0 & 0 & 0 & 0 & -1 & 0 \\ 0 & 0 & 0 & 0 & 0 & 1 \end{pmatrix} \qquad (\sigma'_v) = \begin{pmatrix} 0 & 1 & 0 & 0 & 0 & 0 \\ 1 & 0 & 0 & 0 & 0 & 0 \\ 0 & 0 & 1 & 0 & 0 & 0 \\ 0 & 0 & 0 & -1 & 0 & 0 \\ 0 & 0 & 0 & 0 & 1 & 0 \\ 0 & 0 & 0 & 0 & 0 & 1 \end{pmatrix}$$

5. Appendix 4:Character tables

The character tables of irreducible representations for several groups are given below. The first line of these tables specifies the group operations. Some operations, belonging to the same *class*, are gathered together, as they have the same characters. For instance, in group C_{3v}, $2\,C_3$ is used instead of C_3 and C_3^2. The following rows indicate the symbols of irreducible representations Γ_i and the characters associated with each operation or class. The last two columns indicate the symmetry of linear and quadratic functions of cartesian coordinates x, y, z, centered at an invariant point. These make it easy to obtain the symmetry of atomic orbitals centered at this point.

Each group has a one-dimensional representation, where all characters are equal to 1. This is called the *fully symmetrical* representation.

Group C_1

C_1	E
A	1

Group C_2

C_2	E	C_2		
A	1	1	z	x^2, y^2, z^2, xy
B	1	-1	x, y	yz, xz

Group C_s

C_s	E	σ_h		
A'	1	1	x, y	x^2, y^2, z^2, xy
A''	1	-1	z	yz, xz

Group C_{2v}

C_{2v}	E	C_2	σ_v	σ_v'		
A_1	1	1	1	1	z	x^2, y^2, z^2
A_2	1	1	-1	-1		xy
B_1	1	-1	1	-1	x	xz
B_2	1	-1	-1	1	y	yz

Group C_{3v}

C_{3v}	E	$2C_3$	$3\sigma_v$		
A_1	1	1	1	z	x^2+y^2, z^2
A_2	1	1	-1		
E	2	-1	0	(x, y)	$(x^2-y^2, xy), (xz, yz)$

Group D_{2h}

D_{2h}	E	C_{2z}	C_{2y}	C_{2x}	i	σ_{xy}	σ_{xz}	σ_{yz}		
A_g	1	1	1	1	1	1	1	1		x^2, y^2, z^2
B_{1g}	1	1	-1	-1	1	1	-1	-1		xy
B_{2g}	1	-1	1	-1	1	-1	1	-1		xz
B_{3g}	1	-1	-1	1	1	-1	-1	1		yz
A_u	1	1	1	1	-1	-1	-1	-1		
B_{1u}	1	1	-1	-1	-1	-1	1	1	z	
B_{2u}	1	-1	1	-1	-1	1	-1	1	y	
B_{3u}	1	-1	-1	1	-1	1	1	-1	x	

Group C_6

C_6	E	C_6	C_3	C_2	C_3^2	C_6^5		$\varepsilon = \exp(2\pi i/6)$
A	1	1	1	1	1	1	z	x^2+y^2, z^2
B	1	-1	1	-1	1	-1		
E_1	1	ε	$-\varepsilon^*$	-1	$-\varepsilon$	ε^*	(x, y)	(xz, yz)
	1	ε^*	$-\varepsilon$	-1	$-\varepsilon^*$	ε		
E_2	1	$-\varepsilon^*$	$-\varepsilon$	1	$-\varepsilon^*$	$-\varepsilon$		
	1	$-\varepsilon$	$-\varepsilon^*$	1	$-\varepsilon$	$-\varepsilon^*$		(x^2-y^2, xy)

Group T_d

T_d	E	$8C_3$	$3C_2$	$6S_4$	$6\sigma_d$		
A_1	1	1	1	1	1		$x^2+y^2+z^2$
A_2	1	1	1	-1	-1		
E	2	-1	2	0	0		$(2z^2-x^2-y^2, x^2-y^2)$
T_1	3	0	-1	1	-1		
T_2	3	0	-1	-1	1	(x, y, z)	(xy, xz, yz)

Group R(3) (rotation group)

R(3)	E	∞ C(φ)		
S	1	1		$x^2 + y^2 + z^2$
P	3	$1 + 2\cos\phi$	(x, y, z)	
D	5	$1 + 2\cos\phi + 2\cos 2\phi$		$(2z^2 - x^2 - y^2$
			$x^2 - y^2$	$xy,\ yz,\ xz)$
F	7	...		

Group C∞ᵥ

C∞ᵥ	E	2 C(φ)...	∞ σᵥ		
Σ⁺	1	1 ...	1	z	$x^2 + y^2, z^2$
Σ⁻	1	1 ...	-1		
Π	2	$2\cos\phi$ ...	0	(x, y)	(xz, yz)
Δ	2	$2\cos 2\phi$	0		$(x^2 - y^2, xy)$
Φ	etc.				

6. Appendix 5: Mulliken symbols

Irreducible representations are denoted by Mulliken symbols formed by letters, sometimes bearing subscripts depending on their dimension and symmetry or antisymmetry with respect to certain operations (Table II.II.8).

TABLE II.II.8 — **Key to Mulliken symbols.**

C_n is an axis of highest symmetry, assumed to be vertical, σ_h is an horizontal symmetry plane, σ_v is a vertical symmetry plane containing C_n.

Dimension of representation	Symbols and indices	Symmetry (S) or antisymmetry(A)	Operation
1	A		
1	B	A	C_n
2	E		
3	T		
	1	S	C_2 ($\perp$ à C_n)
	2	A	C_2 ($\perp$ à C_n)
	'	S	σ_h
	"	A	σ_h
	g	S	i
	u	A	i
	+	S	σ_v
		A	σ_v

PART III

TWO COMPLEMENTARY DESCRIPTIONS OF CHEMICAL BONDING

CHAPTER III

Mechanical aspects
of chemical bonding

I. Basics

1. The Born-Oppenheimer approximation and the potential energy of nuclei

The theoretical study of a molecule requires the use of quantum mechanics. Its various stationary states are described as solutions of the eigenvalue equation of the hamiltonian operator H

$$H \Psi = E \Psi \qquad \text{(III.I.1)}$$

This operator is equal to the sum of the kinetic energy operator T and the potential energy operator V

$$H = T + V \qquad \text{(III.I.2)}$$

The operator V corresponds to the multiplication by the potential energy function. [The definition of this function and the form of the potential energy function used in this chapter are given in Appendix 6 (Sec. III.II.4)]. The wave functions Ψ are functions of 4n variables (three space variables and one spin variable for each of the n particles). Thus, the theoretical treatment of a such a molecule is always a complex and difficult task. The aim of this chapter is to show

— When a chemical bond can be considered, from a mechanical point of view, as equivalent to a kind of spring binding the two atoms.

— That the acting forces can then be attributed to the density of electronic charges at each point in space and to charges on the various nuclei.

In order to appreciably reduce the complexity of the equations, we shall begin with the example of the hydrogen molecule. We shall then extend the results to other molecules.

The Hamiltonian operator and wave function of the hydrogen molecule

The hydrogen molecule consists of two protons A and B of charge q $(1.60 \times 10^{-19}$ C) and mass m_p $(1.67 \times 10^{-27}$ kg) and two electrons, 1 and 2, with charge $-q$ and mass m_e $(9.11 \times 10^{-31}$ kg) (Fig.III.I.1).

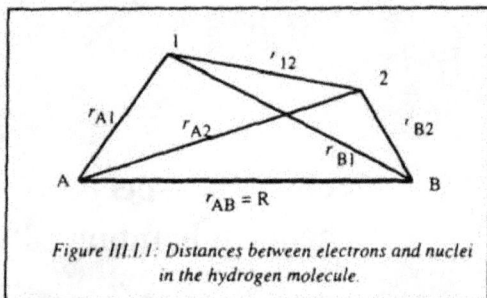

Figure III.I.1: Distances between electrons and nuclei in the hydrogen molecule.

Forces between these charged particles are attractive and repulsive coulombic forces. They can be described as the sum of the potential energy functions associated with each pair (i, j), of charges q_i and q_j (Eq.III.II.19):

$$V_{ij} = q_i\, q_j\, /(4\pi\, \varepsilon_0 r_{ij}) \qquad (\text{III.I.3})$$

As the kinetic energy operator is the sum of the operators for each particle, the hamiltonian operator is

$$H = T_A + T_B + T_1 + T_2$$
$$+ V_{A1} + V_{A2} + V_{B1} + V_{B2} + V_{12} + V_{AB} \qquad (\text{III.I.4})$$

where the operator for the kinetic energy of each particle of mass m_i is

$$T_i = -(\hbar^2/8\pi^2 m_i)(\partial^2/\partial x_i^2 + \partial^2/\partial y_i^2 + \partial^2/\partial z_i^2) \qquad (\text{III.I.5})$$

operator H includes only the kinetic energy of the particles and the coulombic forces, we can predict that bonding between the two atoms does not arise from "mysterious" forces. Nevertheless, the problem, as it is set, is very complex, as the wave function Ψ is a function of space coordinates x, y, z and of the spin coordinate σ of four particles

$$\Psi = \Psi(x_A, y_A, z_A, \sigma_A, x_B, y_B, z_B, \sigma_B, x_1, y_1, z_1, \sigma_1, x_2, y_2, z_2, \sigma_2) \quad (\text{III.I.6})$$

Even if we knew the accurate solution of Eq. III.I.1 above, it would be very difficult to represent. Fortunately, we can obtain an interesting approximate solution using the Born-Oppenheimer approximation.

The Born-Oppenheimer approximation

This approximation can be justified, at least partly, on account of the large value of the proton/electron mass ratio ($m_p/m_e = 1836$). It follows that, if this problem were treated by classical mechanics, the electrons would move much faster than the nuclei (the velocity of the hydrogen atom electron in the smallest of the Bohr orbits is about 2×10^6 ms^{-1} but the average velocity of nuclei corresponding to the vibration of the hydrogen molecule is only 0.7×10^4 ms^{-1} at 0 K). Thus, *the electrons would adjust their motion around the nuclei almost instantaneously with respect to the nuclei* .

The transposition of this remark to quantum mechanics suggests that, in a first approximation, the electron cloud can be studied assuming fixed nuclei. This leads to the omission of operators T_A and T_B associated with the kinetic energy of the nuclei in the hamiltonian H and to a new, simpler operator, called the electron hamiltonian H_{el}, taking only electron motion into account. Let us derive the solutions of the corresponding eigenvalue equation

$$H_{el} \Psi_{el} = E_{el} \Psi_{el} \tag{III.1.7}$$

for all values of the internuclear distance R.

As the potential energy function depends only on internal coordinates, it is independent of the coordinate system. Energy E_{el} is a function of the internuclear distance R and the wave function likewise depends on this distance in addition to the electron coordinates. Thus, functions in the form $\Psi = \Psi_{el} \chi$ (A,B) where χ (A,B) $= \chi(x_A, y_A, z_A, \sigma_A, x_B, y_B, z_B, \sigma_B)$, are generally good, though approximate, solutions of the initial eigenvalue equation of the molecular hamiltonian operator (Eq. III.1.1). This property arises from the fact that

$$H\Psi = (T_A + T_B + H_{el}) \Psi_{el} \chi(A,B) \equiv \Psi_{el} [T_A + T_B + E_{el}(R)] \chi \text{ (A,B)} \tag{III.1.8}$$

when small terms (about 10^{-4} eV) are neglected. This approximation is usually valid for the ground state. Equation III.1.1 is then reduced to

$$[T_A + T_B + E_{el}(R)] \chi(A,B) = E \chi(A,B), \tag{III.1.9}$$

known as the nuclear equation. It is then possible to determine $\chi(A,B)$ and E.

In fact, this is the eigenvalue equation of the hamiltonian operator of a system with two nuclei A and B subject to forces $\mathbf{F}_A$ and $\mathbf{F}_B$ described by the potential energy function

$$U(R) = E_{el}(R \tag{III.1.10}$$

These forces of magnitude $F_A = F_B = |\,dU(R)/dR\,|$ are directed toward each other if $dU(R)/dR$ is positive and in the opposite direction if $dU(R)/dR$ is negative [Appendix 6 (cf., Sec. III.II.4)].

Study of the potential energy function

The function $U(R)$ corresponding to the ground state of the hydrogen molecule is plotted in Fig. III.I.2. For large values of R, it tends toward the energy $(E_A + E_B)$ of two isolated hydrogen atoms. When R decreases, this function decreases, reaches a minimum and then rapidly increases. To the right of minimum, the slope is positive and thus corresponds to attractive forces, which tend to bring the nuclei together. On the left, the slope is negative and forces are repulsive. The distance $R = R_e$, where $dU(R)/dR = 0$ (the minimum of the curve), and the forces vanish, corresponds to an equilibrium position ; it is *the length of the H–H bond* $(R_e = 0.074$ nm). The difference D_e between the energy of two atoms (asymptote) and the energy of the molecule at equilibrium (minimum) is *the bond energy* $(D_e = 4.74$ eV), if the vibration energy is ignored.

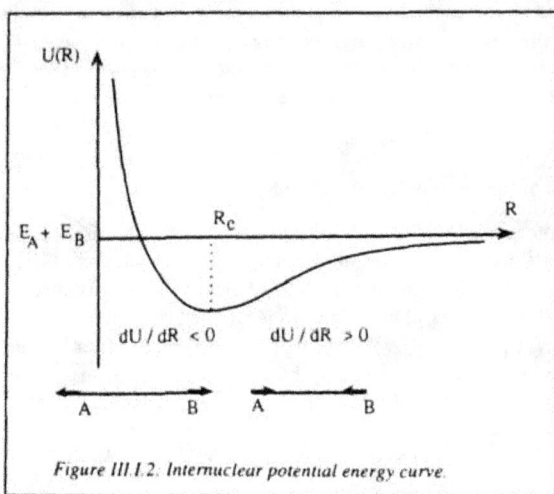

Figure III.I.2. Internuclear potential energy curve.

The expansion of $U(R)$ in the vicinity of $R = R_e$, ignoring terms of order 2 and higher is expressed as

$$U(R) = U(R_e) + (R - R_e)\,(dU/dR)_{R=R_e} + (1/2)(R - R_e)^2\,(d^2U/dR^2)_{R=R_e} \quad \text{(III.I.11)}$$

As dU/dR vanishes at the minimum, this potential is analogous to that of the harmonic oscillator (parabolic potential) with coordinate $x = R - R_e$ and force constant (Appendix 6, Eq. III.II.11):

$$k = d^2U/dR^2 \qquad \text{(III.I.12)}$$

These results lead to the picture of a molecule made up of two atoms, with nuclei vibrating on either side of an average position. If we take into account the other terms of the expansion for $U(R)$, then the picture is that of an anharmonic oscillator. To be more complete, we would have to add rotation about the center of mass and translational motion, that we have left out because they are not essential for the understanding of the chemical bond.

2. Internuclear forces and electron density; the bonding and antibonding Berlin regions

The important point that we want to emphasize in this chapter is the origin of the repulsive and attractive forces that appear in a molecule and act as a spring. It is possible to give a simple electrostatic interpretation of these forces by an alternative calculation as the quantum mean of Coulomb forces acting on the nuclei.

For instance, we shall determine the average of the resultant of forces $\mathbf{f}_A(1)$, $\mathbf{f}_A(2)$ et $\mathbf{f}_A(B)$, due to electrons 1 and 2 and nucleus B respectively, acting on the nucleus A

$$\mathbf{F}_A = \Sigma_{\sigma1}\Sigma_{\sigma2} \int\int \psi_{el}(1,2)^* \{\mathbf{f}_A(1) + \mathbf{f}_A(2) + \mathbf{f}_A(B)\} \, \psi_{el}(1,2) \, dv_1 \, dv_2$$

(III.I.13)

The vector notation replaces the set of three equations corresponding to the vector projections along the three axes $x'x$, $y'y$ and $z'z$; the operators acting on ψ_{el} are thus results of the multiplication by the functions corresponding to the components of these vectors.

The Hellmann-Feynman theorem (Appendix 7, cf., Eq. III.II.5) shows that the result must be the same as that obtained by taking the derivative of the electron energy with respect to R, as we have done previously.

The repulsive force $\mathbf{f}_A(B)$ due to nucleus B does not depend on the integration variables and may be removed from the integral. Taking into account the fact that $\mathbf{f}_A(1)$ depends only on the coordinates of electron 1 and that $\mathbf{f}_A(2)$ depends only on those of electron 2, this force may be expressed as

$$\mathbf{F}_A = \int \rho_1(1) \, \mathbf{f}_A(1) \, dv_1 + \int \rho_2(2) \, \mathbf{f}_A(2) \, dv_2 + \mathbf{f}_A(B) \qquad \text{(III.I.14)}$$

where the function

$$\rho_1(1) = \Sigma_{\sigma1}\Sigma_{\sigma2} \int \psi_{el}(1,2)^* \psi_{el}(1,2) \, dv_2 \qquad \text{(III.I.15)}$$

is the electron density per unit volume for electron 1, averaged over all positions of electron 2. The function

$$\rho_2(2) = \Sigma_{\sigma1}\Sigma_{\sigma2} \int \psi_{el}(1,2)^* \psi_{el}(1,2) \, dv_1, \qquad \text{(III.I.16)}$$

is analogous for electron 2. Because of the antisymmetry properties of wave functions, the two functions ρ_1 and ρ_2 have the same value for each point (x, y, z), and the total electron density due to the two electrons at this point is

$$\rho(x, y, z) = \rho_1(x, y, z) + \rho_2(x, y, z) = 2\rho_1(x, y, z) \qquad \text{(III.I.17)}$$

The total force acting on nucleus A becomes

$$\mathbf{F}_A = \mathbf{f}_A^e + \mathbf{f}_A(B), \qquad \text{(III.I.18)}$$

where f_A^c is the average force due to electron interactions

$$f_A^c = \int \rho(x, y, z)\, f_A\, dv, \qquad\qquad (III.I.19)$$

and f_A is the force that would be exerted on nucleus A due to an electron in the volume element dv at (x, y, z)

$$f_A = u_A\, q^2/(4\pi\varepsilon_0 r_A^2) \qquad\qquad (III.I.20)$$

(u_A is the unit vector directed from A to dv).

The force acting on nucleus B is given by the analogous formula:

$$F_B = f_B^c + f_B(A) \qquad\qquad (III.I.21)$$

$$f_B^c = \int \rho(x, y, z)\, f_B\, dv \qquad\qquad (III.I.22)$$

$$f_B = u_B\, q^2/(4\pi\varepsilon_0 r_B^2) \qquad\qquad (III.I.23)$$

using the same notations as above.

As forces F_A and F_B, and $f_A(B)$ and $f_B(A)$, respectively, are equal in magnitude but of opposite signs, this is also true of forces f_A^c and f_B^c

$$f_A^c = - f_B^c \qquad\qquad (III.I.24)$$

Analysis of electron forces f_A^c et f_B^c arising from electrons

Forces F_A and F_B consist of two terms. Those relative to Coulomb repulsive forces between nuclei, $f_A(B)$ and $f_B(A)$, are easy to calculate. It is interesting, however, to analyze f_A^c and f_B^c in detail as they originate from the electrons. These terms may be interpreted as the vector sum of attractive forces on the nuclei due to the negative charges $dq = - q\rho\, dv$ contained in each elementary volume dv.

As the electron density ρ of the hydrogen molecule revolves symmetrically about the AB axis, the resultant of these forces is directed along that axis and it is therefore sufficient to consider the projections of these forces on the axis. We shall consider as positive the projections from A to B and as negative those from B to A. Thus, after projection, Eq. III.I.19 becomes:

$$f_A^c = \int \rho(x, y, z)\, f_A\, dv \quad \text{with} \quad f_A = \cos\theta_A\, q^2/(4\pi\varepsilon_0 r_A^2) \qquad (III.I.25)$$

(angles θ_A et θ_B are defined in Fig. III.I.3) and Eq. III.I.21 becomes:

$$f_B^c = \int \rho(x, y, z)\, f_B\, dv \quad \text{with} \quad f_B = - \cos\theta_B\, q^2/(4\pi\varepsilon_0 r_B^2) \qquad (III.I.26)$$

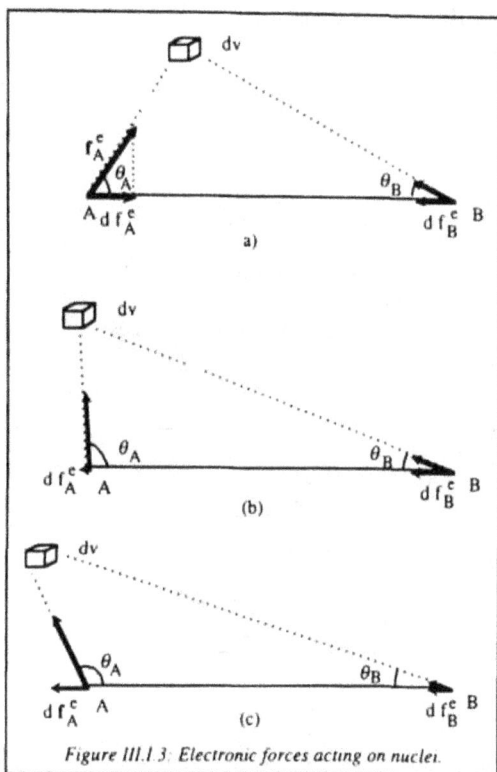

Figure III.1.3: Electronic forces acting on nuclei.

These relationships may be expressed as

$$f_A^e = \int df_A^e \qquad et \qquad f_B^e = \int df_B^e \qquad\qquad (III.I.27)$$

using the projections of the elementary forces acting on the nuclei due to charges in the elementary volume dv

$$df_A^e = \cos \theta_A \, q^2 \rho dv/(4\pi\varepsilon_0 r_A^2) \qquad df_B^e = -\cos \theta_B \, q^2 \rho dv/(4\pi\varepsilon_0 r_B^2) \qquad (III.I.28)$$

Projection df_A^e (or df_B^e) is directed toward the other nucleus if θ_A (or θ_B) is less than $\pi/2$, and in the opposite direction if θ_A (or θ_B) is larger than $\pi/2$. We thus have to consider three cases:

— In Fig. III.1.3a, forces tend to bring the two nuclei together. We conclude that the electron charge in volume element dv has a *bonding effect*.

— In Fig. III.1.3b, force df_A^e acting on nucleus A tends to move it away from nucleus B, but, as $|df_A^e|$ is less than $|df_B^e|$, this brings A nearer to B. We shall therefore consider that the *effect* is still *bonding*.

— In Fig. III.1.3c, where, on the contrary $|df_A^e| > |df_B^e|$, the *effect* will be considered as *antibinding*.

If the two forces are equal with the same direction, the *effect* will be considered as *nonbinding*. The corresponding points are such that

$$(\cos \theta_A)/r_A^2 + (\cos \theta_B)/r_B^2 = 0 \qquad (III.1.29)$$

This set of nonbonding points is separated in space into three regions: A bonding region between nuclei and two antibonding regions on both sides. These regions are similar for all homonuclear diatomic molecules, for which the nuclear charges Z_A et Z_B of the two atoms are equal. These are known as *Berlin regions* [T. BERLIN — Binding regions in diatomic molecules, *J. Chem. Phys.* 19, 208-213 *(1951)*]. (seeFig.III.1.4.)

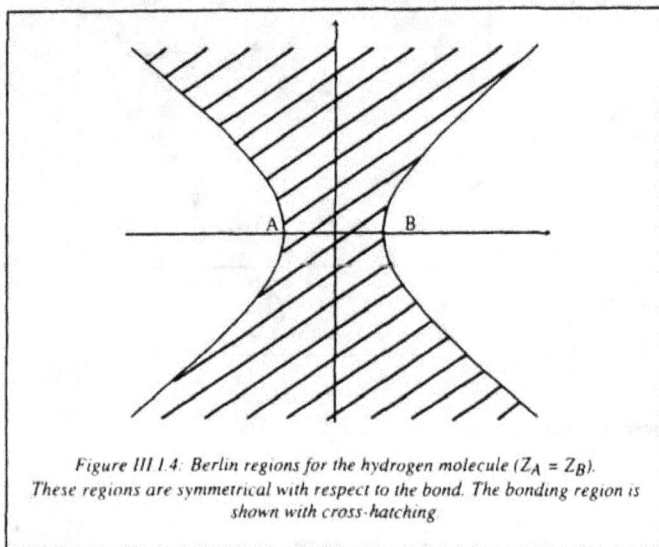

Figure III.1.4: Berlin regions for the hydrogen molecule ($Z_A = Z_B$). These regions are symmetrical with respect to the bond. The bonding region is shown with cross-hatching.

In the case of heteronuclear diatomic molecules, the bonding and antibonding regions proposed by Berlin depend on the charges Z_A et Z_B of the two atoms (Fig. III.1.5). They are separated by surfaces given by

$$Z_A(\cos \theta_A)/r_A^2 + Z_B(\cos \theta_B)/r_B^2 = 0 \qquad (III.1.30)$$

Nevertheless, as shown by Silberbach [H. SILBERBACH, The electron density and chemical bonding: A reinvestigation of Berlin's theorem, *J. Chem. Phys.* 94, 2977-2985 *(1991)*], the representation of electron forces that we have just given is based upon a particular calculation of forces.

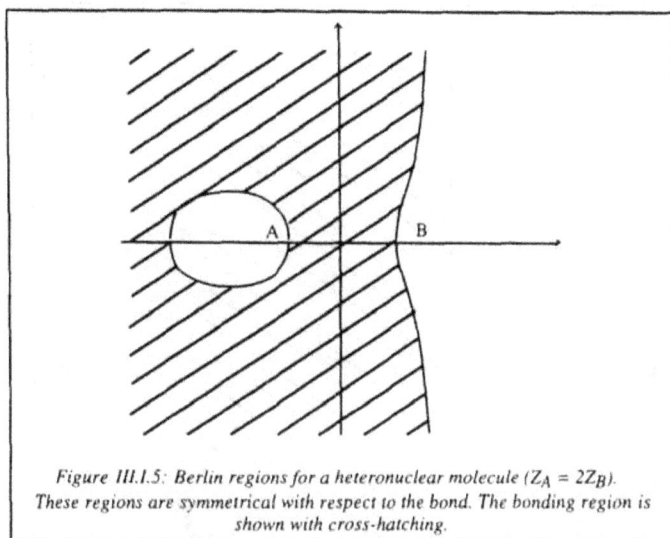

*Figure III.1.5: Berlin regions for a heteronuclear molecule ($Z_A = 2Z_B$).
These regions are symmetrical with respect to the bond. The bonding region is
shown with cross-hatching.*

Indeed, forces f_A^c et f_B^c are equal and of opposite signs (Eq. III.I.24), so that, taking into account Eqs. III.I.25 and III.I.26, it is possible to use the following equation:

$$f_A^c = - f_B^c = \alpha f_A^c - \beta f_B^c = \int \rho(x, y, z)(\alpha f_A - \beta f_B)\, dv \qquad (III.I.31)$$

for any values of α et β as long as $\alpha + \beta = 1$. Since $\rho(x, y, z)$ is positive, the elementary contributions to f_A^c will be bonding if $\alpha f_A - \beta f_B$ is positive and antibonding if this quantity is negative. The bonding and antibonding regions will be separated by the surface (or the surfaces) given by the equation

$$\alpha f_A - \beta f_B = 0 \qquad (III.I.32)$$

Berlin's method is based on $\alpha = \beta = 1/2$, leading to

$$f_A^c = - f_B^c = (f_A^c - f_B^c)/2 = (1/2)\int \rho(x, y, z)(f_A - f_B)\, dv \qquad (III.I.33)$$

assuming that the forces acting on the nuclei are symmetrical. This choice led to the previous definition of bonding and antibonding regions. This choice is, however, arbitrary and we could have used f_A^c only ($\alpha = 1$, $\beta = 0$) or f_B^c only instead ($\alpha = 0$, $\beta = 1$). We would have obtained entirely different bonding and antibonding regions, separated by a plane perpendicular to the bond in A or in B respectively (Fig.III.I.6).

By choosing the values of α and β, it is thus possible to consider each point in space as belonging either to a bonding or to an antibonding region. Concerning heteronuclear molecules, Silberbach has shown that regions are identical to those of homonuclear molecules, provided that proper values are selected for parameters α and β.

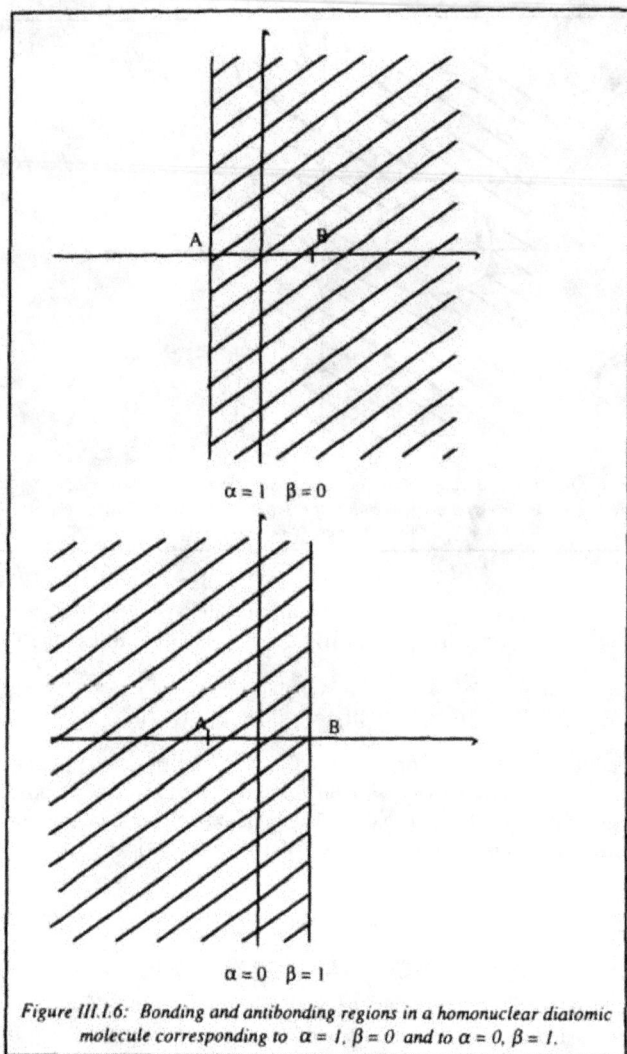

Figure III.I.6: Bonding and antibonding regions in a homonuclear diatomic
molecule corresponding to $\alpha = 1$, $\beta = 0$ and to $\alpha = 0$, $\beta = 1$.

Of course, these remarks weaken the intrinsic meaning of bonding and
antibonding regions, but nevertheless, for a given set of parameters (and why not
those of Berlin which correspond to a rather intuitive image?) the calculation of
forces is correct and it may be interesting to analyze internuclear forces in terms
of these regions. This is the subject of the next section.

II. Applications

1. Diatomic molecules

Following our choice of bonding and antibonding regions (cf., Eq. III.I.2), we may interpret the force acting on a nucleus of a diatomic molecule. This force is considered to be the resultant of three forces: The repulsive force due to other nuclei, the repulsive force due to the electron charges of antibonding regions and the attractive force due to the electron charges of bonding regions. Overall, these forces will be attractive if this last force prevails over the first two.

If we choose the Berlin region model (Fig. III.I.4), the H–H bond may be interpreted as follows: When nuclei are far away from each other, the Coulomb repulsion between nuclei is weak. The total effect will then be attractive if the electron density in the (consequently large) bonding region is high. Thus, we come back to the standard picture of the hydrogen molecule bond, in its ground state, as arising from a large mean electron charge between the two nuclei. On the contrary, if the nuclei are very close, the repulsive value, due mainly to repulsion between the nuclei, increases. It is no longer balanced by the attractive effect corresponding to the bonding region, which is then small.

We then have to consider the electron density between the nuclei. Authors have suggested comparing it with the sum ρ_{atoms} of the electron densities of the two atoms, assumed to be isolated. As these densities are of similar size, it is convenient to use the density difference

$$\Delta\rho = \rho - \rho_{atoms} \qquad \text{(III.II.1)}$$

The contours *of constant density difference* for the hydrogen molecule in a plane containing the bond are calculated using the L.C.A.O. Molecular Orbital method with minimal basis, described in Sec. IV.III.7 (Appendix 9), and are shown in Fig. IV.III.40 for the ground state of H_2. They show that $\Delta\rho$ is positive between the nuclei, which corresponds to the high mean electron density between nuclei necessary for bonding. In the first excited state, on the contrary, $\Delta\rho$ is negative in this region. We can therefore understand why, in that case, the potential energy curve always decreases with R and thus corresponds to repulsive forces for all values of R. [The weak electron density value in the first excited state in the bonding region is related to the antisymmetry of the corresponding wave function with respect to the symmetry of a plane perpendicular to the bond. Indeed, this implies that the wave function vanishes on this plane and that values in the vicinity of this plane (i.e., in the bonding region) are low.]

Similar explanations are valid in the case of other diatomic homonuclear molecules. The differential isodensity contours of some molecules are shown in Fig. III.II.12.

Nevertheless, we should point out that the generalization of these interpretations may be criticized:

— The bonding or antibonding nature of electron forces is strictly equal to the vector sum of all the contributions corresponding to all points in space. We must not restrict our analysis to the positive or negative contributions of one region only, as we have done implicitly. For instance, in the F_2 molecule, for which the charges in the bonding region are relatively weak, bonding is caused by a reduction of the electron density in strongly antibonding regions located away from the F_2 bond.

— The Hellmann-Feynman theorem does not apply to the sum of atomic densities because these densities do not arise from a wave function of the molecule.

— In any event, ρ_{atoms} does not usually correspond to a zero internuclear force, but rather to a repulsive force. Thus, this function is not a fully satisfactory starting point.

2. Other molecules

The calculation of electron density of forces of electron origin acting on the nuclei is easily extended to more complex molecules. Thus, we have at our disposal two methods, in principle entirely equivalent, for the theoretical determination of internuclear forces in a molecule with a given molecular geometry. The first method consists of calculating electron energies, within the Born-Oppenheimer approximation, as a function of the positions of the nuclei

$$E_{el} (..., x_K, y_K, z_K, ...)$$

and deducing the components F_{Ku} of the force on each nucleus K along a given direction Ou ($u = x, y, z$), by taking a derivative of this expression with respect to u_K

$$F_{Ku} = - E_{el} / u_K \tag{III.II.2}$$

The second method is related to the calculation of the electron density $\rho(x, y, z)$ for the given geometry and deduces the force acting on a nucleus K from the expression

$$F_K = f_K^e + \sum_{L \neq K} f_K(L) \tag{III.II.3}$$

where $f_K(L)$ is the Coulomb law force produced by nucleus L on nucleus K and f_K^e is the electron contribution to the total force :

$$f_K^e = \int \rho(x, y, z) \, f_K \, dv, \tag{III.II.4}$$

As previously shown, this may be expressed as a function of total electron density $\rho(x, y, z)$ at each point of space, and of the force f_K which would be produced on nucleus K by an electron located at coordinate point (x, y, z) . The equilibrium geometry arises from the cancellation of the resultant

$$G_N = \sum_{L \neq K} f_k(L)$$

of internuclear forces of electron origin f_K^e (Fig. III.II.1).

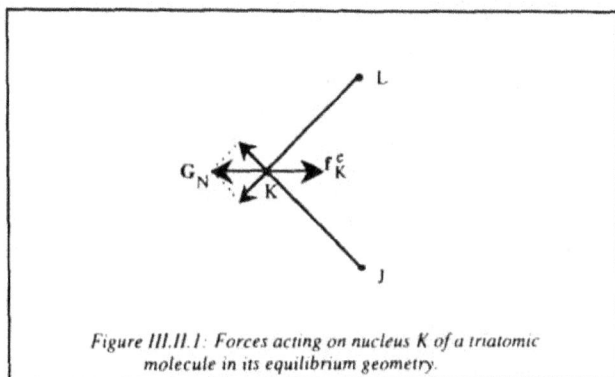

Figure III.II.1: Forces acting on nucleus K of a triatomic molecule in its equilibrium geometry.

In spite of the above criticism and the lack of invariance of bonding and antibonding regions, it may be interesting to discuss electron density and contours of constant density difference in detail.

These electrostatic explanations are of pedagogical value, as they help to understand the origin of chemical bonds, inasmuch as they reduce a chemical bond to an image of the electrostatic interaction between nuclei and an electron cloud of negative charges distributed in normal, three-dimensional space. We may wonder, however, what use these explanations may be for chemists.

We would like to make the following two remarks on this subject:

— The main use of the methods described in this chapter concerns molecular orbitals, as, at this level, the characteristic shapes of atomic orbitals s, p, d, ..., and their signs in the different regions of space lead often to suggestive qualitative explanations of the "strength" of chemical bonds. The success of these explanations is based, of course, upon the possibility of considering, in an initial approximation, some properties as a sum of single-electron contributions. This is not always correct, in particular, when electron correlation is high (cf., Eq. IV.III). Unfortunately, is usually impossible to demonstrate in terms of total electron density which tends to average these contributions. An example will be given relating to the water molecule (cf., Sec. IV.II).

— The two methods we described for calculating forces acting on nuclei in a molecule assume that the eigenvalue equation of the electron hamiltonian was solved beforehand. Meanwhile, it is interesting to note that Hohenberg et Kohn showed in 1964 that electron energy E_{el} was a functional of total electron density $\rho(x, y, z)$. This property raised hopes that it would be possible to realize the theoretical study of a molecule using explicitly only this density in space $\mathbb{R}^3$, instead of the use of a space of 4n variables. But, unfortunately, the form of this functional is unknown. The electron density of a molecule cannot have any form. In particular, it has to satisfy some conditions: It must take into account the coulomb repulsion between electrons (electron correlation) and it must correspond to an antisymmetric wave function with respect to electron exchange, accordingly to the Pauli antisymmetry principle (it is said to be n-representable). In other words, in order to achieve a good representation of a molecule, the electron density must implicitly include the very complex properties necessary for an accurate analysis of chemical bonds. Some methods (*local density method*), derived from the Hohenberg et Kohn theorem, have been used with approximate functionals to take into account the correlation and exchange terms. These methods may give similarly accurate results to those of the Hartree-Fock method. Nevertheless, at present, these methods do not seem to be able to compete with the one derived from the Schrödinger equation, which are by far the most widely used.

3. Computer calculation of the Berlin zones of H_2

Let us take an A-B diatomic molecule (or ion), positioned on the x'Ox axis , as shown in Fig. III.II.2. An elementary volume dv is considered around point M, located by means of distances r_A and r_B and angles θ_A and θ_B. Express conditions for these variables so that the electron charge, $- q\rho dv$, in this volume tends either to push the two nuclei apart or to draw them closer together. We shall call Z_A and Z_B the number of protons in those nuclei (ρ is the volumic electron density at M).

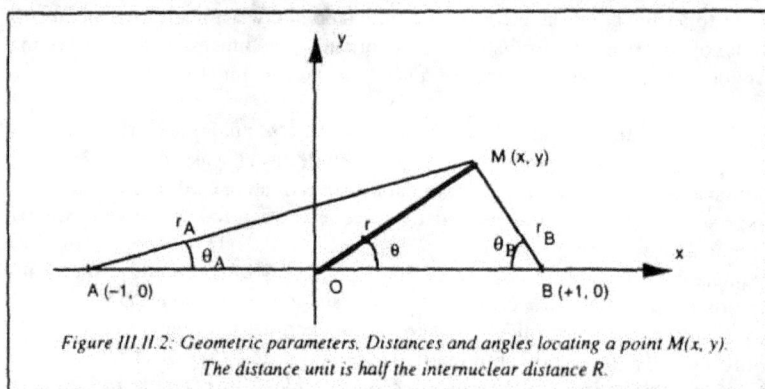

Figure III.II.2: Geometric parameters. Distances and angles locating a point M(x, y). The distance unit is half the internuclear distance R.

As the molecule is symmetrical, the sum of the electron strengths on the nuclei will act along the axis x'Ox. We shall therefore consider only the components along x'Ox for the elementary strengths $d\vec{f}_A$ and $d\vec{f}_B$ acting on the nuclei due to the electrical charge in the elementary volume dv .

The following expressions are given by Coulomb's law:

$$df^e_{A,x} = e^2 Z_A \frac{\cos \theta_A}{r_A^2} \rho dv$$

$$df^e_{B,x} = - e^2 Z_B \frac{\cos \theta_B}{r_B^2} \rho dv$$

As in Sec. II.1.4.4, we have expressed the squared electronic charge as follows $q^2/4\pi\varepsilon_0 = e^2$.

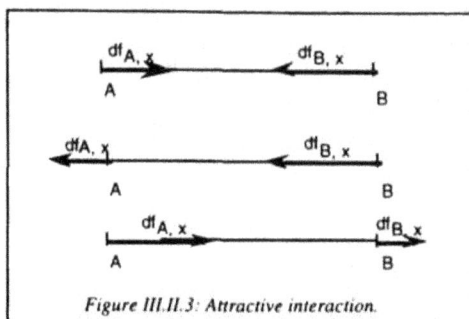

Figure III.II.3: Attractive interaction.

The nuclei are globally submitted to an attractive strength in the cases shown in Fig. III.II.3, corresponding to $df_{A,X} > df_{B,X}$ and leading to

$$Z_A \frac{\cos \theta_A}{r_A^2} + Z_B \frac{\cos \theta_B}{r_B^2} > 0$$

The nuclei are submitted to an overall electron strength which would tend to push the two nuclei apart in the cases shown in Fig. III.II.4, corresponding to $df_{A,X} < df_{B,X}$ and leading to

$$Z_A \frac{\cos \theta_A}{r_A^2} + Z_B \frac{\cos \theta_B}{r_B^2} < 0$$

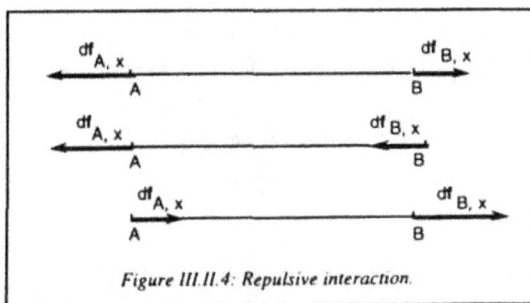

Figure III.II.4: Repulsive interaction.

The equality $Z_A \frac{\cos \theta_A}{r_A} + Z_B \frac{\cos \theta_B}{r_B} = 0$ defines a revolution surface around x'Ox.

We shall only represent its intersection with a plane containing the two atoms.

Expressing the frontier-curve equation with cartesian coordinates in the case of $Z_A = Z_B = 1$ (hydrogen molecule). We shall show that the curve includes both atoms and study the symmetry.

The frontier-curve equation is

$$\frac{\cos \theta_A}{r_A^2} + \frac{\cos \theta_B}{r_B^2} = 0$$

with

$$\cos \theta_A = \frac{1+x}{\left[(1+x)^2 + y^2\right]^{1/2}} \quad ; \cos \theta_B = \frac{1-x}{\left[(1-x)^2 + y^2\right]^{1/2}}$$

So the equation in cartesian coordinates is

$$\frac{1+x}{\left[(1+x)^2 + y^2\right]^{3/2}} + \frac{1-x}{\left[(1-x)^2 + y^2\right]^{3/2}} = 0$$

Let us consider $y = 0$. That leads to

$$\frac{(1+x)(1-x)^3 + (1-x)(1+x)^3}{(1+x)^3(1-x)^3} = 0$$

The final equation

$$1 - x^4 = 0$$

obtained by developing the numerator, has two double solutions $(+1)$ and (-1).

Points A and B are on the graph. Indeed the previous equation is unchanged if $-x$ is put in place of x and $-y$ in place of y. Both Ox and Oy are axes of symmetry.

Let us show that the graph has two asymptotes at an angle $2\theta = 109°28'$.

The frontier-curve equation in polar coordinates (r, θ) is

$$\frac{1 + r\cos \theta}{\left[1 + r^2 + 2r\cos \theta\right]^{3/2}} + \frac{1 - r\cos \theta}{\left[1 + r^2 - 2r\cos \theta\right]^{3/2}} = 0$$

If r tends towards infinity, the previous equation becomes :

$$\frac{1 + r\cos \theta}{\left[1 + \frac{2\cos \theta}{r}\right]^{3/2}} + \frac{1 - r\cos \theta}{\left[1 - \frac{2\cos \theta}{r}\right]^{3/2}} = 0$$

By using the following series expansion, we find

$$\left[1 \pm \frac{2}{r} \cos \theta\right]^{-3/2} \# 1 \mp \frac{3}{r} \cos \theta$$

We obtain

$$\cos \theta = 1/\sqrt{3}$$

i.e.,

$$2\theta = 2 \cos^{-1} \sqrt{3}/3 \# 109°28'$$

Let us draw the graph representing the frontier curve.

The equation is

$$f(x,y) = \frac{1+x}{\left[(1+x)^2 + y^2\right]^{3/2}} + \frac{1-x}{\left[(1-x)^2 + y^2\right]^{3/2}} = 0$$

As the previous equation is implicit, we need to find values of x that fulfill the equation $f(x,y) = 0$ for different values of y $(-\infty < y < +\infty)$.

As (i) there is no solution for $-1 < x < +1$, and (ii) the graph is symmetrical around the Oy axis, we restrict the determination of the x values to those which fulfill the condition $x \geq 1$. These solutions may be obtained by using Newton's approximation and a small computer program. The results are shown in Fig. III.II.5.

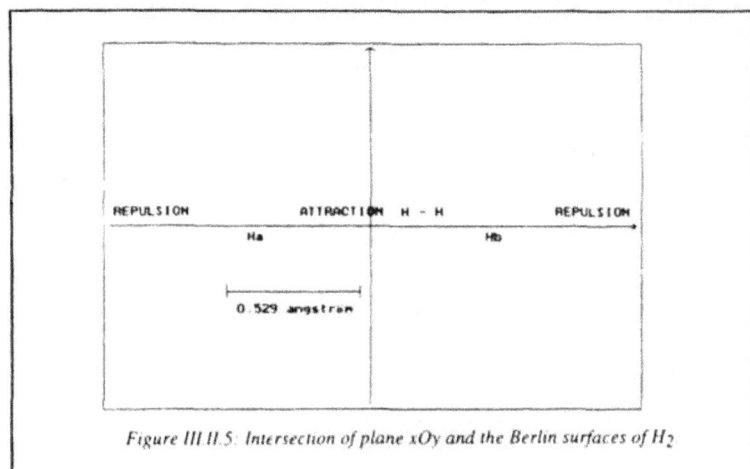

Figure III.II.5: Intersection of plane xOy and the Berlin surfaces of H_2

4. Appendix 6: Forces and potential energies

Elementary standard mechanics lessons use the notion of *forces*. However, it is often more convenient to replace the forces $\mathbf{f}_i$ (f_{xi}, f_{yi}, f_{zi}) acting on each particle i by a function V, called the *potential energy function*. Except for an integration constant, this function is defined by its differential dV

$$dV = -\sum_i \mathbf{f}_i \, d\mathbf{r}_i \qquad\qquad (III.II.5)$$

expressed in Cartesian coordinates

$$dV = -\sum_i (f_{xi} \, dx_i + f_{yi} \, dy_i + f_{zi} \, dz_i) \qquad\qquad (III.II.6)$$

Identifying this expression with that of the function differential

$$dV = -\sum_i ((dV/dx_i) \, dx_i + (dV/dy_i) \, dy_i + (dV/dz_i) \, dz_i). \qquad\qquad (III.II.7)$$

we easily obtain the equations

$$f_{xi} = -(dV/dx_i) \qquad f_{yi} = -(dV/dy_i) \qquad f_{zi} = -(dV/dz_i) \qquad\qquad (III.II.8)$$

which connect this potential function to the forces acting on the various particles.

Example 1: Harmonic oscillator

In the case of a one-dimensional harmonic oscillator in which force f_x is proportional to elongation (Hooke's rule) :

$$f_x = -kx ,\qquad\qquad\qquad\text{(III.II.9)}$$

the potential function is

$$dV = -f_x\,dx = kx\,dx\qquad\qquad\text{(III.II.10)}$$

which gives

$$V = \int kx\,dx = (1/2)kx^2 + constant\qquad\text{(III.II.11)}$$

As the physical properties are independent of the integration constant, this constant is considered to be equal to 0.

Example 2: Potential energy of two interacting charged particles

When representing two forces $f_i(j)$ and $f_j(i)$, of equal magnitudes, acting in opposite directions, arising from the action of a particle i at rest on a particle j at rest and vice-versa, it is convenient to write

$$f_i(j) = -u_{ij}\,f_i = -f_j(i) ,\qquad\qquad\text{(III.II.12)}$$

where f_i is an algebraic quantity, whose absolute value is equal to the modulus of the force, while u_{ij} represents the unit vector associated with vector r_{ij} connecting point i to point j. Taking into account the preceding expression, f_i is positive if the forces are repulsive and negative if the forces are attractive. The increase in potential energy may then be expressed :

$$dV = -f_i(j)\,dr_i - f_j(i)\,dr_j\qquad\qquad\text{(III.II.13)}$$

$$dV = f_i(j)\,(dr_j - dr_i)\qquad\qquad\qquad\text{(III.II.14)}$$
$$= f_i(j)\,dr_{ij} = -f_i\,u_{ij}\,dr_{ij} = -f_i\,dr_{ij}\qquad\text{(III.II.15)}$$

where dr_{ij} is the increment of the distance between the two points.

If V depends only on the distance r_{ij}, we then have

$$f_i = -dV/dr_{ij}\qquad\qquad\qquad\text{(III.II.16)}$$

In view of the definition of f_i (Eq. III.II.12), the derivative dV/dr_{ij} is negative if the forces are repulsive, and positive if the forces are attractive (Fig. III.II.6). This is coherent with the well-known property that forces tend to drive a system toward the state of minimum potential.

Example 3: Coulomb law potential energy of two charged particles

Coulomb forces acting on two particles with electrical charges q_i and q_j are of the preceding type with

$$f_i = q_i\,q_j\,/\,(4\pi\,\varepsilon_0\,r_{ij}^2)\qquad\qquad\text{(III.II.17)}$$

We then have

$$V = V_{ij} = -\int f_i\,dr_{ij},\qquad\qquad\text{(III.II.18)}$$

whence

$$V_{ij} = -(q_i\,q_j\,/\,(4\pi\,\varepsilon_0))\int (1/r_{ij}^2)dr_{ij} = q_i\,q_j\,/(\,4\pi\,\varepsilon_0\,r_{ij}) + C\qquad\text{(III.II.19)}$$

As in the preceding case, the constant is considered to be equal to 0.

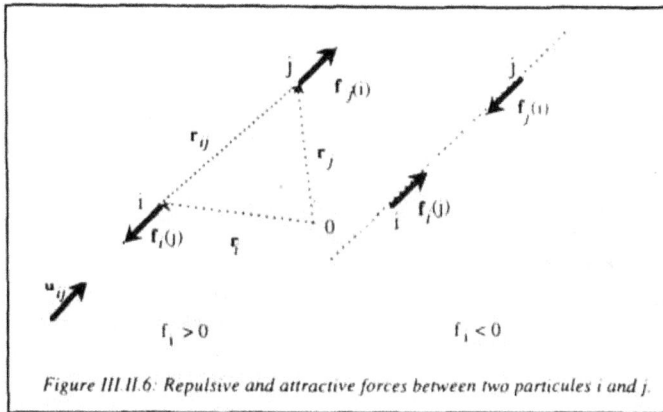

Figure III.II.6: Repulsive and attractive forces between two particles i and j.

5. Appendix 7: The Hellmann-Feynman theorem

In the electron hamiltonian operator H_{el}, nuclear coordinates are assumed to be parameters, *i.e.*, quantities which may take various values but which do not correspond to position variables of moving particles in the system under consideration. If we consider an operator $H(\lambda)$ depending on a parameter of this type, λ, and $\phi(\lambda))$ a normalized eigenfunction of this operator associated with energy $E(\lambda)$. As $E(\lambda)$ is equal to the mean value of $H(\lambda)$, we may write

$$E(\lambda) = \int \phi(\lambda)^* \, H(\lambda) \, \phi(\lambda) \, dv \qquad (III.II.20)$$

$$dE(\lambda)/ \, d\lambda = \int(\partial\phi(\lambda)^*/\partial\lambda) \, H(\lambda) \, \phi(\lambda) \, dv + \int \phi(\lambda)^* \, (\partial H(\lambda) \,/\partial\lambda) \, \phi(\lambda) \, dv$$
$$+ \int \phi(\lambda)^* \, H(\lambda)(\partial\phi(\lambda)/ \, \partial\lambda) \, dv \qquad (III.II.21)$$

[Note that $E(\lambda)$ depends only on parameter λ, while $\phi(\lambda)$ also depends on particle coordinates.] Taking into account the hermiticity of $H(\lambda)$ (cf., Eq. II.I.3), the sum of the first and third integrals is equal to

$$E(\lambda)[\int(\partial\phi(\lambda)^*/\partial \, \lambda) \; \phi(\lambda) \, dv + \int \phi(\lambda)^* \, (\partial \, \phi(\lambda)/ \, \partial\lambda)dv]$$
$$= E(\lambda) \, (\partial \,/ \, \partial\lambda)\int\phi(\lambda)^* \, \phi(\lambda) \, dv \qquad (III.II.22)$$

The left-hand side of this equation is eliminated, as, $\phi(\lambda)$ is assumed to be normalized, so the derivative of the normalization integral is zero, leading to

$$dE(\lambda)/ \, d\lambda = \int\phi(\lambda)^* \, (\partial H(\lambda) \,/\partial\lambda) \, \phi(\lambda) \, dv \qquad (III.II.23)$$

Therefore, the energy derivative with respect to a parameter is equal to the mean value of the derivative of the hamiltonian operator with respect to this parameter: This is the *Hellmann-Feynman theorem*. Its application to coordinate x_A of nucleus A of the hydrogen molecule leads to the expression

$$F_{Ax} = - \partial E_{el}/\partial x_A$$
$$= \Sigma_{\sigma1}\Sigma_{\sigma2} \int \int \psi_{el}(1,2)^* \{ f_{Ax}(1) + f_{Ax}(2) + f_{Ax}(B) \} \; \psi_{el}(1,2) \, dv_1 \, dv_2, \quad (III.II.24)$$

which is the projection along the Ox axis of vector relation (Eq. III.I.13).

Nota bene: Equation III.II.23 is also derived from Eq. III.II.21 if $\phi(\lambda)$ is a variational function corresponding to minimum energy. This theorem therefore also applies to the most frequent case studied in quantum chemistry.

6. Appendix 8 : A didactic adaptation of the mechanical aspects of chemical bonding. A "hands on" explanation of the hydrogen molecule

The following is a teaching program that has been adapted for French students from secondary school to university level

— *Secondary school syllabus:*

Before speaking of chemical bonding, we must be sure that the students have properly understood the nondeterministic nature of the electron and rejected both the planetary model and circular orbits. We have already proposed an elementary presentation for younger students, based on games (see Sec. II.I.8, Appendix 3). We make sure to eliminate all the figures showing circular orbits, preferring those with electron "clouds" (see Fig. III.II.7), ensuring that the word "cloud" has been clearly defined and properly understood.

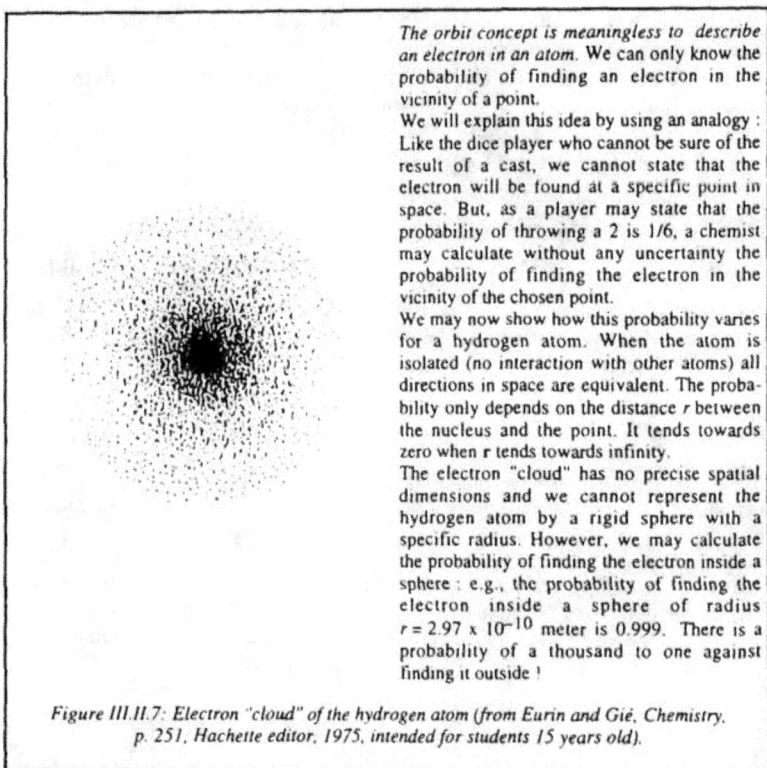

The orbit concept is meaningless to describe an electron in an atom. We can only know the probability of finding an electron in the vicinity of a point.

We will explain this idea by using an analogy : Like the dice player who cannot be sure of the result of a cast, we cannot state that the electron will be found at a specific point in space. But, as a player may state that the probability of throwing a 2 is 1/6, a chemist may calculate without any uncertainty the probability of finding the electron in the vicinity of the chosen point.

We may now show how this probability varies for a hydrogen atom. When the atom is isolated (no interaction with other atoms) all directions in space are equivalent. The probability only depends on the distance r between the nucleus and the point. It tends towards zero when r tends towards infinity.

The electron "cloud" has no precise spatial dimensions and we cannot represent the hydrogen atom by a rigid sphere with a specific radius. However, we may calculate the probability of finding the electron inside a sphere : e.g., the probability of finding the electron inside a sphere of radius $r = 2.97 \times 10^{-10}$ meter is 0.999. There is a probability of a thousand to one against finding it outside !

Figure III.II.7: Electron "cloud" of the hydrogen atom (from Eurin and Gié, Chemistry. p. 251, Hachette editor, 1975, intended for students 15 years old).

Once these notions are well-established, we may go on with the "hands on" presentation of the hydrogen molecule, with the help of Fig. III.II.8.

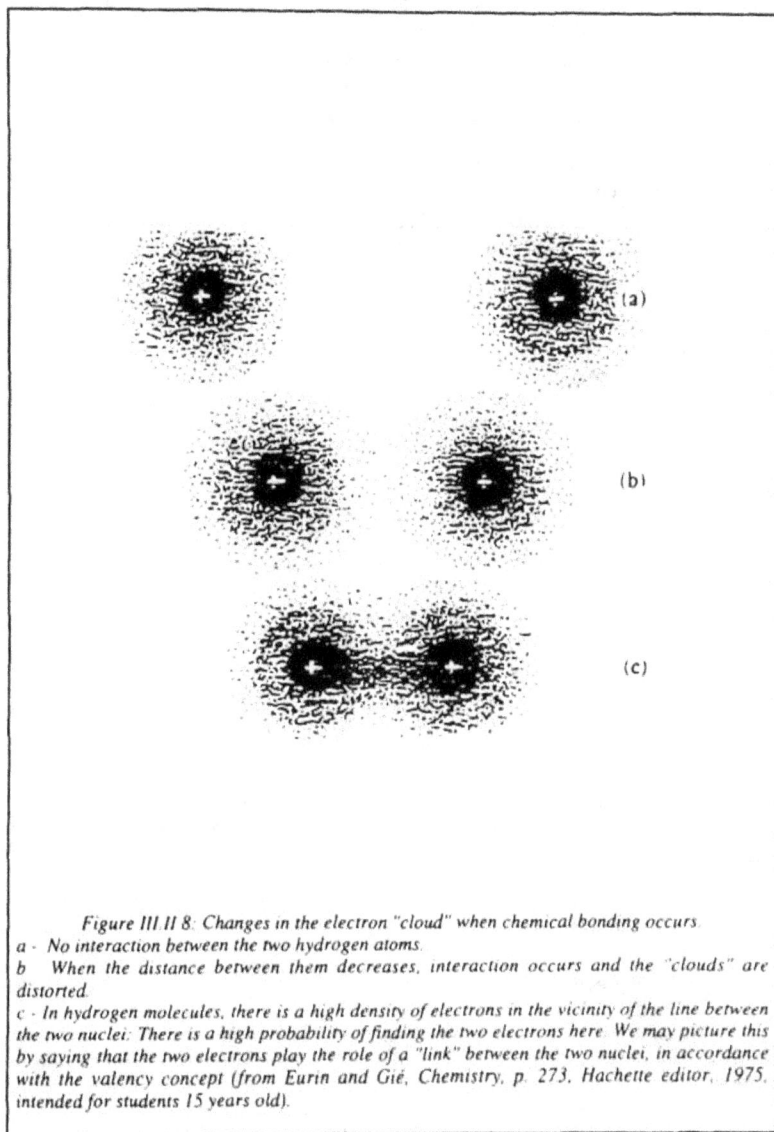

Figure III.II 8: Changes in the electron "cloud" when chemical bonding occurs.
a - No interaction between the two hydrogen atoms.
b   When the distance between them decreases, interaction occurs and the "clouds" are distorted.
c - In hydrogen molecules, there is a high density of electrons in the vicinity of the line between the two nuclei: There is a high probability of finding the two electrons here. We may picture this by saying that the two electrons play the role of a "link" between the two nuclei, in accordance with the valency concept (from Eurin and Gié, Chemistry, p. 273, Hachette editor, 1975, intended for students 15 years old).

The two nuclei, carrying positive electric charges (that we may represent by our two fists) are separated by R. They repel one another with a force

$F = q^2/4\pi\varepsilon_0 R^2$. The density of the electron cloud between the nuclei (fists) produces an attractive force between them. The electron density in the space on the opposite sides of the nuclei produces a repulsive force: A negative charge leads to a larger attraction for the closest nucleus (fist) than for the farthest one.

There is nothing mysterious about these Coulomb electric forces. They depend on the distance R between the nuclei. Mechanical equilibrium is obtained for a particular value R_e and chemical bonding occurs.

Figure III.II.9 is easily understood if the fundamental part played by the electrons has been properly established.

Equilibrium distance between nuclei .

We consider a system consisting of two hydrogen atoms (two protons and two electrons). There are electrical interactions between these charged particles: Each electron is submitted at the same time to the attractive force of the nuclei and to the repulsive force of the second electron. In the same way, each nucleus is submitted to the attractive force of the two electrons and to the repulsive force of the other nucleus.

When one nucleus is very far from the other, the attractive force predominates. In the opposite case, repulsion becomes predominant. Attractive and repulsive forces are equal for $R = R_e$. R_e corresponds to the equilibrium distance between the nuclei (74.1 pm for hydrogen molecule).

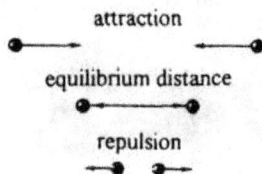

attraction

equilibrium distance

repulsion

Figure III.II.9: Equilibrium distance between two nuclei (from Eurin and Gié, Chemistry, p. 273, Hachette editor, 1975, and for 15 years old students).

We see that the electrons may induce both stabilization and destabilization, and not only stabilization as often assumed.

— *University undergraduate syllabus:*

We must first indicate that we may legitimately assume that the nuclei are practically immobile. The Born-Oppenheimer approximation is presented in its simplest form: As the mass of the electrons is 1836 times smaller than that of the protons, they move very quickly around the protons. A camera attached to a proton would see a static negative electricity in the whole space while, if it were attached to the electron, it would see a fixed positive charge. We may then introduce the expression of the force acting on the nuclei (see Fig. III.II.2), having first specified that we are dealing with the hydrogen molecule in its lowest energy state and ignoring the spin.

For instance, the force acting on nucleus A may be expressed :

$$F_{Ax} = + (q^2 / 4\pi\varepsilon_0) \iiint\limits_{x\ y\ z} \frac{\rho(x,y,z)}{r_A^2} \cos\theta_A \ dx dx dz - \frac{1}{4\pi\varepsilon_0} q^2 / R^2$$

At this level of teaching, ρ represents the probability density of finding one electron (charge -1.6×10^{-19} C) at the point defined by (x, y, z), the other being anywhere in space. ρ may be calculated and its value is very near the experimental data obtained from X-ray diffraction. ρ is nondeterministic and, consequently, the last expression must be considered to be derived from quantum mechanics.

Having discussed the physical interpretation of the graph showing the molecule energy variations in function of R (see Fig. III.I.2), we must clearly indicate that U(R) represents the potential energy from which the forces are derived. The sign of the first derivative dU(R)/dR defines the attractive and repulsive parts of the curve, respectively, and R_e is naturally introduced. We then indicate the part played by the electrons by defining the differential one-electron density given by

$$\Delta\rho = \rho - \sum_{atoms} \rho_a$$

and use diagrams like the one shown in Fig. III.II.1. or computer-assisted systems, as described in Appendix 11 (Sec. IV.III.9).

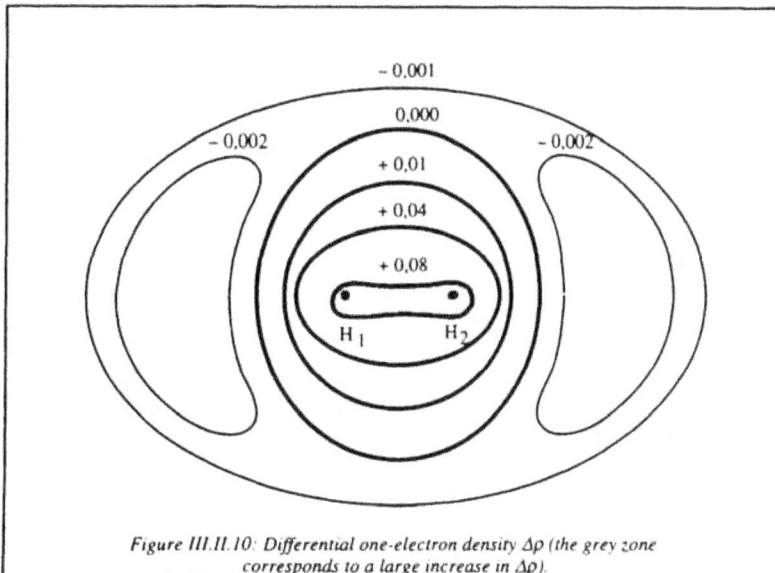

Figure III.II.10: *Differential one-electron density* $\Delta\rho$ *(the grey zone corresponds to a large increase in* $\Delta\rho$*).*

— University post-graduate syllabus:

The whole process will be presented, using documents like the one in Fig.
III.II.11, and computer-assisted systems, used by the students, as outlined in Secs.
III.II.3 and IV.III.7.

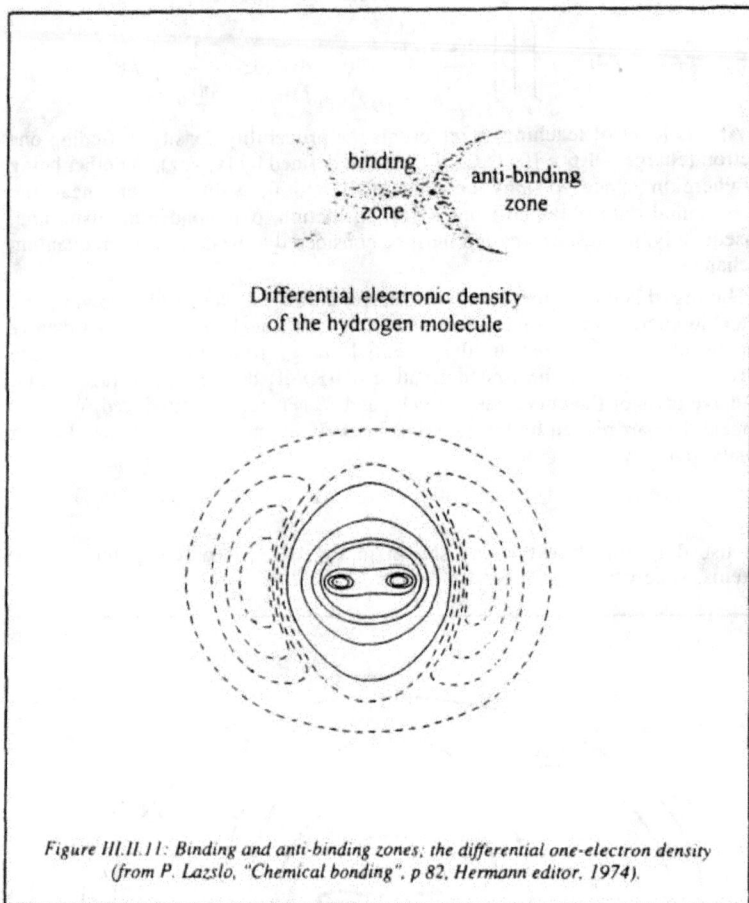

binding zone — anti-binding zone

Differential electronic density
of the hydrogen molecule

*Figure III.II.11: Binding and anti-binding zones; the differential one-electron density
(from P. Lazslo, "Chemical bonding", p 82, Hermann editor, 1974).*

Our own experience has revealed two major difficulties encountered by
students in chemical physics:

— They believe that quantum mechanics postulates that $|\Psi(\mathbf{r}, t)|^2$ represents
the probability density. In fact, this property is really derived from the fourth
postulate (see Sec. II.I.4). The "wave function and probability" section (Sec.
II.I.4.3) must be properly understood by students.

— It would appear to be difficult to progress from $|\Psi_{elec}(1, 2, ..., n)|^2$ to ρ. The physical meaning of these two parameters is often improperly understood. We must clearly show that

$$\sum_{\sigma_1} ... \sum_{\sigma_n} |\Psi_{elec}(1, 2,..., n)|^2$$

represents the n-electron density, i.e., the probability of simultaneously finding electron 1 in the unit volume near r_1, electron 2 in the unit volume near r_2, etc. It is a function of a set of 3n space variables. On the other hand, the one-electron density $\rho(r)$ represents the probability of finding any of the n electrons at point r, with any spin, the others being anywhere in space with undetermined spin. Ψ_{elec} is generally obtained by solving the eigenvalues equation associated with the electron hamiltonian operator.

Finally, we shall present and comment the results obtained for some usual diatomic molecules, as shown in Fig. III.II.12.

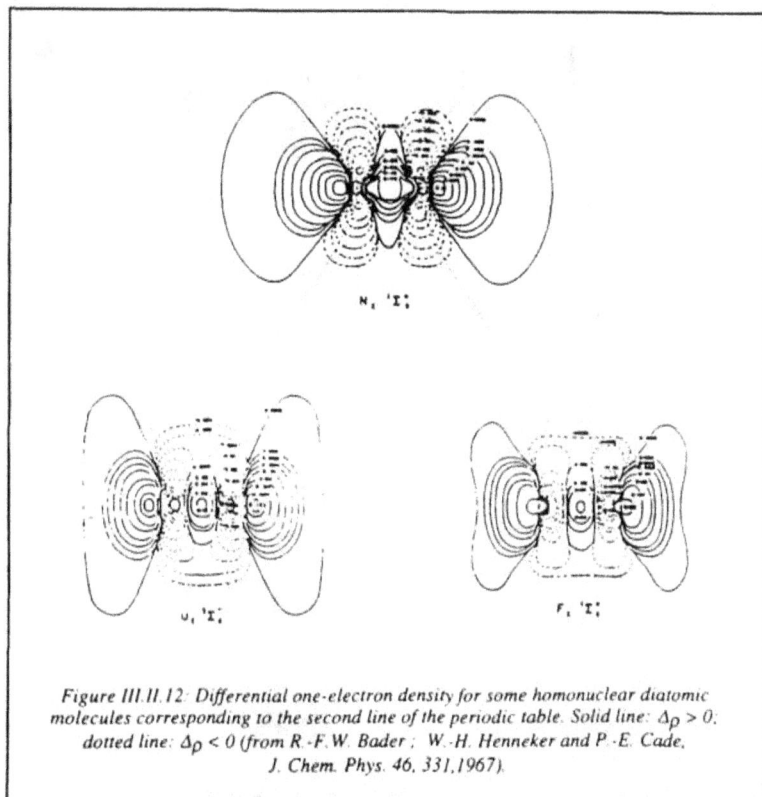

Figure III.II.12: *Differential one-electron density for some homonuclear diatomic molecules corresponding to the second line of the periodic table. Solid line: $\Delta\rho > 0$; dotted line: $\Delta\rho < 0$ (from R.-F.W. Bader ; W.-H. Henneker and P.-E. Cade, J. Chem. Phys. 46, 331,1967).*

CHAPTER IV

Language of orbitals and chemical bonding: applications and limits

This chapter provides a concise presentation of certain major elements of the now well-established methodology for calculating atomic and molecular wave functions. The main objective is to provide the nonspecialist in theoretical chemistry with a vocabulary and basic notions in physical chemistry — a fast-developing field — in order to facilitate understanding of texts on theoretical quantum chemistry. It is vital to understand these concepts and terms, in view of the increasing availability of quantum chemistry calculations and graphics software. We shall attempt, in a few pages, to set out the principles of most of the standard methods (both *ab initio* and semi-empirical) of quantum chemistry, and to outline the limitations of these models. It may be useful to cite a few didactic references, listed in Appendix 9, and which, in addition to an excellent presentation of quantum chemistry, also include a bibliography of related questions (Pilar 1990, Levine 1991, Rivail 1989). A recent, more specialized, work is that of Szabo and Ostlund (Szabo 1989). An excellent, older publication, which has become a classic, is the book by Pauling and Wilson (Pauling 1935). It includes a very lucid presentation of the numerical construction of the atomic self-consistent field (SCF), originally developed by Hartree in 1928. Finally, a fairly old article by Blinder provides a good review of SCF methods (Blinder 1965). In addition to these fairly complete presentations, we wish to cite a few references of pedagogical character with a methodological approach : "Concepts...chimiques", listed under (Ducasse 1985), "Parlez théorique" (Bigot 1984) (Anh 1987) (Ohanessian 1989) and "Faut-il... hybridation" (Chaquin 1984). Numerous articles dealing with quantum chemistry are regularly published in the Journal of Chemical Education.

Section IV.I. presents the average field method attributed to Hartree and Fock. This method plays a major role, not only in quantum chemistry, but also in all fields dealing with the particle interaction, representing, in general, a zero-order approximation to this type of problem. In Sec. IV.II, it is applied to the determination of a wave function acceptable as a representation of the water molecule.

This example will be used to define the concept of molecular orbital (MO) based on atomic orbitals (AO), and to demonstrate the invariant character of the MO concept. An understanding of this method may be used to justify simpler methods for the construction of MO's not requiring a computer (cf., Sec. IV.II.7.2). In addition, molecular orbitals provide a basis for various simple interpretations frequently used in organic and coordination chemistry.

I. One-electron treatment of many-electron systems

1. Introduction to the determination of molecular wave function

In quantum mechanics, the wave function $\psi(q,t)$, a function of particle coordinates and time, describes the state of a given system (1st postulate - cf., Sec. II.1.4.1). The symbol q stands for all the coordinates, i.e., all the spatial and spin coordinates of all particles. The wave function contains all the information about a given system, and its determination therefore represents a major objective. We will first determine it by an "ab initio" approach, as the eigenfunction of a hamiltonian operator (cf., Sec. II.1.4.3), and later use this wave function to analyze the electron structure of a molecular system. We assume the validity of the Born-Oppenheimer (BO) approximation (cf., Sec. III.I.1).

Next the quantum problem concerns the determination of stationary states of a collection of electrons in the field of fixed nuclei. Among all the eigenfunctions of a given system, our interest will be limited to the eigenfunction of the lowest energy, that is to say, to the system ground state. In particular, the determination of eigenstates is complicated by the presence in the hamiltonian of terms describing the interaction among electrons, thus precluding an analytic solution of the eigenvalue equation. The approximate methods used tend to converge towards accurate solutions.

These methods make use of the so-called variational theorem which forms the basis of procedures for wave function optimization.

2. Variational theorem and method

This point was already mentioned in Sec. II.6 and is described here in greater detail. In a quantum system described by hamiltonian H, φ is a convenient function that obeys the boundary conditions of the problem. The ratio W defined below is shown to be at least equal to system ground state energy E_0. the lowest eigenvalue of H. W will only be equal to E if φ is actually the eigenfunction φ_0 of H which describes the ground state in question. We thus have the following inequality:

$$W = \frac{\displaystyle\int_D \varphi^* \, H\varphi \, d\tau}{\displaystyle\int_D \varphi^*\varphi \, d\tau} \geq E_0 \qquad (IV.I.1)$$

The symbol D stands for the domain of integration of the variables characterizing a given problem and must be defined explicitly for each specific case. On the basis of this theorem one may set up a method for wave function determination called variational method. The principle consists of using for φ a function with adjustable parameters, called variational parameters. This function is called the trial function. The idea is to find a collection of parameters which will produce the lowest value of W. The ratio W is also called the variational ratio, or, more simply, variational integral, with reference to the numerator, since the denominator is merely the square of the normalization constant. When the minimum is found, one is sure of having determined the upper limit to the ground state energy. One may also say that the best function was found, in the variational sense, of type φ chosen at the outset.

Remark 1: It should be realized that this theorem concerns only the observable energy; there is no variational theorem for other observables. Therefore it does not follow that a function that yields an energy close to the exact value E is necessarily a good function for some other physical property, such as dipole moment or bond length.

Remark 2: It would seem that the variational theorem concerns only the ground state. Actually there is no a priori assurance that the same procedure will converge for higher energy states, unless, for example, if for symmetry reasons states belong to different irreducible representations. In these cases there may be a variational integral for each symmetry. In this case, our interest concerns only the ground state. Nevertheless, the determination of higher energy states is necessary in spectroscopy or in the analysis of photochemical reactions, where this problem needs to be resolved.

It was shown by McDonald in 1933 (McDonald 1933) that if the trial function φ is developed linearly on the basis of p functions χ_i , each of which obeys the boundary conditions of the problem, i.e., if

$$\varphi = \sum_{i=1}^{p} c_i \chi_i \qquad (IV.I.2)$$

then the variational method yields an upper limit to W_0, W_1,..., W_{p-1} for each of the first p eigenvalues of the hamiltonian. On account of this property, it is crucial to introduce a linear dependence among the variational parameters, as far as possible. This property, related to the linear character of operators, is at the origin of quantum chemistry methods, such as the configuration interaction method mentioned in Sec. IV.II.

3. Choice of zero-order function for a 2N electron system

We have shown (cf., Sec. II.I.4.3) that in a quantum study of stationary states the notion of a particle trajectory disappears. Identical particles, i.e., particles with the same mass m, the same charge q and the same spin s, cannot be distinguished since the only criterion for their identification (the trajectory) vanishes. This indistinguishability must therefore be reflected in the wave function

$$\psi(x_1, y_1, z_1, \sigma_1, x_2, y_2, z_2, \sigma_2, ..., x_{2N}, y_{2N}, z_{2N}, \sigma_{2N})$$

for 2N identical particles which may be expressed more compactly as $\psi(1,2,...,2N)$. Previously x_μ, y_μ, z_μ were the continuous variables in the particle space μ. By convention, σ_μ is the spin variable which, in the case of electrons, may only take the values $- 1/2$ or $+1/2$, i.e., values of the projection of electron spin on the Oz axis in units of $h/2\pi$.

We have shown (cf., Sec. II.1.5) that the application of quantum mechanical postulate 9 requires the wave function ψ to be antisymmetric with respect to binary exchange. This means

$$\psi(1,2,..., 2N) = - \psi(2,1,..., 2N)$$

As already mentioned, the Slater determinant is a convenient but not unique representation of $\psi(1, 2,..., N)$ which reflects this antisymmetry. Slater proposes taking the antisymmetrized products of 2N one-electron functions $f_i(\mu)$, i.e., functions of a single particle:

$$\psi(1,..., 2N) = \frac{1}{\sqrt{(2N)!}} \begin{vmatrix} f_1(1) & f_2(1) & . & f_{2N}(1) \\ f_1(2) & f_2(2) & . & f_{2N}(2) \\ . & . & . & \\ f_1(2N) & f_2(2N) & . & f_{2N}(2N) \end{vmatrix}$$

$f_i(\mu)$ are called spin-orbitals. These are one-electron functions containing both the three continuous spatial variables and the discrete spin variable of particle μ. In general, a spin-orbital $f_i(\mu)$ is expressed as the product of a purely spatial one-electron function $\varphi_i(\mu)$, (the orbital) by a one-electron spin function, $\alpha(\mu)$ or $\beta(\mu)$, which represents the spin state of electron μ. In a system with paired electrons, each orbital appears twice in the determinant, once with spin α and once with spin β.

As an example, let us write the Slater determinant corresponding to an acceptable description of the ground state of the helium atom, where two paired electrons appear in orbital 1s:

$$\Phi^0 = \frac{1}{\sqrt{2!}} \begin{vmatrix} f(1) & g(1) \\ f(2) & g(2) \end{vmatrix} = \frac{1}{\sqrt{2!}} \begin{vmatrix} 1s(1)\,\alpha(1) & 1s(1)\,\beta(1) \\ 1s(2)\,\alpha(2) & 1s(2)\,\beta(2) \end{vmatrix} \qquad (IV.I.3)$$

The function ψ(1, 2,..., 2N) would be an exact wave function if there were no two-electron interaction terms in the hamiltonian. Therefore there is no way it can represent the exact wave function for the ground state of a 2N particle system. Nevertheless, this function is often acceptable, and is useful for the understanding of chemical bonding. In addition, it is the starting point of various techniques for the construction of better wave functions.

The determinant of $\psi(1,..., 2N)$ is often expressed more compactly as follows: The normalization constant is taken for granted and only the diagonal terms of the determinant are shown between two vertical bars, the convention being that, for spin orbitals associated with spin function α , only the spatial part is shown, and that spin orbitals associated with spin function β are shown with a horizontal bar above the spatial function symbol (cf., Sec. I.I). Thus the determinant $\psi(1,..., 2N)$ is expressed in the form

$$\psi(1,..., 2N) = \left| \varphi_1(1) \, \overline{\varphi_1}(2) \, \varphi_2(3) \, \overline{\varphi_2}(4) \,.........\, \varphi_N(2N-1) \, \overline{\varphi_N}(2N) \right|$$

$$= \left| \varphi_1 \, \overline{\varphi_1} \, \varphi_2 \, \overline{\varphi_2} \,.............\, \varphi_N \, \overline{\varphi_N} \right| \qquad (IV.I.4)$$

To simplify further, the labels of electrons associated with spin orbitals may be left out, since the functions that appear on the diagonal are labeled in ascending order; thus the helium determinant of order two defined above may be expressed compactly in the form

$$\Phi^0 = \left| 1s(1) \, \overline{1s}(2) \right| = \left| 1s \, \overline{1s} \right| \qquad (IV.I.5)$$

4. The Hartree-Fock method (1928-1930)

4.1 Principle and objective of the method

Starting with a Slater determinant $\psi(1, 2,..., 2N)$ for a system of 2N paired electrons

$$\psi(1,..., 2N) = \frac{1}{\sqrt{(2N)!}}$$

$$\times \begin{vmatrix} \varphi_1(1)\alpha(1) & \varphi_1(1)\beta(1) & . & \varphi_N(1)\beta(1) \\ \varphi_1(2)\alpha(2) & \varphi_1(2)\beta(2) & & \varphi_N(2)\beta(2) \\ & . & . & . \\ \varphi_1(2N)\alpha(2N) & \varphi_1(2N)\beta(2N) & & \varphi_N(2N)\beta(2N) \end{vmatrix}$$

the Hartree-Fock (H-F) method consists of finding the best spatial functions $\varphi_i(\mu)$ that minimize W

$$W = \frac{\displaystyle\int_D \psi^* H \psi \, d\tau}{\displaystyle\int_D \psi^* \psi \, d\tau} \geq E_c$$

The hamiltonian operator H of such a system of 2N electrons and K nuclei has the following form:

$$H = \sum_{\mu=1}^{2N} h(\mu) + \sum_{\mu=1}^{2N} \sum_{v > \mu}^{2N} g(\mu, v) + V_{NN} \qquad (IV.I.6)$$

In this expression the one-electron operators $h(\mu)$ are defined by

$$h(\mu) = \frac{\hbar^2}{2m_e} \Delta_\mu - \sum_{\alpha=1}^{K} k \, Z \, q^2 \, \frac{1}{r_{\mu\alpha}} \qquad (IV.I.7)$$

where $k = 9 \times 10^9$ in the MKSA system, $Z_\alpha \, q$ is the charge on nucleus α and K is the number of nuclei,

— The two-electron operators $g(\mu, v)$ have the form:

$$g(\mu, v) = k \, q^2 \, \frac{1}{r_{\mu v}} \qquad (IV.I.8)$$

— The constant term V_{NN} represents the pair-wise repulsion between the nuclei:

$$V_{NN} = \sum_{\alpha=1}^{K} \sum_{\beta > \alpha}^{K} \frac{k \, Z_\alpha Z_\beta \, q^2}{r_{\alpha\beta}} \qquad (IV.I.9)$$

The fact that V_{NN} is constant [inasmuch as, in the spirit of the BO approximation, the function $\psi(1,..., 2N)$ is determined for fixed positions of the nuclei] means that this term may be ignored, at least temporarily. To this end, a simplified hamiltonian $H - V_{NN}$ is substituted for H. The variational ratio $W' = W - V_{NN}$ is then minimized, so that, in order to obtain W, it is sufficient to add V_{NN} to W'.

The formal development for W' is rather complicated, considering the expression for the determinant, and the same is true for the minimization of W'. In order to facilitate understanding of the physical content of the H-F method, we will start by examining the elements of a simple problem, namely the Hartree-Fock (H-F) treatment of the ground state of helium.

4.2 H-F treatment of ground state of the helium atom

The hamiltonian operator for the helium atom (assuming fixed nucleus) is

$$H_{el} = h(1) + h(2) + g(12) \qquad \text{(IV.I.10)}$$

$$\text{with} \quad h(\mu) = -\frac{\hbar^2}{2m_e}\Delta_\mu - k\frac{Zq^2}{r_\mu} \quad \text{and} \quad g(1,2) = k\frac{q^2}{r_{12}} \qquad \text{(IV.I.11)}$$

Consider now the 2-electron Slater determinant Φ^0, for which we have to find the best function φ, according to the variational theorem. However, instead of formally setting up the equations to minimize the variational ratio, we shall try to find a better φ function by a more easily-interpreted physical approach.

For this reason, let us start with an approximate function $\varphi^{(0)}$ (e.g., the 1s function of the hydrogen atom). The probability density relative to electron μ is given by $|\varphi^{(0)}(\mu)|^2$ and the charge density is

$$\rho_\mu = q\,|\,\varphi^{(0)}(\mu)\,|^2 \qquad \text{(IV.I.12)}$$

The electrostatic interaction energy between two charges q_1 and q_2 separated by distance r_{12} is $kq_1 q_2/r_{12}$. Between a point charge q_1 and a charge with spatial distribution ρ_2 this interaction energy is equal to

$$k\,q_1 \int_D \frac{\rho_2}{r_{12}}\,dv_2$$

Consequently, we may proceed as follows: Replace electron 2 by a static charge distribution associated with function Φ^0. The average electrostatic interaction between electron 1 and this charge distribution is

$$k\,q^2 \int_D \frac{|\varphi^{(0)}(2)|^2}{r_{12}}\,dv_2$$

The interaction potential energy $V^{(0)}(1)$ between electron 1 and the "electron 2 + nucleus" set is equal to

$$V^{(0)}(1) = -k\frac{Zq^2}{r_1} + k\,q^2 \int_D \frac{|\varphi^{(0)}(2)|^2}{r_{12}}\,dv_2 \qquad \text{(IV.I.13)}$$

With the help of $V^{(0)}(1)$ we build a one-electron eigenvalue equation in which the operator acting on $\varphi^{(1)}$ is called the Fock operator:

$$\left[-\frac{\hbar^2}{2m_e}\Delta_1 + V^{(0)}(1)\right]\varphi^{(1)}(1) = \varepsilon^{(1)}\,\varphi^{(1)}(1) \qquad \text{(IV.I.14)}$$

The function $\varphi^{(1)}$, which is the solution of this eigenvalue equation, represents an "improvement" with respect to $\varphi^{(0)}$ which was used to construct $V^{(0)}(1)$. We now construct a new average field for electron 1 using the same approach, by writing the interaction energy in the form:

$$k \ q^2 \int_D \frac{\left|\varphi^{(1)}(2)\right|^2}{r_{12}} \ dv_2$$

The new potential $V^{(1)}(1)$ felt by electron 1 is

$$V^{(1)}(1) = -k \ \frac{Z \, q^2}{r_1} + k \ q^2 \int_D \frac{\left|\varphi^{(1)}(2)\right|^2}{r_{12}} \ dv_2$$

enabling us to write a new one-electron eigenvalue equation

$$\left[-\frac{\hbar^2}{2m_e} \Delta_1 + V^{(1)}(1) \right] \varphi^{(2)}(1) = \varepsilon^{(2)} \ \varphi^{(2)}(1) \qquad \text{(IV.I.15)}$$

The process is repeated until, after n iterations, the average field created by electron 2 no longer changes ; we then say that we have reached *self-consistency*. The nth iteration result $\varepsilon^{(n)}$, denoted simply ε, is called the energy of orbital φ.

This procedure, which progressively builds up the average field felt by each electron, is exactly the same as that produced by the mathematical procedure of *minimizing the variational ratio W, in which one seeks the best function φ for the* determinant representing the helium wave function. In fact, the minimization of W leads to the solution of a pseudo-eigenvalue equation of a one-electron operator F:

$$F(1) \ \varphi(1) = \varepsilon \ \varphi(1) \qquad \text{(IV.I.16)}$$

The Fock operator F has the following form:

$$\begin{aligned}
F(1) &= -\frac{\hbar^2}{2m_e} \Delta_1 - \frac{k \ Z \ q^2}{r_1} + k \ e^2 \int_D \frac{\left|\varphi(2)\right|^2}{r_{12}} \ dv_2 \\
&= h(1) + k \ q^2 \int_D \frac{\left|\varphi(2)\right|^2}{r_{12}} \ dv_2
\end{aligned}$$

where $h(1)$ is the hydrogen-like part of F. The operator F is in all respects similar to the operator mentioned above. It depends on the function φ to be determined. The preceding equation is only a pseudo-eigenvalue equation which requires iteration adapted to this type of equation.

An expression for the orbital energy ε may be obtained from the preceding equation:

$$\varepsilon = \int_D \varphi^*(1)\, F(1)\varphi(1)\, dv_1$$

$$= \int_D \varphi^*(1)\, h(1)\varphi(1)\, dv_1 + k\, q^2 \int \int_D \frac{|\varphi(1)|^2 |\varphi(2)|^2}{r_{12}}\, dv_1\, dv_2 = I + J$$

where I and J are one- and two-electron integrals, respectively. The quantity W_{min}, reached at self-consistency, is directly related to the orbital energy ε by

$$W_{min} = 2\varepsilon - k\, q^2 \int \int_D \frac{|\varphi(1)|^2 |\varphi(2)|^2}{r_{12}}\, dv_1\, dv_2$$

as the average mutual repulsion of the electrons is included twice in 2ε. Indeed, considering the expressions for the determinant ψ and the hamiltonian operator (Eqs. IV.I.10 and IV.I.11), the ratio W is

$$W = \int_D \varphi^*(1)\, h(1)\varphi(1)\, dv_1 + \int_D \varphi^*(2)\, h(2)\varphi(2)\, dv_2$$

$$+ k\, e^2 \int \int_D \frac{|\varphi(1)|^2 |\varphi(2)|^2}{r_{12}}\, dv_1\, dv_2 = 2I + J \qquad \text{(IV.I.17)}$$

Given that $\varepsilon = I + J$, it follows that:

$$W_{min} = 2\varepsilon - J \qquad \text{(IV.I.18)}$$

W_{min}, called E_{HF}, is the "Hartree-Fock" limit of the electron problem under consideration. In the case of He, one finds $E_{HF} = -77.8$ eV, compared with the exact value -79.0 eV (obtained as the sum of the two ionization energies of He). The difference is thus 1.2eV (115.8 kJ/mol).

4.3 The general case

We will give here only the final results of the developments. A reader interested in a more detailed presentation may consult the various quantum chemistry texts mentioned in Appendix 9. The minimization procedure of W' leads to a system of coupled one-electron integro-differential equations, called Fock equations:

$$F(1)\, \varphi_i(1) = \varepsilon_i\, \varphi_i(1) \quad \text{avec } i = 1, 2,..., N \qquad \text{(IV.I.19)}$$

In this equation system, the ε_i are energies of orbitals $\varphi_i(1)$ at convergence. The hermitian operator $F(1)$ is the Fock operator. It is an effective one-electron hamiltonian, identical for each electron. As expressed, it depends on the spatial coordinates of electron 1. In fact, we could have given it any other label, as the labeling of electrons is immaterial, given that they are indistinguishable. This operator $F = F[\varphi_i]$ depends in turn on the spatial functions φ_i to be determined. The mathematical form is

$$F(1) = h(1) + V_{average}(1) \tag{IV.I.20}$$

The operator $h(1)$ (Eq. IV.I.11) is a one-electron operator containing the kinetic energy operator of electron 1 as well as the sum of interactions between electron 1 and the K nuclei. The average potential $V_{average}$ (1) is defined by its effect on function $\varphi_i(1)$:

$$V_{average}(1)\, \varphi_i(1) = k\, q^2 \sum_{j=1}^{N} \varphi_j^*(2) \frac{2 - P_{ij}}{r_{12}} \varphi_j(2)\, \varphi_i(1)\, dv_2 \tag{IV.I.21}$$

The $V_{average}$ (1) potential is a function of the coordinates of electron 1 only, as integration involves a definite integral over the coordinates of electron 2. $V_{average}$ (1) is an integral operator containing the permutation operator P_{ij}, leading to the appearance of exchange integrals as a result of the antisymmetric nature of the wave function. It should be noted that exchange integrals do not appear in the treatment of two-electron systems like He. In such cases there is only one spin orbital of spin α and another of spin β. No exchange is therefore possible, because it may take place only in orbitals associated with functions of the same spin. As a result of this functional dependence between F and φ_j, the preceding equations are only pseudo-eigenvalue equations. The method for solving the preceding system (Eq. IV.I.19) is an iterative approach to the final solution. At iteration zero, take an initial set of functions $[\varphi_1^{(0)}(1), \varphi_2^{(0)}(1), ..., \varphi_N^{(0)}(1)]$. This set must not be too far from the final solution if the convergence is to be efficient and fast. This set is used to determine the initial operator denoted $F^{(0)}$.

The eigenfunctions of $F^{(0)}$, which we will call $[\varphi_i^{(1)}(1)]$, make it possible to determine a new operator $F^{(1)}$ and its eigenfunctions $[\varphi_i^{(2)}(1)]$, and so on. The process is repeated until the difference between the nth iteration eigenfunctions $[\varphi_i^{(n)}(1)]$ and those of iteration n–1 is less than a predetermined value. At this point the process has reached self-consistency. Thus the method consists of replacing 2N electrons in mutual interaction by a system of 2N quasi-independent particles. The one-electron operator F, substituted for the many-electron hamiltonian operator H, describes the movement of an electron μ in the field of the nuclei, represented by operator $h(\mu)$, and in the average field $V_{average(\mu)}$ induced by the remaining 2N – 1 electrons. This average field replaces the individual electron repulsion terms $1/r_{\mu\nu}$ in H and is defined by the functions $\varphi_i(\mu)$ to be determined. These iterations gradually build up the average field. At self-consistency we may say that we have created a *self-consistent field*, hence the name of the method (SCF). The orbital functions $\varphi_i(\mu)$ are thus the best coordinate-space molecular functions, according to the variational method. This is based on a procedure originally proposed by Hartree (Hartree 1928) for a wave function in the form of a simple product of orbitals, later extended by Fock (Fock 1930) and Slater (Slater 1930) to an antisymmetrized product of determinants.

Although each molecular orbital φ_j contains two electrons with antiparallel spins, energy W' is not twice the energy ε_i

$$W'_{min} \neq 2 \sum_{i=1}^{N} \varepsilon_i \qquad (IV.I.22)$$

because this sum includes the electron repulsion twice. It is therefore necessary to subtract the electron repulsion once. We thus obtain

$$W'_{min} = 2 \sum_{i=1}^{N} \varepsilon_i - \sum_{i=1}^{N} \int_{D} \varphi^*_i(1) V_{moyen}(1) \varphi_i(1) dv_1 \qquad (IV.I.23)$$

where these repulsions are evaluated at convergence by taking the sum of the quantum average of the operators $V_{average}(\mu)$. The quantity $W_{min} = W'_{min} + V_{NN}$ is the "best" electron energy for this particular choice of basis functions, i.e., the Slater determinant. The energy $W_{min} = E_{HF}$ is called the Hartree-Fock energy or the Hartree limit. The term "limit" is justified below. This quantity, of course, does not have any strict physical significance. However it serves as a reference point for the calculation of atomic or molecular wave functions because it is mathematically well-defined as the extremum of a function. According to the variational theorem, W_{min} is higher than the real energy value because the real wave function of a many-electron interacting system cannot be put in the form of a single Slater determinant. We shall now give a practical procedure for obtaining the Hartree-Fock (H-F) solution.

5. The Roothaan method (1951)

5.1 Overview of the method

This method was put forward by Roothaan in 1951 (Roothaan 1951). When applied to the study of a molecular system, it is also called the molecular orbital (MO) method, making use of the "Linear Combination of Atomic Orbitals" (LCAO) method. The basic idea of this useful method consists of taking a linear combination of the MO's φ_i using a basis of Q one-electron functions χ_p (which, in general, play the role of atomic orbitals of the constituent atoms), according to

$$\varphi_i(1) = \sum_{p=1}^{Q} c_{ip} \chi_p(1)$$

This transforms a system of integro-differential equations into a system of equations with Q unknowns (the c_{iq}) of the form

$$\sum_{q=1}^{Q} c_{iq} (F_{pq} - \varepsilon_i S_{pq}) = 0 \quad \text{avec } p = 1,\ldots, Q \qquad (IV.I.24)$$

where the terms F_{pq} are the elements of a matrix that represents the operator $F(1)$ in the chosen base set

$$F_{pq} = \int_D \chi^*_p (1) F(1) \chi_q (1) dv_1 \qquad (IV.I.25)$$

and the terms S_{pq} are the overlapping integrals of the base functions

$$S_{pq} = \int_D \chi^*_p (1) \chi_q (1) dv_1 \qquad (IV.I.26)$$

Since the operator F depends on the functions φ_i, the matrix elements of F are functions of the coefficients c_{ip} which define the MO. However, this system is linearized by temporarily keeping constant the coefficients that appear in F_{pq}. The above system thus becomes a system of homogeneous linear equations. The condition for a non-trivial solution of such a system is that the following determinant must be equal to zero:

$$|F_{pq} - \varepsilon_i S_{pq}| = 0 \quad \text{with} \quad p \text{ and } q = 1, 2,\ldots, Q$$

We thus obtain a polynomial equation in ε_i ($i=1,2,..Q$) of degree Q with Q real roots. For each root ε_i, we determine the collection of coefficients c_{ip}, taking into account the normalization of the MO's. These coefficients are then used to construct a new operator F, i.e., a new matrix (F). This iterative procedure is repeated until convergence of the energy or the coefficients of the MO's. The implementation of this procedure requires an initial set of coefficients to construct the first matrix (F).

As a result of the construction of the self-consistent field, the MO's are *delocalized over all the base functions*. The energies ε_i obtained in this way are arranged in order of increasing energy. Each MO is occupied by at most two electrons. We thus obtain for energy a *shell structure* analogous to that of atoms.

5.2 Choice of base functions for molecular calculations

How many base functions χ_p should one take? *A priori*, an infinite number. This si the prerequisite for obtaining a Hartree-Fock solution of the system under study, whence the expression "Hartree-Fock limit" mentioned above. As it is impossible to take an infinite number of functions, it is necessary to cut the development off at some point. The energy W_{min} obtained in this way is larger than the Hartree-Fock limit, denoted W_{SCF}. It is a function of Q and depends on the nature of the selected base. Thus

$$W_{min} = W_{SCF} (Q < \infty) > E_{HF} = W_{SCF} (Q = \infty)$$

If we use atomic orbitals (AO) as the base set, centered on the constituent atoms of the molecule under study, the Hartree-Fock limit may be reached with a fairly small Q.

Various types of functions have been proposed over the past twenty years. For reasons of speed of execution, gaussian base functions, originally proposed by Boys in 1950, are now mainly used. However they have a few "pathological" drawbacks at long and short distances (relative to equilibrium bond lengths). A gaussian function is a one-electron function consisting of a radial part R(r) and an angular part $Y(\Theta, \varphi)$ due to the spherical symmetry of the atom. Many gaussian base functions have been proposed by various authors, including Huzinaga, van Duijneveldt, Pople, etc. (e.g., Huzinaga 1965, van Duijneveldt 1971, Dunning 1970 and 1977, Hehre 1986).

Why choose gaussian functions? In the exact quantum treatment of the hydrogen atom H, the radial part appears as an exponential in ($-r$), multiplied by a polynomial in r of degree $n-1$, where n is the principal quantum number characterizing the energy levels of H. In the case of many-electron atoms, the H-F radial part is no longer hydrogen-like due to repulsion among the electrons. However, Slater has shown an approximate description in a form similar to that of H may be useful for further developments. In this description, the radial part consists of an exponential in ($-r$) multiplied by r^{n-1}, the highest degree of the equivalent hydrogen-like orbital. This type of orbital is called the *Slater orbital* (Zener 1930, Slater 1930b). Therefore, an H-F atomic orbital may be represented either by a Slater orbital, or by a linear combination of Slater orbitals.

In the case of molecular systems this type of hydrogen-like or Slater-like function in $\exp(-r)$, is not very well-suited to calculating two-electron integrals when the atomic functions are centered on different atoms, as is often the case in the study of molecular wave functions. A convenient way of resolving this difficulty is to replace $\exp(-r)$ by $\exp(-r^2)$. This type of function, associated with an angular part $Y(\theta, \varphi)$, is called the atomic gaussian. It is obvious that the behavior of this type of function at the nucleus ($r = 0$) is quite different from that of the function $\exp(-r)$ [the first derivative is zero for the gaussian and negative for $\exp(-r)$]. Furthermore, at large distances, the gaussian $\exp(-r^2)$ decreases faster than $\exp(-r)$. These shortcomings of the gaussian function are partially compensated by using a predominantly gaussian base set. The additional numerical effort this represents is largely compensated by easier calculation of gaussian base integrals.

The minimum number of atomic functions corresponds to the sum of the number of internal and valence orbitals, forming the minimum base set. For example, to represent atoms in the second row of the periodic system requires one orbital 1s, one orbital 2s and three orbitals 2p, i.e., five orbitals per atom. By contrast, for an atom in the next row, nine valence orbitals must be added to the five core orbitals, giving a total of fourteen orbitals. If greater precision is required, that is, an energy closer to the H-F limit, it is necessary to increase the number of functions per atom.

Remark: Before giving an example of this approach, an important point concerning this type of calculation should be emphasized. In studying a system of 14 electrons, taking 20 atomic functions as a basis, the linearized system is of dimension 20. We thus obtain 20 real roots, i.e., 20 molecular orbitals in order of increasing energy. As the system has only 14 electrons, only the 7 lowest MO's will be occupied in the ground state. The Slater determinant is of dimension 14 x 14, but only includes 7 MO's, each associated once with spin α and once with spin β. There are thus 13 virtual MO's which are not used in the calculation of H-F energy but which may be used to describe many-electron systems beyond the Hartree-Fock approximation, as we shall see in Sec. IV.III.

5.3 Application to a demonstration problem: The helium atom using basis of dimension two

By way of illustration, it is of interest to perform an SCF calculation of the ground state of the helium atom with the atomic orbital φ developed in term of two base functions χ_1 and χ_2:

$$\varphi(1) = c_1\,\chi_1(1) + c_2\,\chi_2(1) \qquad\qquad \text{(IV.I.27)}$$

with $\chi_p(1) = \sqrt{\alpha_p^3/\pi}\; e^{-\alpha_p r_1}$

The development of the matrix F according to Roothaan yields

$$F_{pq} = I_{pq} + \sum_{r=1}^{2}\sum_{s=1}^{2} c_r\,c_s\,(pq\,|\,rs) \qquad\qquad \text{(IV.I.28)}$$

$$\text{where } I_{pq} = \int_D \chi_p^*(1)h(1)\chi_q(1)\,dv_1$$

$$\text{and } (pq\,|\,rs) = \iint_D \chi_p^*(1)\chi_q(1)\frac{ke^2}{r_{12}}\chi_r^*(2)\chi_s(2)\,dv_1dv_2$$

and the energy W is given by

$$W = \sum_{p=1}^{2}\sum_{q=1}^{2} c_p^*\,c_q\,(I_{pq} + F_{pq}) \qquad\qquad \text{(IV.I.29)}$$

In view of the analytic expressions for the base functions χ_p, it is possible to find convenient forms of the various integrals S_{pq}, I_{pq} and $(pq|rs)$ as functions of the exponents α_1 and α_2, as cited by Ducasse (Ducasse 1985). The exponents α_p that yield the lowest energy are $\alpha_1 = 1.45$ and $\alpha_2 = 2.91$. These values are used to calculate the SCF energy W_{min}

$$W_{min} = -2.8617 \text{ Hartree} = -77.87 \text{ eV}$$

and identify the atomic orbital φ:

$$\varphi(1) = 0.842\,\chi_1(1) + 0.183\,\chi_2(1)$$

of energy $\varepsilon = -0.9183$ Hartree. The 2×2 linear system also yields a virtual atomic orbital $\varphi'(1)$ of the form:

$$\varphi'(1) = 1.620\, \chi_1\,(1) - 1.816\, \chi_2\,(1)$$

with energy $\varepsilon' = 2.047$ Hartree. The results of this simple calculation of the SCF energy W_{min} are fairly close to the H-F limit, $E_{HF} = -2.8616799$ Hartree, obtained by Roetti and Clementi (Roetti 1974) on the basis of 5 Slater functions of the type used here.

II. Chemical bonding in terms of MO language

We now propose to apply the Hartree-Fock method to the study of simple molecular systems like water molecules.

1. Canonical orbitals of the water molecule

Starting with a minimal AO base set, i.e., the set $1s(O)$, $2s(O)$, $2p(O)$, $2p(O)$, $2p(O)$, $1s(H_1)$ and $1s(H_2)$, each MO may be described by an expansion of the form

$$\varphi_i = c_1\ 1s(O) + c_2\ 2s(O) + c_3\ 2p_x(O) + c_4\ 2p_y(O) + c_5\ 2p_z(O)$$
$$+ c_6\ 1s(H_1) + c_7\ 1s(H_2) \qquad \text{(IV.II.1)}$$

For a value of the HOH angle in the neighborhood of $100.3°$ and O-H distances equal to 0.99 Å, and allowing for the C_{2v} symmetry of the molecule, we obtain the following MO's (Fig. IV.II.1).

These results were obtained by Pitzer and Merrifield in 1970 using a minimal base set of Slater AO's (Pitzer 1970). Each MO belongs to the irreducible representation of the C_{2v} group, and as a result, some zero coefficients appear in the general expansion of the φ_i according to the irreducible representation considered. Indeed, as we have seen in Secs. II.II.4.3 and II.II.4.4, as the oxygen atom is situated at the invariant point of the C_{2v} group, its AO's are the symmetry orbitals for this group: $1s(O)$, $2s(O)$ and $2p_z(O)$ each form a basis for the a_1 representation, $2p_x(O)$ belongs to b_1, and $2p_y(O)$ to b_2. As for the combinations including the $1s$ AO's of the two hydrogens, the combination $[1s(H_1) + 1s(H_2)]$ belongs to a_1, while $[1s(H_1) - 1s(H_2)]$ belongs to b_2. It is therefore logical that MO's of symmetry "a" should be a linear combination of the AO's $1s(O)$, $2s(O)$, $2p(O)$ and $[1s(H_1) + 1s(H_2)]$. The same is also true for other symmetries. We may therefore build a maximum of four orbitals of symmetry a_1, as there are four AO's — or combinations of AO's — which belong to this symmetry, two MO's of symmetry b_1 and only one MO of symmetry b_2.

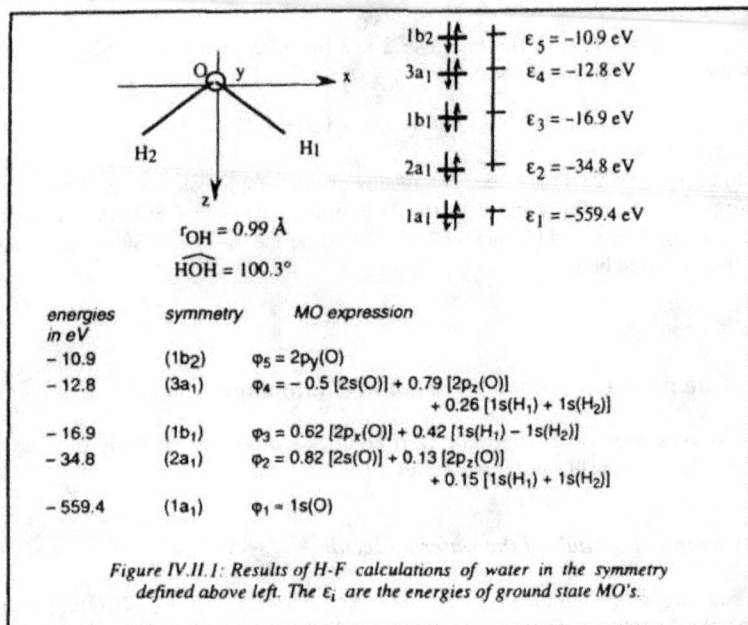

energies in eV	symmetry	MO expression
− 10.9	(1b$_2$)	$\varphi_5 = 2p_y(O)$
− 12.8	(3a$_1$)	$\varphi_4 = -0.5\,[2s(O)] + 0.79\,[2p_z(O)]$ $+ 0.26\,[1s(H_1) + 1s(H_2)]$
− 16.9	(1b$_1$)	$\varphi_3 = 0.62\,[2p_x(O)] + 0.42\,[1s(H_1) - 1s(H_2)]$
− 34.8	(2a$_1$)	$\varphi_2 = 0.82\,[2s(O)] + 0.13\,[2p_z(O)]$ $+ 0.15\,[1s(H_1) + 1s(H_2)]$
− 559.4	(1a$_1$)	$\varphi_1 \approx 1s(O)$

Figure IV.II.1: Results of H-F calculations of water in the symmetry defined above left. The ε_i are the energies of ground state MO's.

Each MO "steers" two electrons of opposite spin in accordance with the 9th postulate of quantum mechanics. The orbital energies calculated by the H-F method from the base under consideration are shown in the above diagram. Note the considerable difference between the energy of φ_1, the first MO, and the other four orbitals. This MO φ_1, of symmetry a$_1$, consists mainly of the core orbital 1s(O) of oxygen. In fact, it also contains nonzero coefficients on the AO's 2s(O), 2p (O) and [1s(H$_1$) + 1s(H$_2$)] as each of these atomic orbitals — or their combinations — belong to the representation a$_1$. But these coefficients are so small compared to that of orbital 1s(O) that the corresponding contributions are negligible. By contrast, the other four MO's consist mainly of combinations of valence AO's. Here also, 1s(O) contributes a priori to each MO of symmetry "a", but this contribution is so small that it has been ignored. This absence of mixing is due to the large energy difference between the 1s(O) AO and the valence AO's of oxygen and hydrogen. The resulting large energy difference then gives rise to an electron structure distributed in shells analogous to those found in atoms. Neither the total energy of a molecule [as was mentioned previously in Eq. IV.I.22], nor the bond energy may be calculated directly from these orbital energies.

The highest occupied orbital is fairly closely related to the first ionization energy of the molecule (see Koopmans' theorem - Section III.1).

Analysis of the MO's from φ_1 to φ_5 illustrates the characteristics described below. Figs. IV.II.2 to IV.II.6 are based on bonding orbitals of the water molecule of a given geometry, as shown in Fig. IV.II.1, and represent contour curves (cf., Sec. II.I.4.4) in the xOz plane (Figs. IV.II.2 to IV.II.5) and the perpendicular plane yOz (Fig. IV.II.6). Continuous and dotted curves distinguish curves of different algebraic signs. They were obtained on a Silicon Graphics workstation at the Laboratoire de Physicochimie Théorique, Université de Bordeaux 1 (URA 503) using software developed by P. Halvick.

– MO φ_1, of symmetry a_1, is formed almost exclusively from the 1s function of the oxygen atom. This core orbital function does not contribute in any way to the cohesion of the molecular structure. Its energy, equal to – 560 eV, is highly negative. Therefore, this core orbital does not couple with hydrogen AO's of energy –13.6 eV (Fig. IV.II.2).

– MO φ_2, of symmetry a_1, is formed by constructive fusion (overlap of lobes with the same sign) of the hybrid combination [2s(O), 2p (O)], which points toward the positive part of the Oz axis, and the sum [1s(H$_1$) + 1s(H$_2$)]. Hence it is a bonding orbital delocalized on the three atoms (Fig. IV.II.3).

– MO φ_3, of symmetry b_1, is formed by constructive fusion of the AO 2p (O) and the combination [1s(H$_1$) – 1s(H$_2$)]. As the function 2p (O) is odd, it may combine only with the antisymmetric combination of the two hydrogen AO's. The resulting MO, delocalized on the three atoms, is of bonding character on the two O-H bonds, and has yOz as the nodal plane (Fig.IV.II.4).

– MO φ_4, of symmetry a_1, contributes only modestly to chemical bonding as the spatial extension of hybrid [2s(O), 2p (O)] mainly points towards the negative side of the Oz axis. Hence the overlap between the combination [1s(H$_1$) + 1s(H$_2$)] and the positive part of this hybrid, with limited extension towards positive z, is not significant (Fig. IV.II.5). We may therefore consider that this MO corresponds to a quasi-free electron pair oriented toward the negative side of the Oz axis. In view of its weak bonding capacity, this orbital has a high energy in the energy diagram in Fig. IV.II.1.

— Finally, as the 2p$_y$ (O) orbital is the only function of symmetry b_2, the φ_5 MO consists of the electron lone-pair of the oxygen atom in this 2p$_y$ (O) AO, oriented perpendicularly to the xOz molecular plane (Fig. IV.II.6). This MO is of strictly nonbonding character and its Hartree-Fock energy corresponds satisfactorily to the first ionization energy of water (on the order of 11 eV). (The molecular ion H_2O^+ has one electron less than H_2O, so there is one non-bonding electron). This property is directly related to Koopmans' theorem (Koopmans 1934), stating that the absolute value of the ionization energy of a many-electron system is given by the energy of the orbital from which the electron is removed (ignoring the change of self-consistent field from the molecule to the ion).

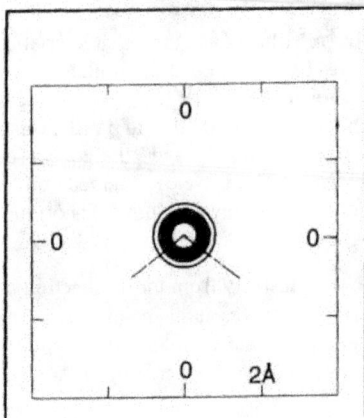

Figure IV.11.2: Contour curves representing the MO φ_1 of symmetry $1a_1$. This orbital is formed almost exclusively of the 1s orbital of oxygen. Contours at intervals of 0.05 a.u. from 0.025 a.u. to 0.975 a.u.

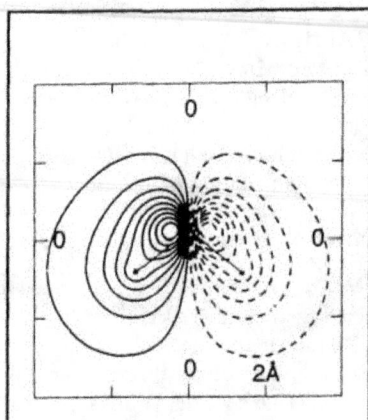

Figure IV.11.4: Contour curves representing the bonding MO φ_3 of symmetry b_1. Contours at intervals of 0.05 a.u. from -0.425 a.u. to 0.420 a.u.

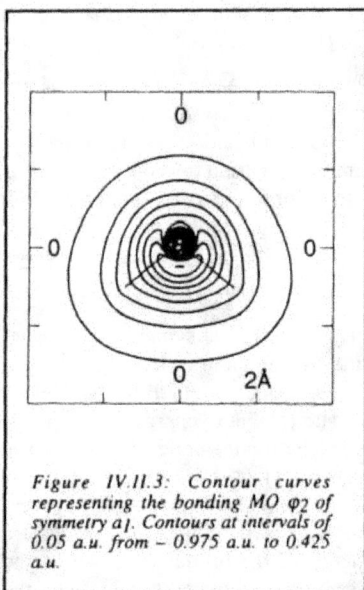

Figure IV.11.3: Contour curves representing the bonding MO φ_2 of symmetry a_1. Contours at intervals of 0.05 a.u. from -0.975 a.u. to 0.425 a.u.

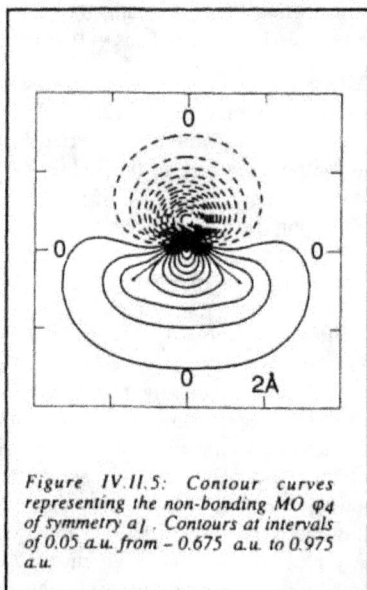

Figure IV.11.5: Contour curves representing the non-bonding MO φ_4 of symmetry a_1. Contours at intervals of 0.05 a.u. from -0.675 a.u. to 0.975 a.u.

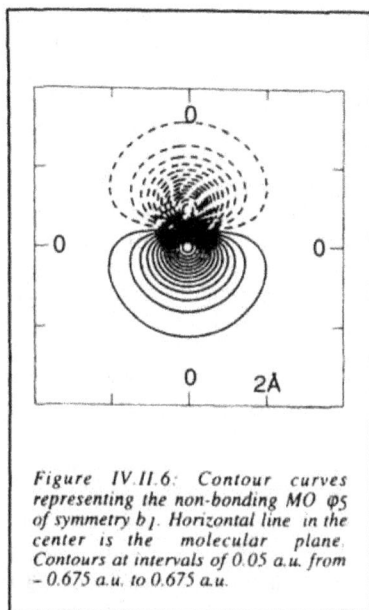

Figure IV.11.6: Contour curves representing the non-bonding MO φ_5 of symmetry b_1. Horizontal line in the center is the molecular plane. Contours at intervals of 0.05 a.u. from – 0.675 a.u. to 0.675 a.u.

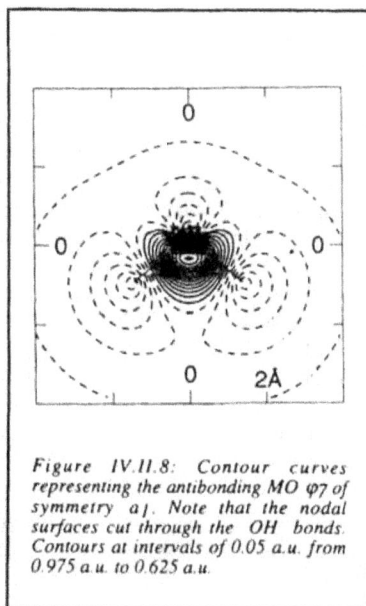

Figure IV.11.8: Contour curves representing the antibonding MO φ_7 of symmetry a_1. Note that the nodal surfaces cut through the OH bonds. Contours at intervals of 0.05 a.u. from 0.975 a.u. to 0.625 a.u.

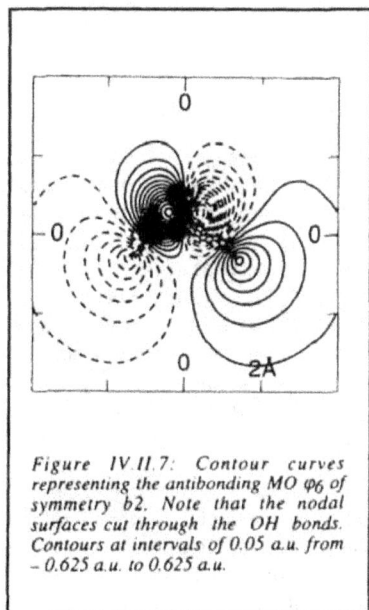

Figure IV.11.7: Contour curves representing the antibonding MO φ_6 of symmetry b_2. Note that the nodal surfaces cut through the OH bonds. Contours at intervals of 0.05 a.u. from – 0.625 a.u. to 0.625 a.u.

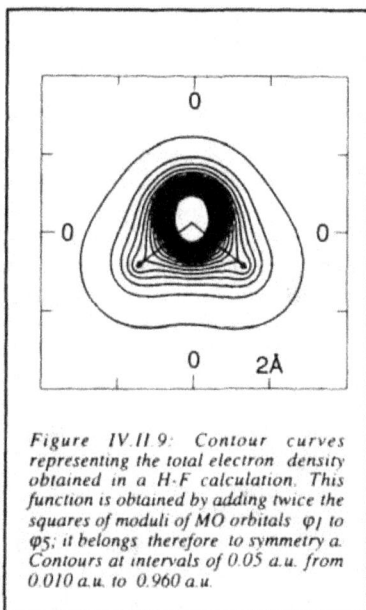

Figure IV.11.9: Contour curves representing the total electron density obtained in a H-F calculation. This function is obtained by adding twice the squares of moduli of MO orbitals φ_1 to φ_5; it belongs therefore to symmetry a. Contours at intervals of 0.05 a.u. from 0.010 a.u. to 0.960 a.u.

It is important to note that, as the minimum AO base set is composed of 7 functions, calculation yields 7 MO's of which only the first five, in order of increasing energy, are occupied by the 10 electrons of the molecule in the ground state. These five MO's are therefore included in the 10 x 10 Slater determinant describing the ground state of the water molecule. Hence the virtual space is composed here of only two MO's, φ_6 of symmetry a_1 and φ_7 of symmetry b_1. The φ_6 MO is the antibonding counterpart of the φ_2 MO with a node on each OH bond. These nodes are described by the destructive combination between the hybrid $[2s(O), 2p(O)]$ and the sum $[1s(H_1) + 1s(H_2)]$. The φ_7 MO is the antibonding counterpart of the φ_3 MO, and also has a node on each OH bond. It is given by the destructive combination of $2p_x(O)$ AO and the combination $[1s(H_1) - 1s(H_2)]$ (Figs. IV.II.7 and IV.II.8).

We thus obtain a description of the water molecule in terms of delocalized molecular orbitals on the atoms. The orbitals are also called canonical orbitals of water. The delocalized character is related directly to the gradual build-up of the average electron field, the basis of the Hartree-Fock method.

Figure IV.II.9 shows the total electron density of the water molecule as determined by an H-F calculation. It is obtained by summing the squares of the moduli of MO orbitals occupied in the ground state (φ_1 to φ_5) and multiplying by two, as each MO "steers" two electrons. This figure gives only general information on the distribution of electrons in the molecule. It would seem that there is more information in the MO's, taken individually, than in the total electron density. However, the following analysis demonstrates that prudence is in order when using H- F orbitals.

2. Localized molecular orbital description of the water molecule

The concept of a bond between a pair of atoms, essential in chemistry, is supported by much experimental evidence. Thus the notion of bond energy provides a very satisfactory estimation of the formation heat of a molecule by adding its bond energies. Other molecular properties (e.g., magnetic susceptibility, dipole moment) may also be analyzed in terms of contributions of individual bonds and lone pairs. Examination of the infrared spectrum of compounds containing O-H bonds shows the presence of a characteristic band in the neighborhood of 3600 cm^{-1} This band is associated with the elongation of the O-H oscillator and appears in the same spectral region in compounds as diverse as water H_2O, hydrogen hypochlorite HOCl or methanol CH_3OH. In addition, O-H bond length is practically equal to 0.96 Å in any compound with an O-H bond.

The description of the electron structure of water in terms of delocalized molecular orbitals, presented in the preceding paragraph, seems, a priori, incompatible with the existence of individual bonds in H_2O, as the construction of canonical MO's is delocalized over the entire molecule. Furthermore, if we compare the canonical MO's of H_2O with those of HOCl and CH_3OH, we see that these MO's have no common feature. In particular, knowing the MO's of H_2O does not tell us anything about those of CH_3OH.

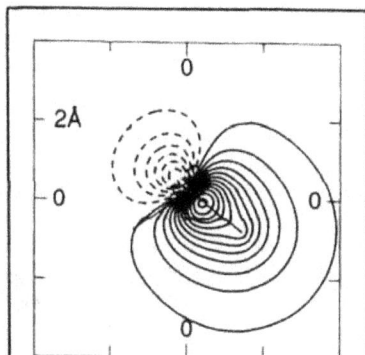

Figure IV.11.10: Contour curves representing the MO of the OH_1 bond in water. Note that this function does not belong to the irreducible representation of the C_{2v} group. The banana-like shape mentioned in the text is barely visible. Contours at intervals of 0.05 a.u. from -0.625 a.u. to 0.575 a.u.

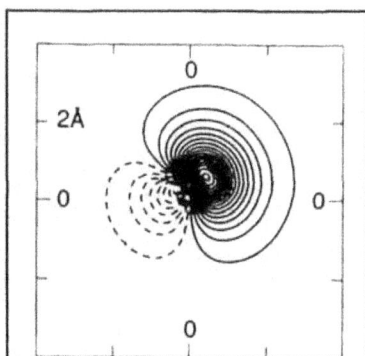

Figure IV.11.12: Contour curves of one of the free electron pairs of oxygen. The central line represents the molecular axis. Contours at intervals of 0.05 a.u. from -0.975 a.u. to 0.775 a.u.

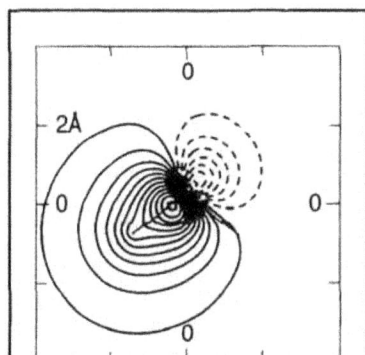

Figure IV.11.11: Same as in Fig. IV.11.10 but representing the OH_2 bond in water.

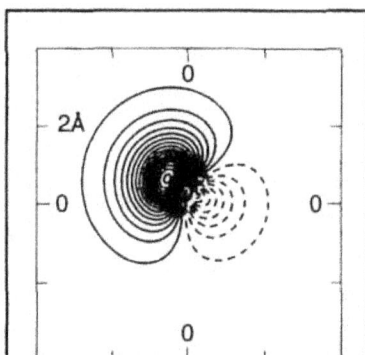

Figure IV.11.13: Same as in Fig. IV.11.12 but representing the other lone electron pair in oxygen.

This apparent incoherence is due to the fact that the canonical MO's represent just one particular solution of the Hartree-Fock problem among an infinite number of possibilities. However all solutions lead to the same wave function.

What is the origin of this infinite number of solutions? From a mathematical point of view, these are inherent in the very structure of the Hartree-Fock equations. It was shown by Roothaan that a general solution of the H-F problem is given by

$$F(1)\, \varphi_i(1) = \sum_{i=1}^{N} \varepsilon_{ij}\, \varphi_j(1) \text{ with } i=1,...,N \qquad \text{(IV.II.2)}$$

where ε_{ij} is the element of matrix (ε) representing the operator F on the basis of functions $\{\varphi_L\}$. The completely delocalized description corresponds to eigenfunctions of operator F, i.e., to functions φ_i which diagonalize matrix (ε). This is therefore just one solution. The result is that the canonical MO's, as solutions of the Hartree-Fock problem, are defined only to within an *orthogonal transformation*, that is to say, a transformation that conserves the norm of these vectors and their relative angles. In fact, energy and electron density are both invariant with respect to this transformation. This property may easily be demonstrated in the case of a determinant of order two. Consider for example the determinant

$$D = \begin{vmatrix} f(1) & g(1) \\ f(2) & g(2) \end{vmatrix} = f(1)\, g(2) - f(2)\, g(1) \qquad \text{(IV.II.3)}$$

If we apply an orthogonal transformation $(\mathbb{U})$ to the column vector with components f and g such that

$$\begin{bmatrix} f' \\ g' \end{bmatrix} = \begin{pmatrix} \cos\theta & -\sin\theta \\ \sin\theta & \cos\theta \end{pmatrix} \begin{bmatrix} f \\ g \end{bmatrix} = (\mathbb{U}) \begin{bmatrix} f \\ g \end{bmatrix}$$

the determinant D' constructed from f' and g' is

$$D' = \begin{vmatrix} f'(1) & g'(1) \\ f'(2) & g'(2) \end{vmatrix} = f'(1)\, g'(2) - f'(2)\, g'(1)$$

$$D' = [f(1) \cos\theta - g(1) \sin\theta]\, [f(2) \sin\theta + g(2) \cos\theta]$$
$$- [f(1) \sin\theta + g(1) \cos\theta]\, [f(2) \cos\theta - g(2) \sin\theta]$$
$$\qquad \text{(IV.II.4)}$$

If we develop this expression, we obtain $D' = D$.

The energy calculated from D is the same as that calculated from D', and the same is true of the total electron density, which is proportional to the square of D. Without entering into the details of the calculation, we may give an illustration of the localization procedure based on the mathematical property of determinant invariance. We shall now consider determinant D, constructed from functions f and g. We know that the value of a determinant is not changed if all the elements in a row (or column) are replaced by the same linear combination including corresponding elements of many rows (or columns). Also, multiplying the elements of a column (or row) by a number b, multiplies the entire determinant by b. These properties easily justify the following sequence of operations:

$$D = \begin{vmatrix} f(1) & g(1) \\ f(2) & g(2) \end{vmatrix} = \begin{vmatrix} f(1) + g(1) & g(1) \\ f(2) + g(2) & g(2) \end{vmatrix}$$

$$= -\frac{1}{2} \begin{vmatrix} f(1) + g(1) & -2 g(1) \\ f(2) + g(2) & -2 g(2) \end{vmatrix} = -\frac{1}{2} \begin{vmatrix} f(1) + g(1) & f(1) + g(1) - 2 g(1) \\ f(2) + g(2) & f(2) + g(2) - 2 g(2) \end{vmatrix}$$

$$D = -\frac{1}{2} \begin{vmatrix} f(1) + g(1) & f(1) - g(1) \\ f(2) + g(2) & f(2) - g(2) \end{vmatrix} \tag{IV.II.5}$$

This sequence may be used for the transition from delocalized to localized MO's. If we suppose that the two functions f and g correspond to MO's delocalized on the two bonds b_1 and b_2 of a molecule, as is the case with the MO's φ_2 and φ_3 of water

$$f = b_1 + b_2 \quad \text{and} \quad g = b_1 - b_2$$

we then have

$$f + g = 2b_1 \quad \text{and} \quad f - g = 2b_2$$

Introducing these relations into the initial and final forms of D, we have

$$D = \begin{vmatrix} b_1(1) + b_2(1) & b_1(1) - b_2(1) \\ b_1(2) + b_2(2) & b_1(2) - b_2(2) \end{vmatrix} = -\frac{1}{2} \begin{vmatrix} 2b_1(1) & 2b_2(1) \\ 2b_2(2) & 2b_2(2) \end{vmatrix} \tag{IV.II.6}$$

The expression on the left-hand side is the delocalized description, while the localized description is on the right-hand side.

We may now look for the type of orthogonal transformation that will produce localized orbitals, in order to come back to the classical concept of chemical bonding. Various localization criteria have been proposed (Edminston-Rudenberg, Boys, etc.; see Levine 1991), all of which lead to the same picture. Thus Liang and Taylor (Liang 1970) obtained for water:

$$i(O) \quad = 1s(O)$$
$$b(OH_1) = 0.5 \, [1s(H_1)] + 0.25 \, [2s(O)] + 0.41 \, [2p_z(O)] + 0.44 \, [2p_x(O)]$$
$$b(OH_2) = 0.5 \, [1s(H_2)] + 0.25 \, [2s(O)] + 0.41 \, [2p_z(O)] - 0.44 \, [2p_x(O)]$$
$$l_1(O) \quad = 0.63 \, [2s(O)] - 0.39 \, [2p_z(O)] - 0.71 \, [2p_y(O)]$$
$$l_2(O) \quad = 0.63 \, [2s(O)] - 0.39 \, [2p_z(O)] + 0.71 \, [2p_y(O)]$$

The two orbitals, denoted $b(OH_1)$ and $b(OH_2)$, correspond to functions localized in the direction of the bonds between the oxygen atom and hydrogen atoms H_1 and H_2, respectively, (Figs. IV.II.10 and IV.II.11). They are formed essentially by the fusion of a 1s orbital of the hydrogen atom H_1 (or H_2) with a hybrid orbital χ_1 (or χ_2) of oxygen and defined by the following two combinations (not normalized):

$$\chi_1 = 2s(O) + \frac{0.41}{0.25} \, [2p_z(O)] + \frac{0.44}{0.25} \, [2p_x(O)]$$

$$\chi_2 = 2s(O) + \frac{0.41}{0.25} \, [2p_z(O)] - \frac{0.44}{0.25} \, [2p_x(O)] \tag{IV.II.7}$$

The direction of the hybrids χ_1 (or χ_2) is given by the combination of oxygen orbitals $2p_z$ and $2p_x$, i.e., by the ratio of their coefficients. The angle γ defined in Fig. IV.II.14 is given by

$$\gamma = tg^{-1} \frac{0.44}{0.41} = 47°$$

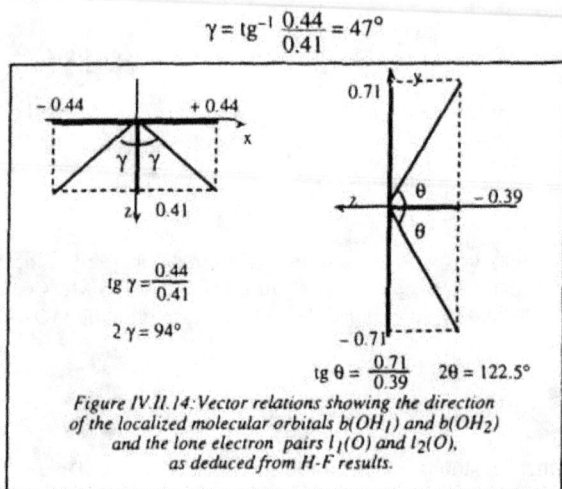

$$tg\,\gamma = \frac{0.44}{0.41}$$

$$2\,\gamma = 94°$$

$$tg\,\theta = \frac{0.71}{0.39} \qquad 2\theta = 122.5°$$

Figure IV.II.14: Vector relations showing the direction
of the localized molecular orbitals b(OH₁) and b(OH₂)
and the lone electron pairs l₁(O) and l₂(O),
as deduced from H-F results.

The angle 2γ (94°) between the two directions χ_1 and χ_2 is apparently smaller than the angle of OH bonds in water (experimental value: 104.5°, optimized theoretical value of Pitzer and Merrifield: 100.3°). This description leads to localized banana-shaped orbitals, oriented toward the inside of the angle H_1OH_2. In fact, calculations using an extended atomic base set (von Niessen 1973) do not have as pronounced a banana character as the angle 2γ comes out to 103°, whereas optimization of the equilibrium geometry of H_2O yields the value 104.5° for the angle H_1OH_2 (consistent with experimental data). It would appear that the banana character is due to the minimal base set used by Pitzer and Merrifield in their calculations.

The two lone electron pairs are described by the functions $l_1(O)$ and $l_2(O)$. They are located on either side of the molecular plane and in the xOz plane (Fig.IV.II.14). The angle 2θ between the directions of these two lone pairs is given by

$$2\theta = tg^{-1} (0.71/0.39) = 122.5°$$

This value seems to be slightly overestimated when compared with the value of 114° obtained by von Niessen (von Niessen 1973) using a more precise approach than that of Liang and Taylor (Liang 1970). Figures IV.II.12 and IV.II.13 show the contour curves for the two lone pairs $l_1(O)$ and $l_2(O)$ in the yOz plane.

These two descriptions of the electron structure of the water molecule (localized or delocalized) are just two of an infinite number of possibilities, as there is an infinite number of unitary transformations between any two sets of molecular orbitals.

We have thus arrived at the curious situation where, among the infinite number of solutions of the H-F problem, we have managed to construct molecular orbitals which give a description of bonding familiar to chemists.

Several questions come to mind :

— Do localized MO's have other virtues?
— What are the uses for delocalized canonical MO's?
— Which representation to choose? Is one representation better than another?
— The H-F method is only approximate. What are its shortcomings? How can they be corrected? What happens to the orbital concept if we go beyond H-F?

We shall try to sketch out a few answers to these questions.

III. Beyond the one-electron description

1. The noninvariant concept of the molecular orbital: Delocalized and localized description

We have just made the point that there are an infinite number of solutions to the Hartree-Fock problem; "delocalized MO's" and "localized MO's" are merely two particular solutions. All the solutions are equivalent as they all lead to the same total energy value and the same total electron density (as well as any quantity related to electron density such as dipole moment), the observables of the system under study, unlike molecular orbitals which are not physical observables of the system. The concept of MO is therefore not invariant. As a result, one representation is chosen rather than another for reasons of convenience and not on the basis of physically valid criteria. All these choices are equivalent insofar as the observables of the system are concerned.

1.1 The interest of delocalized MO's.

We have shown that the delocalized MO's do not suggest the traditional notion of a chemical bond, whereby the orbital electron density (the square of the modulus of MO) is concentrated principally at the midpoint between two atoms. Apart from this disadvantage, which may be corrected by localization, delocalized MO's nevertheless have certain useful properties for chemists.

Access to orbital energies by photoelectron spectroscopy

The delocalized representation leads to orbitals which are eigenfunctions of the Fock operator with eigenenergies, especially the first few, that correspond roughly to energies of successive ionizations of the molecule (Koopmans' theorem, cf., Sec. IV.I).

Photoelectron spectroscopy, providing access to energies close to these eigenvalues, may be considered as a tool for the discovery of delocalized molecular orbitals (although strictly speaking this is not quite correct). Let us recall at this point a few characteristics of this highly specialized type of spectroscopy. Readers may wish to consult Levine 1975, Allan 1987, and, for greater detail, Turner 1970. In general, "traditional" spectroscopy measures the energy of photons emitted or absorbed by molecules. Photoelectron spectroscopy measures the kinetic energy $T(e^-)$ of electrons emitted by molecules photoionized by very intense electromagnetic radiation (UV or X).

The process involved is $M + h\nu \rightarrow M^+ + e^-$

where M represents a molecule. The energy balance is

$$T(e^-) = h\nu + E(M) - E(M^+)$$

If the energy of the source is known (in general this is a helium lamp emitting intense radiation at 584 Å, corresponding to 21.11 eV, or at 304 Å, corresponding to 40.8 eV), the determination of $T(e^-)$ yields the energy difference between M and M^+. The ion M^+ is produced in its ground electron state, or in one of its excited states, depending on the nature of the MO of M from which the electron is removed. Therefore, intense transitions are observed at energies closely related to those of the MO's of M. In fact, a line structure is recorded for each transition.

This structure is related to the fact that ion M^+ is produced in various vibration states of the electron state considered. At 300 K, the molecule M is most often in its electron and vibration ground state. It then follows that the vibration structure observed must only be that of M^+. The line structure makes it possible to observe changes in vibration frequencies of M^+ with respect to those of M. This provides additional information concerning the bonding or anti-bonding nature of the MO of M from which the electron was removed.

Delocalized MO's obey molecular symmetry

Since delocalized orbitals are eigenfunctions of the Fock operator, they obey the symmetry of a given molecular system, i.e., they belong to irreducible representations of the molecular symmetry group. We have seen an example of this in the H-F calculation of water.

The fact that symmetry properties are maintained has been widely used by quantum chemists for qualitative interpretation of reactivity in organic chemistry.

1.2 The interest of localized MO's

The familiar description of chemical bonding

The advantage of a localized solution is, above all, that it reflects the usual image of a chemical bond, as stated in Sec. IV.I.

Transferability of orbital wave functions

Unlike canonical MO's, these bond orbitals are approximately transferable among similar compounds. Thus a localized orbital associated with a C–H bond in methane is transferable for the   description of a C–H bond in ethane or methanol (Rothenberg 1969-71). As a result, it is possible to construct large-scale wave functions from orbital functions, or groups of orbitals, obtained in simpler systems. These localized orbitals are no longer eigenfunctions of the Fock operator, and cannot, therefore, be associated with the notion of eigenenergy. It is important to be aware that it is not possible to obtain both orbital energies and transferable localized MO's at the same time, as these two notions are mutually exclusive.

Return to hybridation

Let us return to the well-known chemistry problem of the hybridation of the oxygen atom in the water molecule (or of the carbon atom in methane, ethylene or acetylene). The calculation described in Sec. IV.I does not make use of any hybrid state of the central atom. The atomic base functions of the quantum calculation contain orbitals 2s and 2p of the oxygen atom of spherical and axial symmetry, respectively. The principle of the LCAO method is to take into account the "mixture" of atomic orbitals, but this is only a convenient approximate mathematical "device" for dealing with the problem of chemical bonding. The localized description of H_2O leads to the notion of bonding hybrids χ_1 and χ_2 (cf., Sec. IV.I) between the 2s and 2p AO's of oxygen, each pointing in the direction of a hydrogen atom, so as to create the two bond orbitals $b(OH_1)$ and $b(OH_2)$. Thus hybridation has no physical basis and is merely a convenient mathematical model for certain types of description.

2. The merits of the Hartree-Fock model

2.1 Mathematical advantages

For a given many-electron system (fixed geometry and fixed number of electrons), the Hartree-Fock limit corresponds to an extremum of a given type of trial function: The best function of the Slater determinant type, according to the variational theorem. It is therefore a well-defined mathematical limit. Even though it has no experimental counterpart, this limit has a clear numerical relevance, on the basis of which we may measure, for example, the quality of atomic base functions. From the start, Hartree realized that computing time for many-electron wave functions will rise exponentially with the number of variables in a given problem. Suppose, for example, that we wish to calculate the wave function of a 10-particle system for 10 values of each variable (3 variables per particle). This will require the calculation of 10^{30} points. In the Hartree-Fock method, which consists of minimizing the (antisymmetrized) product of one-electron functions, the calculation of a 30-variable function (for a 10-electron system) is replaced by the calculation of 10 functions with 3 variables each. As a result, the calculation is reduced to 10^4 points, which represents a significant numerical advantage, regardless of any physical consideration.

2.2 Acceptable descriptions of one-electron properties

The H-F method leads to a description of atomic and molecular systems in terms of internal shells and valence shells, and thus the reactivity of atoms and molecules may be interpreted qualitatively in terms of valence orbitals. This description is justified mainly by the fact that it leads to predictions of real behavior which are often correct.

One justification of this model may be obtained by comparing the radial electron density of the argon atom, as calculated by the H-F method, with the radial density deduced indirectly from measurements of electron diffraction intensity by the intermediate determination of the structure factor (Bartell 1953, Malli 1966). Figure IV.III.1 shows the maxima of radial density associated with the K, L and M shells, corresponding to increasing distance from the nucleus (i.e., to energies decreasing in absolute value).

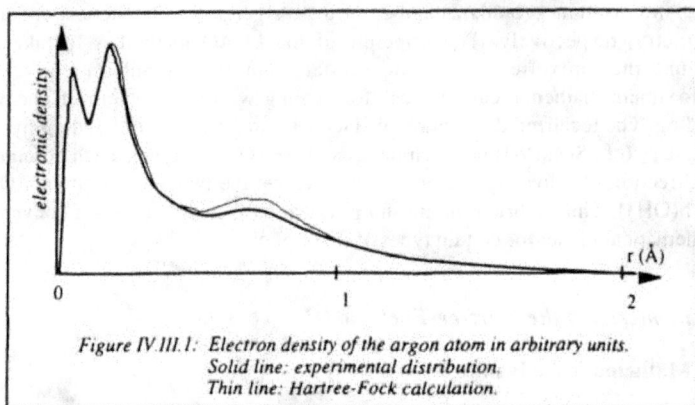

Figure IV.III.1: Electron density of the argon atom in arbitrary units. Solid line: experimental distribution. Thin line: Hartree-Fock calculation.

In general, as a consequence of the fact that electron densities are acceptable, one-electron properties, like the dipole moment, are also satisfactorily reproduced. For example, the H-F calculation of the dipole moment of LiH yields a value of 6.0 D (Debye) (experimental value : 5.83 D); while the calculation for NaCl yields 9.18 D (experimental value : 9.02 D). Average errors are on the order of 0.2 D. However, the "theory vs. experiment" agreement is valid only if absolute values are relatively large. Thus for CO, where the experimental dipole moment is 0.11 D (with polarity C^-O^+), the H-F calculation yields 0.27 D (with polarity inversed to C^+O^-).

2.3 Utility of the H-F model in chemistry

The H-F method accounts reasonably well for equilibrium geometries. Thus in the case of H_2O, from a minimal base set of Slater functions, Pitzer and Merrifield (see preceding chapter) obtained equilibrium OH distances on the order of 0.99 Å and the HOH plane angle in the neighborhood of 100.3° . In the case of larger base sets, this is a fairly "cheap" way of obtaining a geometry quite close to the experimental equilibrium geometry. Equilibrium geometries are among the most easily accessible observables in quantum chemistry.

Concerning the predictive possibilities of MO's in chemical reactivity, it is known that considerations of symmetry, coupled with a numerical determination of modest precision, enabled Woodward and Hoffman (Anh 1970) to propose general rules of behavior for chemical systems. We shall illustrate these rules in connection with the problem of cyclization of butadiene (cf., Sec. IV.III.6.4).

3. The limitations of the Hartree-Fock model

3.1 Electron correlation and the fluctuation potential

Let us analyze the contents of the many-electron H-F function $\Psi(1,2,...,N)$. This represents a product of one-electron functions, with the additional constraint of antisymmetry with respect to binary exchange. It is thus an approximate function, in which the movement of electron μ is virtually independent of the movement of other electrons. The movements appear to be uncorrelated. Actually this virtual independence of movement is partially limited. The antisymmetry of $\Psi(1,2,...,N)$ implies a correlation among electrons with the same spin, as, regardless of any electrostatic considerations, there is zero probability of finding electrons of the same spin next to each other. On the other hand, electrons of opposite spins remain uncorrelated in $\Psi(1,2,...,N)$. This correlation shortcoming produces an energy overestimate, as these situations are poorly described by the H-F function. The H-F energy E exceeds the exact energy E (nonrelativistic) by an amount called the correlation energy E_{corr}.

Thus

$$E_{corr} = E_{HF} - E_{NR} > 0$$

We would like to emphasize that the exact nonrelativistic energy E_{NR} is the exact eigenenergy of a hamiltonian operator containing only Coulomb-like kinetic and electrostatic potential terms. The exact, nonrelativistic, energy E is thus a purely theoretical quantity, differing from exact experimental energy only by a small amount due to relativistic effects (variation of electron mass with velocity, various couplings involving spin, non-local character of coulombic potential, etc.). In the case of water, all of these represent only 1/10 th of the correlation energy. The relativistic effects become more marked higher in the periodic system, i.e., in heavy atoms. In the case of light atoms, we can neglect these effects and identify the nonrelativistic energy E_{NR} with exact experimental energy E, without committing a significant error. The relative importance of correlation energy with respect to exact energy is on the order of 1%. Thus the H-F model accounts for 99% of the exact energy E. Unfortunately, E happens to be numerically very large — the zero of potential energy in atomic and molecular calculations corresponds to infinite separation of electrons (and of nuclei in the case of molecules) — so that 1% of E is far from negligible compared with the energies involved in chemical reactions. It is essential to go beyond the H-F limit to account with acceptable precision for bond breaking or bond formation.

These processes involve energies of about 100 to 400 kJ, one order of magnitude smaller than the correlation energy. In fact, the situation is not quite as catastrophic as this analysis makes it appear, as actual calculations show that in some cases this correlation energy may be split into contributions from electron pairs. E_{corr} is estimated to be equal, on the average, to 1.2 eV for each pair of electrons, independently of the nuclear charge to which they are subjected. There is thus an apparent additivity of the contribution to correlation energy from each pair of electrons, so that the variation of correlation energy may rapidly be estimated. In support of this additivity rule, note that the energy of isodesmic reactions, i.e., bond-conserving reactions, is correctly described on the H-F level.

This correlation shortcoming, inherent in the H-F function, may be illustrated by means of the concept of fluctuation potential V_F defined by Sinanoglu (Sinanoglu 1961). It is the difference between the exact two-electron repulsion potential and the average inter-electron potential that appears in the expression for the Fock operator F

$$V_F = \sum_{\mu=1}^{2N} \sum_{v>\mu}^{2N} g(\mu, v) - \sum_{\mu=1}^{N} V_{moyen}(\mu) \qquad \text{(IV.III.1)}$$

This fluctuation potential may easily be evaluated by the SCF calculation of helium, where the AO is developed linearly on the basis of two Slater functions

$$\varphi(1) = c_1 \chi_1(1) + c_2 \chi_2(1) \qquad \text{(IV.III.2)}$$

$$\text{with} \quad \chi_p(1) = \left(\sqrt{\alpha_p^3}/\sqrt{\pi}\right) e^{-\alpha_p r_1}$$

where the values of the exponents α_1 and α_2 are 1.45 and 2.91, respectively. Define by $J(1)$ the following one-electron operator, called the Coulomb operator:

$$J(1) = k\, q^2 \int_D \frac{|\varphi(2)|^2}{r_{12}}\, dv_2 \qquad \text{(IV.III.3)}$$

According to this definition, the Fock operator $F(1)$ is

$$F(1) = h(1) + J(1) \qquad \text{(IV.III.4)}$$

The fluctuation operator V_F is by definition

$$V_F = H - H^0 \qquad \text{(IV.III.5)}$$

where $H = h(1) + h(2) + kx/r_{12}$ is the exact system hamiltonian and $H = F(1) + F(2)$ is the sum of two one-electron Fock operators relative to each He electron.

Thus

$$V_F = H - H^0 = \frac{k\,q^2}{r_{12}} - J(1) - J(2) \qquad \text{(IV.III.6)}$$

If the position of electron 2 is fixed in space [for example, at a distance of 1 Bohr from the nucleus, which amounts to taking the operator $J(2)$ as a constant equal to 0.89355 Hartree], it is possible to calculate the dependence of V_F on x (such that $r_1 = |x|$) along the Ox axis, passing through the nucleus O and electron 2. The distance r is the algebraic difference $|r_2 - r_1|$. The problem then amounts to calculating the integral $J(1)$ of Eq. IV.III.3 from coefficients c_1 and c_2 of orbital Ī so that $J(1)$ is

$$J(1) = k\,q^2 \sum_{p=1}^{2} \sum_{q=1}^{2} c_p^* c_q \int_D \frac{\chi_p^*(2)\,\chi_q(2)}{r_{12}}\,dv_2 \qquad\text{(IV.III.7)}$$

Figure IV.III.2 shows the x-dependence of potential V_F as well as of operators $1/r_{12}$ and $J(1)$: The mean electron potential, calculated by the H-F method, is not infinite when the two electrons are in the same place ($r = 0$), whereas it must be in the case of exact two-electron potential. In the Hartree-Fock description, therefore, two electrons of opposite spin may occupy the same place, which is a real defect of the method.

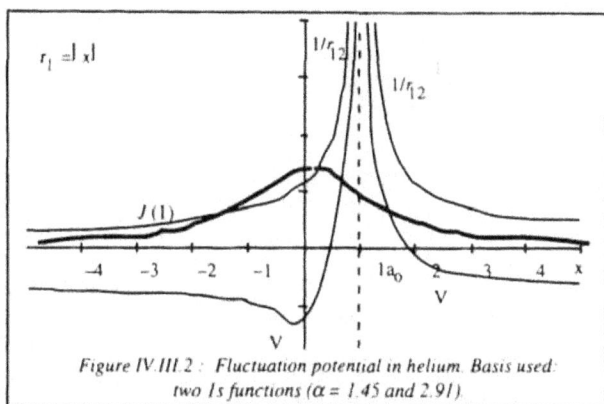

Figure IV.III.2 : Fluctuation potential in helium. Basis used: two 1s functions ($\alpha = 1.45$ and 2.91).

3.2 Study of bond energies and bond ruptures at the H-F level

Let us analyze the behavior of Hartree-Fock functions in the case of bond energies and bond ruptures, that is, in situations where there may be a wide variation in correlation energy

Bond energies

In the H-F method, the dissociation energy D is the difference between the H-F energy of the molecule and the sum of H-F fragment energies. The H-F energy of the nitrogen molecule N_2 is 5.3 eV, whereas the experimental value is 9.9 eV. In the case of the water molecule, the experimental atomization energy value is 10.1 eV, of which about 5 eV may be attributed to each O-H bond. A minimal base calculation of Pitzer and Merrifield (Pitzer 1970) yields an atomization energy of 4.5 eV, also using a minimal base for fragments, as well as optimized exponents. At the quasi-H-F level, Dunning, Pitzer and Aung find an atomization energy of 6.9 eV (Dunning 1972). For the fluorine molecule F_2, the H-F calculation yields the negative value $D = -1.4$ eV, while the experimental value is + 1.65 eV. The situation where the two fluorine atoms are at infinity is more stable than when they are bound. In other words, the molecule F_2 does not exist in the H-F model!

Bond rupture

Insofar as the bond rupture problem is concerned, the H-F function of a molecule such as hydrogen H_2 produces unacceptable behavior when the internuclear distance H-H tends to infinity. This behavior may be analyzed using the Slater determinant for the ground state of H_2, constructed from a minimal base bonding MO (not normalized) $\Psi = 1s_A + 1s_B$

$$\Phi_0 = \frac{1}{\sqrt{2}} \begin{vmatrix} \psi(1)\,\alpha(1) & \psi(1)\,\beta(1) \\ \psi(2)\,\alpha(2) & \psi(2)\,\beta(2) \end{vmatrix} = \psi(1)\,\psi(2)\,\frac{1}{\sqrt{2}}\left[\alpha(1)\,\beta(2) - \beta(1)\,\alpha(2)\right]$$

Extracting the coordinate part gives

$$\psi(1)\,\psi(2) = [1s_A(1) + 1s_B(1)][1s_A(2) + 1s_B(2)] \tag{IV.III.8}$$

$$= 1s_A(1)\,1s_B(2) + 1s_B(1)\,1s_A(2) + 1s_A(1)\,1s_A(2) + 1s_B(1)\,1s_B(2)$$

The first two terms correspond to a so-called covalent wave function, denoted H-H, as the two electrons involved in the bonding are "steered" by two functions centered on two different atoms. The presence of these two terms is the result of particle indiscernability. The next two products describe so-called ionic situations, denoted H^-H^+ and H^+H^- because these two electrons are each "steered" by two functions centered on the same atom. As these four terms appear with the same multiplicative constant, a covalent situation, where one electron is attached to each nucleus, is just as probable as that where both electrons are attached to the "left" atom (ionic form H^-H^+) or the "right" atom (ionic form H^+H^-). The situation is even more surprising in a system with a triple bond such as N_2. The H-F function of the molecule N_2 contains, with equal probability, chemically acceptable forms with three electrons "steered" by orbitals p_x, p_y and p_z of the "left" nitrogen, and three electrons "steered" by equivalent orbitals of the "right" nitrogen (covalent form N≡N), as well as highly unlikely forms where all the six electrons are either on the "left" atom (form $N^{3-}-N^{3+}$), or on the "right" atom (form $N^{3+}-N^{3-}$). This development thus shows the weakness of the H-F description of equilibrium geometries of nitrogen and hydrogen. With increasing internuclear distance R, this shortcoming of the H-F function is even more apparent. All ionic forms should gradually disappear, as these two molecules fragment into two neutral atoms. However the H-F function prevents this: The relative weight of ionic and covalent forms remains constant, independently of distance R. In the case of hydrogen, it may be shown that, as R increases continuously, the H-F total energy converges to a value 7 eV higher than the exact value –27.2 eV, corresponding to the energy of two separate hydrogen atoms (twice –13.6 eV). The nonsensical H-F energy is in fact the sum of the energies of two neutral hydrogen atoms, plus one-half the energy of ion H^- (one-half of -14.35 eV), so that the H-F function corresponding to this energy has no physical meaning.

These examples, among others, show the inadequacy of the H-F function owing to the lack of a description of electron correlation. By its construction, a Slater determinant consists of an algebraic sum of terms (products of one-electron functions), the contribution of each being of equal weight.

4. General remarks concerning electron correlation

4.1 Early contributions of Hylleraas

According to the analysis of Sinanoglu, the H-F function may be improved by including the fluctuation potential in the wave function. This potential creates what quantum chemists call a "correlation hole", so that there is little probability that the electrons will be close to each other. The objective of all practical methods beyond the H-F approximation is precisely to create this "correlation hole".

One of the first quantum chemists to be aware of the electron correlation problem was without doubt Hylleraas who, as early as 1928, had the idea of introducing a correlation between the movements of the two helium electrons into the trial function itself (Hylleraas 1928). Starting with a function consisting of the product of two 1s-type one-electron functions with an effective exponent ζ, he introduced an additional term which varied linearly with the distance r_{12} between two electrons:

$$\Phi(1, 2) = N \; e^{-\zeta r_1 / a_0} \; e^{-\zeta r_2 / a_0} \; [1 + b \, r_{12}] \qquad \text{(IV.III.9)}$$

The optimization of parameters ζ and b, according to the variational theorem, leads to $\zeta = 1.849$ and $b = 0.364$, yielding the energy $W_{min} = -78.7$ eV, quite close to the experimental value -79.0 eV. The H-F limit for He is -77.86 eV. Hylleraas has thus accounted for two thirds of the correlation energy. Introduction of the r_{12}-dependent part with a positive constant b favors the situations where r_{12} is large at the expense of those where r_{12} is small. In fact, the square of the modulus of F represents the probability of such situations.

Hylleraas improved his function by introducing a sum of terms, each of which is a function of r_1, r_2 and r_{12} (instead of a single term r_{12}) of the form:

$$\Phi'(1, 2) = N' \; e^{-\zeta r_1 / a_0} \; e^{-\zeta r_2 / a_0} \; \sum_{i,j,k} C_{ijk} \, (r_1 + r_2)^i \, (r_1 - r_2)^j \, r_{12}^k \qquad \text{(IV.III.10)}$$

where i,j,k are positive integers. In addition, j must be even so that the coordinate space function Φ' is symmetric with respect to binary exchange, inasmuch as it is ultimately multiplied by a spin function antisymmetric with respect to this exchange. Hylleraas used a six-term function of this type and obtained a value for the ground state energy of the He atom only 0.01 eV below the experimental

224 Orbital description of chemical bonding

value. In 1959, Pekeris used this type of function again, with 1078 terms and a computer (Pekeris 1959), and obtained the value of -2.903724375 Hartree, within 2×10^{-9} Hartree of the relativistic value.

A similar electron correlation approach for the H_2 molecule had already been used in 1933 by James and Coolidge (James 1933). They obtained a theoretical dissociation energy $D_e = 4.72$ eV, compared with the theoretical value 4.746 eV, using a function very similar to that used by Hylleraas. It contains a sum of 13 terms comprising the elements $(r_{A1} + r_{B1})$ and $(r_{A2} + r_{B2})$, $(r_{A1} - r_{B1})$ and $(r_{A2} - r_{B2})$, as well as the distance r_{12}. Each of these terms appears to an integral power (r_{Ni} is the distance of electron i from nucleus N). In the 60's, this treatment was extended by Kolos, Roothaan and Wolniewicz (Kolos 1960 - 1964 -1968) by means of a function similar to that of James and Coolidge, including about a hundred terms. Taking into account relativistic corrections, these authors obtained a value for the dissociation energy D within 4 cm^{-1} of Herzberg's experimental determination. On re-examination, Herzberg later corrected his value by the same minuscule 4 cm^{-1}, thus proving the exactness of the theoretical approach (Herzberg 1970).

Despite its incontestable elegance, this correlation treatment is difficult to apply to an arbitrary many-electron system, although this has recently been achieved (Dulieu 1987a – 1987b) (Henriet 1984-1987) (Moumeni 1990). However, in research work concerned primarily with practical rather than methodological aspects, it is preferable to use a different approach, more or less derived from the perturbative concept of Sinanoglu. In this approach, the H-F function is considered as a zero-order function obtained from a zero-order hamiltonian $H^0 = F(1) + F(2)$, and the fluctuation potential V_F is treated as an additive perturbation term in order to obtain the exact non-relativistic system hamiltonian H^0. From the second order onwards, this perturbation treatment involves mixing the unperturbed state (the H-F determinant) with functions that describe excited states with respect to the H-F function (actually only doubly excited functions are necessary for the second order). In general, starting from the original electron distribution in the ground state, it is necessary to take all singly-, doubly-, triply- etc. excited functions in order to construct a more realistic function. This perturbation approach had already been proposed in 1934 by Møller and Plesset (Mﬂler 1934) and is called the MP method. It belongs to the general class of perturbation methods developed for the study of systems of interacting particles, known generically as the "Many-Body Perturbation Theory" (MBPT). MBPT was first applied to practical molecular problems in the U.S. around 1975 by Pople and Bartlett (Hehre 1986 Bartlett 1981). When the perturbation is taken to order 2, 3 or 4, the method is called MP2, MP3 or MP4. Efficient quantum chemistry software represents practical applications of the various perturbation methods.

In France, Malrieu et al. (Huron 1973) applied a perturbation approach to a reference function comprising several determinants. In this way it was possible to estimate the correlation energy, not only in the vicinity of the equilibrium geometry, but also at various points on the potential energy surface. The many-electron function often involves only a single determinant in the study of systems in equilibrium geometries, in particular systems with closed shells, so that the MP approach is sufficient. In general, however, the perturbation approach of Malrieu et al., based on the many-determinant function is very useful for the study of reactive processes.

In addition to these perturbation approaches, there is also a variational-type procedure for evaluating the correlation energy, called *"configuration interaction"* *(CI)*. Although it has some drawbacks compared with the perturbation approach (slow convergence and nonrespect of a property called "size consistency"; see Szabo 1989 or Levine 1991), configuration interaction is without doubt the technique used most often to take account (in general partially) of electron correlation. Also its presentation facilitates the understanding of calculation methods for electron correlation.

4.2. Configuration interaction (CI)

Description of the method

We have seen that the H-F method, when applied to a system of 2N electrons with Q one-electron base functions, leads to Q orbitals (Q > N) among which N of the lowest energy are occupied by two electrons each with antiparallel spins. The (Q – N) orbitals of higher energy constitute virtual orbitals in the field of the N occupied ones.

Although CI is a variational and not a perturbative method, it takes into account the various electron excitations using the complete set of H-F orbital levels (occupied and virtual orbitals). The virtual orbitals play a crucial role in the make-up of electron correlation. These orbitals are antibonding, and therefore they contain nodes, i.e., regions of space where the square of their modulus, that is to say, their orbital density, is zero. These regions of zero orbital density are often localized on bonds, hence the name "anti-bonding orbital". As a result, these orbitals favor situations where the electrons are far from each other, so that their role in configuration interaction will be to privilege instantaneous dynamic situations where electron repulsions are at a minimum (Coffey 1974). As the nature and the number of these virtual MO's depends on the choice of the AO basis, this initial choice is critically important for the success of CI.

Let Δ_1 be the closed-shell determinant (paired spins) built up from N AO's. It is possible to build M determinants Δ_i (M is usually quite large) in which one, two, three, four, etc. atomic spin-orbitals (ASO's), occupied in the ground state of the system under study, are replaced by one, two, three, four, etc. virtual ASO's. We may then say that we have built singly- (M_i), doubly- (D_i), triply-(T_i), quadruply-(Q_i), etc. excited determinants. The total wave function of the system is expressed as a linear development in the base set of these M determinants:

$$\Phi (1, 2,..., 2N) = \sum_{k = 1}^{M} a_k \Delta_k \qquad (IV.III.11)$$

The problem is, therefore, to find the coefficients which will produce the lowest variational energy W

$$W = \frac{\displaystyle\int_D \Phi^* H\Phi \, d\tau}{\displaystyle\int_D \Phi^* \Phi \, d\tau}$$

This amounts to solving a system of M homogeneous linear equations with M unknowns (the a_k), of the form

$$\sum_{k=1}^{M} a_k (H_{ik} - W S_{ik}) = 0 \quad \text{with } i = 1, \dots M \qquad \text{(IV.III.12)}$$

where

$$H_{ik} = \int_D \Delta_i^* H \Delta_k d\tau \quad \text{and} \quad S_{ik} = \int_D \Delta_i^* \Delta_k d\tau$$

This sort of system has nontrivial solutions only for values of W which satisfy the secular Eq.of degree M :

$$| H_{ik} - W S_{ik} | = 0 \qquad \text{(IV.III.13)}$$

The smallest root W_0 is an upper limit of the ground state energy,. Its quality depends on the choice and number of determinants. There will also be $(M - 1)$ real roots W_n $(n > 0)$, larger than W_0 , which, according to Mc Donald's theorem (Mc Donald 1933) constitute the upper energy limits of excited states. From the equation system and the normalization condition of the functions, it is possible to determine the coefficients a_k which define the composition in terms of the determinants Δ_k of the state linked to each root.

Remark 1: Note that if the AO's constituting the determinants Δ_k are orthonormalized, the overlap matrix S of the determinants is a unit matrix, i.e., we have $S_{ik} = \delta_{ik}$. The problem reduces strictly to the diagonalization of the matrix that represents hamiltonian H on the basis of determinants.

Remark 2 : The off-diagonal term H_{ik} of H is zero if the determinants Δ_i and Δ_k belong to different symmetries, i.e., to different irreducible representations of the symmetry group of the system under study. It is thus possible to avoid having to manipulate a matrix H of large dimensions if interaction is only permitted between functions of the same symmetry. The size of matrix H increases rapidly, proportional to M^N, where N is the number of electrons and M the number of determinants. If all possible determinants, from singly- to N-excited, are taken into account, this produces what is called *complete CI for the base used*. In fact, the total number of determinants is finite if one uses a base of AO's constituted of a finite number of functions, as the SCF calculation leads to as many MO's as there are AO's in the base.

Remark 3: It should be emphasized that the SCF calculation is a redundant stage in the case of complete finite CI. Since the CI base is complete, any set of base functions produces the same result. In particular, it is possible to use an set of AO's directly to construct a complete set of determinants singly- to N-excited.

Results obtained from a complete finite CI for the energy of each state do not correspond to exact nonrelativistic energies. Exact results may only be obtained from an infinite number of determinants built up from an infinite number of AO's or MO's, which is, of course, not feasible. Even a complete finite CI is unattainable, except for small systems and those with a very small number of base functions. It is therefore important to know which electron excitations, built from the original determinant (in general it is the closed-shell determinant Δ_1), play the most important role. It may be shown that it is necessary above all to account for *double excitations* ($\Delta_k = D_k$) which contain all the first-order corrections to the H-F function. At the same time, however, it is equally important to account for single excitations ($\Delta_k = M_k$) in order to represent correctly one-electron properties of the molecule under study, like the dipole moment (Shavitt 1977). It may be shown that the single excitations M_k do not interact directly with the fundamental closed-shell determinant. This is because the elements H_{1k} of the matrix representing the hamiltonian in the base formed by the Δ_1 (SCF solutions of closed-shell system) and M_k are zero:

$$H_{1k} = \int_D \Delta_1 {}^* H\, M_k\, d\tau = 0 \qquad \text{(IV.III.14)}$$

This property is called "Theorem of Brillouin" Actually, the interaction between Δ_1 and M_k is indirect and mediated by functions of determinants corresponding to higher excitations, such as doubly-excited determinants D_k. The inclusion of the all single and double excitations is usually called SDCI (Single and Double Configuration Interaction). Comparison between SDCI and the MP-type perturbation approach shows that the SDCI technique does not produce all the second-order corrections to the H-F function. To obtain all these corrections, it is necessary to include determinants corresponding to triple (T_k) and quadruple (Q_k) excitations, in addition to single (M_k) and double (D_k) excitations. It should be pointed out that only the CI approach converges monotonically to an upper energy limit, due to the variational theorem. In perturbation methods, convergence to exact energy is approached either from below or from above, so that, in any given case, it is impossible to tell if one is below or above the exact value. Most molecular calculations focus on properties in which only valence electrons play a role. It is therefore possible to limit the calculation to valence excitations, thus reducing the number of functions to be considered. Note that the choice of base functions is crucial in the CI calculation, as the set of virtual MO's is responsible for the quality of the correlation calculation. For this reason, an extended basis of atomic functions is normally used, giving the largest possible degree of freedom to the electrons so that they "avoid each other as much as possible". This favors deformations of the electron cloud in all spatial directions.

An important practical problem that has to be resolved in the treatment of correlation by the CI method is that of achieving fast convergence to the nonrelativistic limit with the smallest possible number of determinants. It may be shown that using canonical SCF MO's produces relatively slow convergence, so that many determinants are necessary for acceptable accuracy. There are a number of acceleration techniques, mentioned below in the paragraph entitled "Other quantum approaches". Suffice it to mention here that using localized molecular orbitals instead of canonical orbitals speeds up the convergence of CI considerably. Thus Saebø and Pulay (Saebø 1985) have shown that using localized MO's (instead of canonical MO's) in the CI treatment of trans-butadiene reduces computing time by a factor of almost thirty.

Application to the study of the helium atom ground state

For the purpose of illustration, we shall apply the CI method to a more accurate determination of the ground state energy of He relative to a simple H-F calculation with base (χ_1, χ_2) (χ_1 and χ_2 are Slater functions with exponents $\alpha_1 = 1.45$, $\alpha_2 = 2.91$). We have two AO's, one of which (φ) is occupied by two electrons in ground state, and the other (φ') is a virtual orbital (the expression for this was given previously).

We may thus form at most M = 6 Slater determinants Δ_i:

$$\Delta_1 = \left| \varphi \; \bar{\varphi} \right|$$
$$\Delta_2 = \left| \varphi' \; \bar{\varphi'} \right|$$
$$\Delta_3 = \left| \varphi \; \bar{\varphi'} \right|$$
$$\Delta_4 = \left| \bar{\varphi} \; \varphi' \right| \qquad\qquad (IV.III.15)$$
$$\Delta_5 = \left| \varphi \; \varphi' \right|$$
$$\Delta_6 = \left| \bar{\varphi} \; \bar{\varphi'} \right|$$

A complete CI is performed in the basis (χ_1, χ_2) or (φ, φ'). The matrix that represents the Hamiltonian H in this base is a 6×6 matrix, whose elements are integrals expressed in terms of the orthonormalized MO's φ and φ'. The molecular integrals which appear in these expressions are calculated from atomic integrals of the SCF problem. The diagonalization of H yields six eigenvalues W_0, W_1, W_2, W_3, W_4 and W_5. The smallest of these, W_0, is below the previously obtained SCF energy, hence better. The ground state energy of He is thus

$$W_0 = -2.876 \text{ Hartree} = -78.25 \text{ eV},$$

compared with -2.8617 Hartree $= -77.87$ eV for the SCF result with base (χ_1, χ_2).

In fact, it turns out that the construction of the 6×6 matrix is not really necessary because considerations related to the total spin of the system cause this matrix to be block-diagonal, in the form of one 4×4 block and two 1×1 blocks:

$$(H) = \begin{pmatrix} H_{11} & H_{12} & H_{13} & H_{14} & 0 & 0 \\ H_{21} & H_{22} & H_{23} & H_{24} & 0 & 0 \\ H_{31} & H_{32} & H_{33} & H_{34} & 0 & 0 \\ H_{41} & H_{42} & H_{43} & H_{44} & 0 & 0 \\ 0 & 0 & 0 & 0 & H_{55} & 0 \\ 0 & 0 & 0 & 0 & 0 & H_{66} \end{pmatrix}$$ (IV.III.16)

Diagonalization concerns exclusively the 4 x 4 block. This characteristic, represents a considerable advantage for numerical calculations. It leads to energy ordering of states that is very useful in spectroscopy, and is the consequence of the fact that eigenstates of the nonrelativistic hamiltonian are both eigenstates of the square S^2 of the total spin operator S of the system and its projection S_z on a given axis Oz. The eigenvalues of these two spin operators are proportional to $S(S+1)$ and M, respectively. S is either a positive integer (including zero), or a positive half-integer:

$$S = 0, 1/2, 1, 3/2, 2, 5/2, 3, ..., \text{etc}..$$

and M is an algebraic integer such that $|M| \leq S$, with unit difference between two adjacent values:

$$M = -S, -S+1, -S+2,..., S-2, S-1, S$$

Thus there are $(2S+1)$ values of M. The spin state multiplicity of a many-electron system is defined by the integer $(2S+1)$.

Examples: if $S = 2$, $\Rightarrow M = -2, -1, 0, 1, 2$
if $S = 3/2$, $\Rightarrow M = -3/2, -1/2, 1/2, 3/2$

It is customary to label the multiplicity of a spin state as a *singlet* (S = 0; 2S+1 = 1), *doublet* (S = 1/2; 2S+1 = 2), *triplet* (S = 1; 2S+1 = 3), *quartet* (S = 3/2; 2S+1 = 4), *quintet* (S = 2; 2S+1=5), etc. The result is that on this level of relativistic description, distinct states of total spin S and/or projections of M on different Oz axes do not couple via the hamiltonian operator which contains no spin variables (at this level of description). Among the preceding six determinants, Δ_1, Δ_2, Δ_3 and Δ_4 are associated with M = 0 (M is obtained by taking the algebraic sum of spin projections of each spin-orbital). By the same procedure, it may be seen that Δ_5 and Δ_6 correspond to M = +1 and M = -1, respectively. This characteristic thus justifies the separation of the H matrix into three blocks. For a given value of S, the number of values of M (i.e., spin multiplicity) yields the degeneracy of the energy level corresponding to the non-relativistic hamiltonian. We may thus continue the analysis, as the eigenvalues $W_4 = H_{55}$ and $W_5 = H_{66}$, which are equal according to the proposition above, represent the energy ^3E of the lowest triplet state of He (in this approximate description). This common value ^3E must appear for the third time in the diagonalization of (H) as this is a triplet. One of the eigenvalues of the remaining 4 x 4 block of (H) must therefore contain the eigenvalue ^3E.

The diagonalization of (H) yields eigenfunctions which are also eigenfunctions of S^2. Therefore, four functions Φ_1, Φ_2, Φ_3 and Φ_4 are produced, linear combinations of the determinants, of which one (Φ_4), associated with ^3E, is a triplet; and the other three are singlets.

We obtain

$$\Phi_1 = 0.999\ \Delta_1 - 0.047\ \Delta_2 - 0.018\ \Delta_3 + 0.018\ \Delta_4 \qquad \text{with } W_o = -2.876\ \text{eV}$$
$$\Phi_2 = 0.0056\ \Delta_1 - 0.063\ \Delta_2 - 0.706\ \Delta_3 + 0.706\ \Delta_4 \qquad \text{with } W_1 = -0.392\ \text{eV}$$
$$\Phi_3 = 0.047\ \Delta_1 - 0.997\ \Delta_2 - 0.045\ \Delta_3 + 0.045\ \Delta_4 \qquad \text{with } W_2 = -2.876\ \text{eV}$$
$$\Phi_4 = (\Delta_3 + \Delta_4)/\sqrt{2} \qquad \text{with } W_3 = {}^3E = -2.876\ \text{eV}$$

The triplet function cannot be used to improve the ground state helium function which is a singlet. If we had used the combination $\Omega = (\Delta_3 - \Delta_4)\ \sqrt{2}$ of the determinants Δ_3 and Δ_4 directly ,a singlet eigenfunction of S^2 , as a basis for CI, instead of the determinants Δ_3 and Δ_4 as such, we would have reduced the 4 x 4 block to one 3 x 3 block containing only singlet functions, and one 1x1 block. The CI base functions would then have been Δ_1 , Δ_2 and Ω. This result is, of course, identical to that obtained by diagonalization of the 4 x 4 block (without the triplet function), but it is represented by three eigenfunctions (Δ_1, Δ_2 and Ω) of S^2. Note that the preceding calculation only gives a mediocre description of the first two excited states of He, as the AO base contains no orbital capable of accounting for the presence of an electron in the region of space far from the nucleus, such as the functions 2s or 2p.

From a practical point of view, it is preferable in a CI calculation to choose an eigenfunction base of S^2 and S_z , formed by a linear combination of determinants, rather than the determinants themselves. These combinations are often called CSF (Configuration State Function). This makes it possible to work with smaller matrices which are easier to diagonalize. Tables of linear combinations of determinants, eigenfunctions of S^2, may be found in the literature (Pauncz 1979).

CI study of bond rupture in the hydrogen molecule

Let us return to the minimal base set ($1s_A$, $1s_B$) used previously for the H-F study of the H-H bond rupture in ground state hydrogen molecule. A complete CI is possible with (nonnormalized) molecular orbitals $\psi = 1s_A + 1s_B$ and $\psi' = 1s_A - 1s_B$, the first being a bonding MO and the second an anti-bonding MO. The latter is the minimal-base SCF virtual orbital of hydrogen H_2 . It is therefore possible to construct six determinants from these two MO's, but the CI calculation just presented has shown that it is pointless to work with a 6 x 6 matrix as the H_2 ground state is a singlet. A priori, three base functions, analogous to those we have denoted Δ_1, Δ_2 and Ω in the case of helium, will be sufficient. However inasmuch as orbital ψ' is antisymmetric with respect to the midpoint of the H-H bond, while ψ is symmetric, the symmetry of the Ω-type function is different from those of Δ_1 and Δ_2. It is therefore possible to limit the development to two base functions of the type D_1 and D_2 (see Remark 2 of paragraph 2). Let us call Φ_0 the closed-shell determinant of type Δ_1 constructed from ψ:

$$\Phi_0 = \frac{1}{\sqrt{2}} \begin{vmatrix} \psi(1)\,\alpha(1) & \psi(1)\,\beta(1) \\ \psi(2)\,\alpha(2) & \psi(2)\,\beta(2) \end{vmatrix} = \psi(1)\,\psi(2)\,\frac{1}{\sqrt{2}}\left[\alpha(1)\,\beta(2) - \beta(1)\,\alpha(2)\right]$$

and Φ_1 the closed-shell determinant of type Δ_2 constructed from ψ':

$$\Phi_1 = \frac{1}{\sqrt{2}} \begin{vmatrix} \psi'(1)\,\alpha(1) & \psi'(1)\,\beta(1) \\ \psi'(2)\,\alpha(2) & \psi'(2)\,\beta(2) \end{vmatrix} = \psi'(1)\,\psi'(2)\,\frac{1}{\sqrt{2}}\left[\alpha(1)\,\beta(2) - \beta(1)\,\alpha(2)\right]$$

The wave function ψ_1 describing the H_2 ground state is the (non-normalized) combination of the type:

$$\Psi_1 = \Phi_0 + \lambda\,\Phi_1 \tag{IV.III.17}$$

By analyzing the contents of the spatial part of ψ_1, we find

$$\tag{IV.III.18}$$

$$\psi(1)\,\psi(2) + \lambda\,\psi'(1)\,\psi'(2)$$
$$= [1s_A(1)\,1s_B(2) + 1s_B(1)\,1s_A(2)]\,(1-\lambda) +$$
$$[1s_A(1)\,1s_A(2) + 1s_B(1)\,1s_B(2)]\,(1+\lambda)$$

The coefficient λ is a function of the distance R between the two protons. At the equilibrium distance $R_e = 0.74$ Å we find $\lambda(R) = -0.59$. These numbers are the result of a calculation with optimized exponents of the two atomic functions $1s_A$ and $1s_B$ ($\zeta = 1.19$), obtained from the determination of equilibrium distance at potential minimum. We obtain a description of the H-H bond where the ionic component constitutes only one-fourth ($0.41/1.59 = 0.26$) of the covalent component. This situation is certainly more acceptable than the SCF result, obtained from a MO ψ that neglects the overlap of functions $1s_A$ and $1s_B$, where the ionic part has the same weight as the covalent part. This improvement also results in better energy values, as the SCF dissociation energy is $D_e = 3.48$ eV (with $\zeta = 1.19$), whereas CI yields $D_e = 4.02$ eV and the experimental value is 4.746 eV.

As R increases constantly, $\lambda(R)$ tends towards -1. The ionic forms disappear. This simple CI, using only the ground state determinant and a determinant describing the double excitation of ψ to ψ', provides a correct description of the dissociation of the hydrogen molecule into two neutral atoms. A complete review of quantum calculations on the hydrogen molecule is available in the literature (McLean 1960).

Other quantum approaches

We have previously raised the important issue of the speed of CI convergence. By a method called the *method of multi- configurational SCF (MCSF)* it is possible, for a given set of MO's, to calculate a larger proportion of the correlation energy than by simple SCF-CI. In this method, the total wave function of a many-electron system is expressed as a linear combination of determinants (or CSF's). Using the variational theorem, it is then possible to determine the best coefficients a of the base of determinants (or CSF's), and the best AO coefficients c_p in the expression for the MO's at the same time. This variational approach involves two linear parameters. The optimal MCSCF orbitals are obtained by an iterative procedure analogous to the traditional SCF method (Levy 1968a-1968b-1971)(Wahl 1977) (Shepard 1987). Actual results obtained with a more limited number of determinants (or of CSF's) are better than those obtained using CI, at considerably higher computing costs.

The method of Complete Active Space SCF (CASSCF) developed by Roos (Roos 1987) makes use of the MCSCF method and includes all excitations from 1 to N using only so-called *active* MO's. Doubly occupied, or inactive, MO's are ignored. The set of valence MO's of the system under study is a reasonable choice for the active MO's, as the inactive MO's are core orbitals centered on the atoms. This procedure yields acceptable dissociation energies D_e, equilibrium distances R_e and wave numbers of molecular vibrations (Wahl 1977) (Chavy 1991). This is a preferred method for the study of bond rupture, i.e., for the theoretical study of elementary reactions. In addition, it is the starting point of methods that come closer still to the nonrelativistic energy limit (see Szabo 1989 and Levine 1991).

Remark: One very interesting, promising approach has not yet been analyzed in this chapter (but was rapidly presented in Chapter I). This is the so-called *valence bond method* (VB), originally introduced by Heitler and London in 1927 and developed by Slater and Pauling (see summary in Pilar 1990). This method consists of writing the wave function of a molecular system in the form of a linear combination of determinants, built from system AO's. It describes covalent or ionic structures similar to those invoked in connection with the H molecule. The great merit of this method is that it provides a description of molecular systems close to the chemist's traditional image. Updated around 1970 into a so-called "generalized" form by Goddard III and al. (Bobrowicz 1977), it is presently in full development (Cooper 1987-1991). We will note that it is possible to construct good VB functions from MCSCF-type developments (Levasseur 1991), without having to construct VB functions from a nonorthogonal base. From the chemist's point of view, the VB approach yields a very promising interpretation of molecular structure and reactivity. It forms the basis of useful models such as the Woodward-Hoffmann rules, expressed in terms of MO orbitals (Shaik 1991).

This brief analysis of alternative theoretical approaches to the calculation of electron correlation is far from exhaustive. We hope, nevertheless, that it will arouse the reader's curiosity. There is little doubt that correct treatment of electron correlation is currently a very active field of research. There are numerous problems relative to the convergence of the calculations. Readers wishing to learn more about this question are invited to embark on the "initiation route" (to use the terminology of the authors) proposed by Malrieu and Maynau (Malrieu 1978).

5. A few thoughts on the exact wave function (by way of conclusion)

In the last part of this chapter, we have given a critical analysis of the Hartree-Fock method, as well as a brief account of a few methods for evaluating correlation energy. In particular, we have shown that complete configuration interaction leads to exact nonrelativistic energy values. From the function associated with this exact nonrelativistic energy, which is thus the exact function in the nonrelativistic approximation, we may calculate the exact electron density which should agree with that obtained from X-ray diffraction experiments.

Let us consider the contents of this exact function. In addition to total electron density, this function provides electron pair probability, i.e., the probability of simultaneously finding two electrons μ and ν in two given regions of space $\Delta\nu_\mu$ and $\Delta\nu_\nu$ with given projections of spin $s_{z\mu}$ and $s_{z\nu}$. We thus find, in the water molecule, for example, that there is high probability of finding two electrons with antiparallel spins in a more-or-less banana-shaped "loge" (Daudel 1971) surrounding each of the O-H bonds. We also find that it is very probable to have electron pairs in "loges" situated symmetrically on either side of the molecular plane of water, and in a plane perpendicular to the latter. This brings us back to the chemist's familiar image, although, as it is deduced from the exact wave function, all the shortcomings of the Hartree-Fock method have now been eliminated.

It is important to keep in mind that the success of Pauling's theory of hybridation is due to the fact that it is based on a description that takes into account this distribution of electron pairs and their spins.

The same comment may be made concerning the principle of Valence Shell Electron Pair Repulsion (VSEPR) method (Gillespie 1963). In this method, described in Sec. IV.III.6.1, it is possible to predict molecular geometries of a system $AX_n E_m$ — where A is a central atom, X a ligand bound to this atom and E a lone electron pair — simply by counting the electron pairs (lone and bound) in the vicinity of the central atom A. Equilibrium geometry is obtained when there is minimum repulsion among the electron pairs. Structures may be in the shape of a star, triangle, tetrahedron, etc., depending on the number of pairs involved in the structure.

6. Problems and computer-assisted visualization

6.1 Molecular geometry determination using the VSEPR method

Introduction

When using the quantum mechanical approach, one usually needs to know beforehand the relative locations of the atoms. However it is possible to use this method to obtain the geometry of molecules or of polyatomic ions by calculating the energy corresponding to the various structures and choosing the one with the lowest energy. We have seen in Sec. IV.I that the H.F.- L.C.A.O. method usually leads to correct bond lengths and angles, even with semiempirical formulae, but that the necessary computing power is always large. Methods making it possible to determine the structures or main characteristics of chemical species are therefore of high value for chemists (e.g., why is the triatomic molecule CO_2 linear while H_2O is bent?). Among these methods, the V.S.E.P.R. theory (Valence Shell Electron Pair Repulsion) developed by R.J. Gillespie in 1960, from an idea published by Sidgwick and Powell in 1940, has become extremely popular [R.J. Gillespie, The Valence-Shell Electron-Pair Repulsion Theory (V.S.E.P.R.) of Directed Valency, Journal of Chemical Education, **40**, 295(1963)].

The basic idea behind this theory is quite simple: Most molecules possess electron pairs, as shown in the Lewis formula. The locations of these pairs in space around the central atom determine the geometry of the molecule.

AX_n molecules with single bonds only

Let us first consider molecules of the type AX_n in which the central atom A is linked to n atoms X by single bonds (σ bonds). Besides the n pairs corresponding to the A—X bonds, atom A may possess n' free pairs (nonbonding). The V.S.E.P.R. theory states that these $p = n + n'$ pairs occupy localized orbitals oriented in space around A in such a way as to maximize their average distance so that the repulsion energy between them is minimized.

Let us consider, *as a first approximation*, that bonding and nonbonding pairs are equivalent. We may thus depict them as radii of a sphere with center A.

When $n' = 0$ (i.e., $p = n$), we obtain the structures, shown in Fig. IV.III.3, corresponding to the largest angles between pairs. When $n' \neq 0$, a bonding pair is replaced with a nonbonding pair.

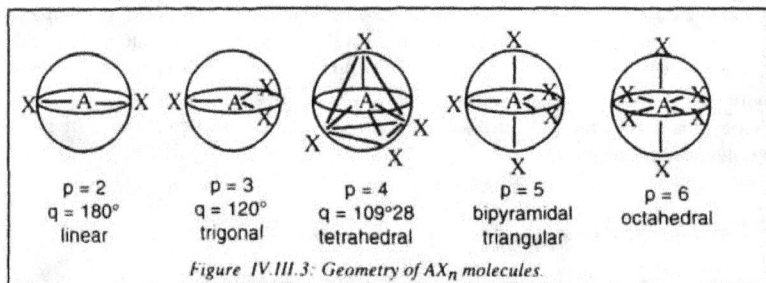

$p = 2$	$p = 3$	$p = 4$	$p = 5$	$p = 6$
$q = 180°$	$q = 120°$	$q = 109°28$	bipyramidal	octahedral
linear	trigonal	tetrahedral	triangular	

Figure IV.III.3: Geometry of AX_n molecules.

Let us determine the geometry of a molecule AX_n. In the first stage, we must count the number of bonding and nonbonding pairs surrounding A. We thus need to know about the electron structure of atom A and, in particular, the number N of electrons occupying the outermost shell. The n ligands X will involve n bonding pairs of electrons and thus n electrons of atom A, if we assume that atoms A and

X contribute one electron each to the A—X bond. In that case, $(N - n)$ electrons are left with A and $n' = (N - n)/2$ nonbonding pairs. We may then calculate:

— The total number of pairs: $p = n + n'$.

— The structure of the molecule.

— Its spatial representation.

— The Lewis structure of the molecule in which the pair, in a single A—X bond, is represented by a line between atoms and the nonbonding pair by: A $|$.

Example 1: H_2O

In the case of the oxygen atom we have

$Z = 8$
electron configuration: $1s^2\ 2s^2\ 2p^4$
valence shell quantum number: 2
number of electrons in outermost shell: $N = 6$

The two hydrogen atoms provide $n = 2$ *bonding pairs*. There are thus $n' = (N - n)/2 = 2$ *nonbonding pairs* left. Therefore $p = n + n' = 4$ and H_2O has a tetrahedral structure. The H_2O formula of water will then be completed by representing the nonbonding pairs by E: H_2O will thus be reexpressed as OH_2E_2, which should have, on this level of approximation, the same geometry as AX_2E_2 or AX_4 (tetrahedron). The spatial representation and the Lewis formula are given in Fig. IV.III.4.

H_2O is indeed bent as shown experimentally, but the HOH angle ($\approx 105°$) is slightly different from that of the proper tetrahedral structure ($\theta = 109°28'$). We suggest below an explanation for that discrepancy.

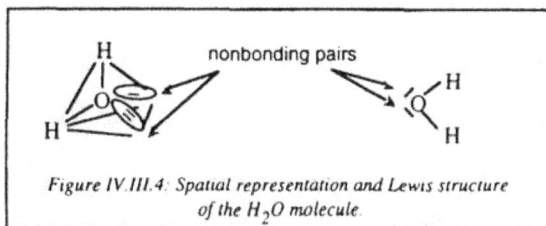

Figure IV.III.4: Spatial representation and Lewis structure of the H_2O molecule.

Example 2: SF_6

For the sulphur atom we have

$Z = 16$
electron configuration: $1s^2\ 2s^2\ 2p^6\ 3s^2\ 3p^4$
valence shell quantum number: 3
number of electrons in outermost shell: $N = 6$

The six fluorine atoms correspond to $n = 6$ *bonding pairs*. There are $n' = (N - n)/2 = 0$ *nonbonding pairs* left and the value of p is: $p = n + n' = 6$, leading to an octahedral structure. The geometry of the molecule is shown in Fig. IV.III.5.

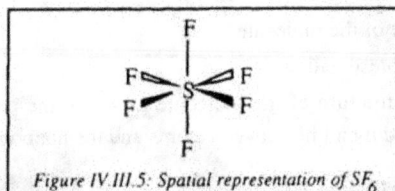

Figure IV.III.5: Spatial representation of SF_6

Other examples: The same method may be used to determine the geometry of the following molecules:

BeH_2, $BeCl_2$, $HgCl_2$, CdI_2, MgF_2, BF_3, CH_4, NH_3, H_2S, PCl_5, IF_5, SiH_4, $TeCl_4$, ClF_3, PH_3, SCl_2, $SnCl_2$, CCl_4, $SbCl_5$, XeF_4, ClF_5, SF_4, ClF_3, $SnCl_2$. Results are shown in Table IV.III.1.

Molecules with multiple bonds

In double or triple bonds $(A = Y, A \equiv Z)$ there are two or three electron pairs involved but their electron cloud is not much larger than that of a single bond. Therefore we may, as a first order approximation, replace multiple bonds with single bonds, the structure of the molecule $AX_n Y_m E_{n'}$ being similar to that of $AX_{n+m+n'}$. In counting the pairs, however, the number of pairs involved in multiple bonds must be taken into account.

Example 1: CO_2 (O=C=O)

For the carbon atom, we have

$Z = 6$

electron configuration: $1s^2 \, 2s^2 \, 2p^2$

number of electrons in outermost shell: $N = 4$

Between the two oxygen atoms there are $m = 2$ *double bonds*. Thus there $n' = (N - 2m)/2 = 0$ *nonbonding pairs* left on the carbon atom and $p = 2$, as the number of electrons of the carbon atom involved in double bonds is $2m$. The CO_2 molecule corresponds to the AY_2 structure and has a geometry similar to that of a AX_2 molecule. It is linear as shown in Fig. IV.III.6.

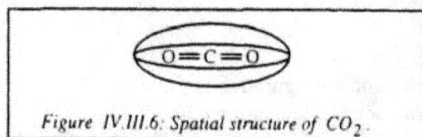

Figure IV.III.6: Spatial structure of CO_2.

TABLE IV III I — **Geometry of some simple molecules and ions.**

p	structure	structure	single bonds molecules	multiple bonds molecules	ions
2	linear	AX_2	$HgCl_2$ BeH_2 $BeCl_2$ CdI_2	CS_2	
3	triangular	AX_3	BF_3	SO_3	CH_3^+ NO_3
		AX_2E	$SnCl_2$	$ClNO$ O_3 SO_2	
4	tetraedral	AX_4	CH_4 CCl_4 SiH_4	SO_2Cl_2 $OPCl_3$	NH_4^+ ClO_4
		AX_3E	NH_3 PH_3	$SOCl_2$	H_3O^+ ClO_3
		AX_2E_2	H_2O H_2S SCl_2		ClO_2^-
5	bipyramidal triangular	AX_5	PCl_5	SOF_4	
		AX_4E	SF_4	XeO_2F_2 $XeOF_4$	
		AX_3E_2	ClF_3		
		AX_2E_3			ICl_2^- I_3^-
6	octaedral	AX_6	SF_6		
		AX_5E	IF_5 ClF_5		

Example 2: $COCl_2$

The two chlorine atoms correspond to $n = 2$ *single bonds*. The oxygen atom gives $m = 1$ *double bonds*. Thus there are $n' = (N - n - 2m)/2 = 0$ *nonbonding pairs* left. The geometry of $COCl_2$ is of the AX_2Y type. It is triangular, like the AX_3 molecule (Fig. IV.III.7).

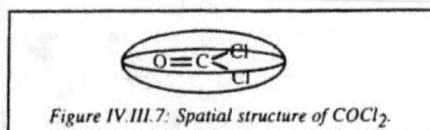

Figure IV.III.7: Spatial structure of $COCl_2$.

Example 3: HCN

In the HCN molecule, the hydrogen atom is linked by $n = 1$ *single bonds* and the nitrogen atom by $m = 1$ *triple bonds*. Thus there are $n' = (N - n - 3m)/2 = 0$ *nonbonding pairs*. The HCN molecule has a geometry of the type AYX, similar to the linear AX_2 molecule. The spatial structure is thus $H - C \equiv N$.

Other examples: The structure of the following molecules is shown in Table IV.III.1: CS_2, SO_3, ClNO, SO_2, SO_2Cl_2, $POCl_3$, $SOCl_2$, SOF_4, XeO_2F_2, $XeOF_4$, and O_3.

Polyatomic ions of the type AX_n^{a-} or AX_n^{b+}

The method is the same as used for the AX_n molecules but, in counting the N electrons of the central atom A, the excess or lack of electrons must be taken into account : In the case of AX_n^{a-} one must add a electrons while for AX_n^{b+} one must subtract b from the number of electrons in the valence shell of A.

Example 1: H_3O^+

For the O^+ ion, we have

> $Z = 8$
> electron configuration: $1s^2 2s^2 2p^3$
> quantum number of the valence shell: 2
> number of electrons in outermost shell: $N = 5$

There are five electrons involved in bonding and nonbonding pairs. The three hydrogen atoms give $n = 3$ bonding pairs. There is $n' = (5 - 3)/2 = 1$ nonbonding pair left. We then have $p = n + n' = 4$ and hence a tetrahedral geometry and a structure expressed as OH_3E^+, corresponding to a tetrahedral geometry of the AX_3E type, similar to AX_4.

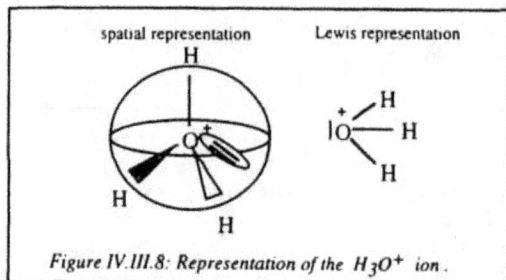

Figure IV.III.8: Representation of the H_3O^+ ion.

Other examples: The structures of the following ions are given in Table IV.III.1: CH_3^+, NH_4^+, I_3^-, ICl_2^-, $PtCl_6^{2-}$, FeF_6^{3-}, $Fe(CN)_6^{4-}$, ICl_4^-, NO_3^-, ClO_2^-, ClO_3^-, ClO_4^-

Alterations of bond angles due to nonbonding pairs and multiple bonds

So far we have assumed that, in a first approximation, there was an equivalence between bonding and nonbonding and between bonding pairs involved in single and double bonds. However the theoretical angle values of 180°, 120°, 109°28' and 90° are not truly obtained unless all pairs are equivalent, which is not always the case: If the electron clouds are not identical, their mutual repulsion is modified as well as the angles between the pairs.

Lack of equivalence between bonding and nonbonding pairs

This phenomenon is due to the fact that the electrons in a nonbonding pair (attached to a single atom) occupy a larger volume than a bonding pair of electrons, which are located in a smaller volume around the axis between the two atoms. This effect is shown in Fig. IV.III.9.

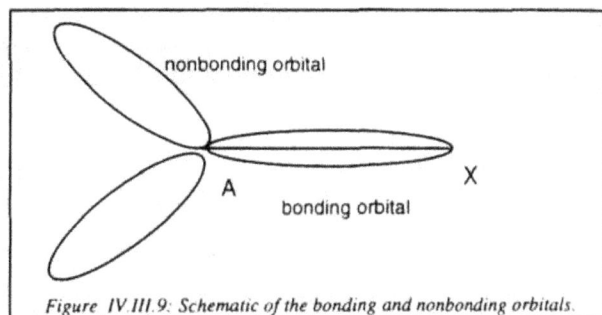

Figure IV.III.9: Schematic of the bonding and nonbonding orbitals.

The repulsion between nonbonding and bonding pairs is thus at its maximum around atom A. This contributes towards a decrease in the angle between the bonds.

Examples:

— The geometry of the CH_4, NH_3, and H_2O molecules corresponds to the tetrahedral organization of the four electron pairs. Angles of 109°28' could be expected, as in a regular tetrahedron, but the presence of nonbonding pairs in nitrogen and oxygen atoms leads to smaller values of these angles as indicated in Fig. IV.III.10.

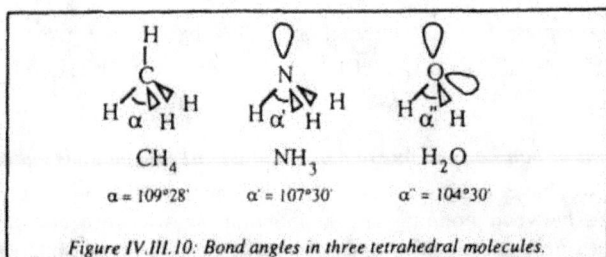

Figure IV.III.10: Bond angles in three tetrahedral molecules.

— The peculiar "umbrella" shape of the ClF_5 molecule may be justified in spite of the predicted octahedral structure:

In the case of the chlorine atom, we have

$Z = 17$
electron configuration: $1s^2\ 2s^2\ 2p^6\ 3s^2\ 3p^5$
quantum number of the valence shell: 3
number of electrons in outermost shell: $N = 7$
$n = 5$ bonding pairs
$n' = (N - n)/2 = 1$ nonbonding pair

The molecule with structure ClF_5E should correspond to an *octahedral* geometry, like AX_6, but the free pair is in the axial position which is freer than the equatorial ones. There is a repulsion between this pair and the bonding pairs. This leads to a folding of the bonds attached to the four fluorine atoms located in the plane shown in Fig. IV.III.11 and to the "umbrella" shape.

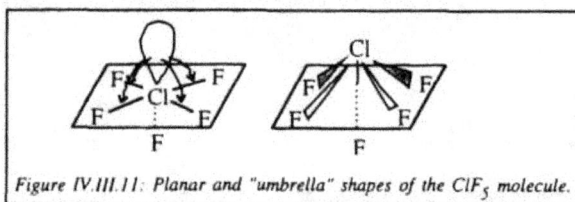

Figure IV.III.11: Planar and "umbrella" shapes of the ClF_5 molecule.

Non equivalence between single and multiple bonds

A multiple bond consists of two or three pairs of electrons shared by two atoms. It is therefore larger in volume than a single bond. Its higher electron density leads to a stronger repulsion than that of the single bonds in the following order:

$$A{\equiv}Y > A{=}Y > A{-}Y.$$

This is why the angles between multiple bonds are larger than those of single bonds around the same central atom, as shown in Fig. IV.III.12.

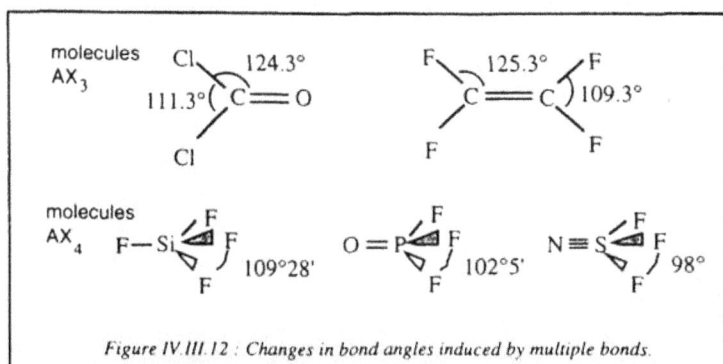

Figure IV.III.12 : Changes in bond angles induced by multiple bonds.

Role of ligand electronegativity

The shape of the volume occupied by a bonding pair depends on the difference in electronegativity between the two atoms. In the case of two atoms with similar electronegativities, the shape is almost symmetrical. When the electronegativities are different, the shape is distorted and the more strongly electronegative atom attracts the bonding pair.

The electron cloud gives a qualitative picture of the phenomenon as shown in Fig. IV.III.13.

For atom A linked to a more electronegative atom X, the volume occupied by the bonding pair is smaller around A leading to a decrease in the repulsion of the pair and thus to a decrease in the angles. The opposite is true if A is linked to a more electropositive atom.

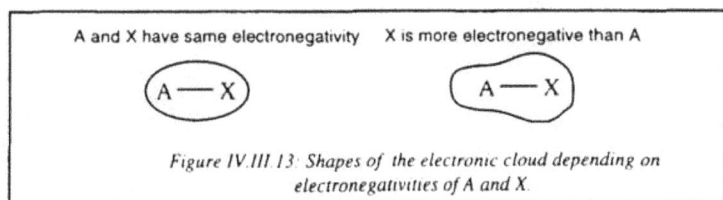

Figure IV.III.13: Shapes of the electronic cloud depending on electronegativities of A and X.

Examples: Molecules of the type AX_3E and AX_2E_2

In type AX_3E molecules, the tetrahedral structure is distorted because of the presence of a nonbonding pair and the difference in electronegativity of the ligands. Following the increasing electronegativity of halogen atoms from iodine to fluorine, the angle α between A—X bonds decreases in the following series of molecules:

molecule	PI_3	PBr_3	PCl_3	PF_3
α	102°	101.5°	100.3°	97.3°

The tetrahedral structure of AX_2E_2 type molecules is distorted by the presence of two nonbonding pairs and the different electronegativity of the central atom. In contrast to the preceding example, this leads to an increase in the angle α as a function of electronegativity. This is shown in Table IV.III.2, given that electronegativity increases from selenium to oxygen.

TABLE IV.III.2 — **Variation of angle α with electronegativity.**

molecule	H_2Se	H_2S	H_2O
α	90°	92.3°	104.5°

Conclusion

The V.S.E.P.R. method thus leads to qualitative predictions of the geometry of molecules and polyatomic ions with a *central atom*. However this method cannot be used to determine the geometry of more complex molecules: In the case of the ethylene molecule $CH_2=CH_2$, it is possible to predict the triangular environment of each of the carbon atoms, including alterations due to the multiple bond, but this method cannot help predict how the two pieces fit together (see Fig. IV.III.14).

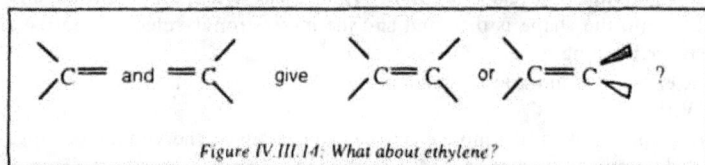

Figure IV.III.14: What about ethylene?

Moreover, this method does not determine the geometry of molecules or ions with a central atom in the transition element series, because, in that case, electrons from the d level may play a role in bonding.

In conclusion, it should be remembered that this method may, in some cases, predict an erroneous geometry. For example, V.S.E.P.R predicts a linear structure for the BaF_2 molecule (as for the four AX_2 molecules shown in Table IV.III.1) whereas the molecule is, in fact, bent. However, the ease of use of this method makes it extremely valuable for the physical chemist, as a starting point for more elaborate calculations.

6.2 "Manual" construction of a molecular orbital (MO) diagram

We have previously given (cf., Sec. IV.II) the canonical MO's of water, from which we deduced their representation. In this exercise we shall show that, using a simple, logical procedure, it is possible to obtain the main qualitative characteristics of these orbitals by coupling the valence AO's of the atoms concerned. The rules are as follows:

— MO's are obtained by coupling the AO's.
— The number of MO's is equal to the number of the original AO's.

— Coupling among the AO's is possible only if they belong to adjacent energy levels.

— It is convenient to use AO's, or their linear combinations, which are functions of a symmetry adapted to molecular symmetry. To put it more precisely, we say that these functions will form the basis of the irreducible representation of the group. For coupling to be possible under these conditions, it is necessary for these AO's, or their linear combinations, to belong to the same irreducible representation. The same holds for the MO thus formed (cf., Sec. II.II.2).

— The coupling of these AO's, or their linear combinations, will produce an A–B bond of the following character:

Constructive, if, in the MO obtained in this way, the contributions of the two AO's, centered on A and B, are of the same sign at mid-point between A and B (in graphical terms, if the two "lobes" have the same sign). This type of coupling tends to produce a bonding orbital with energy below that of the original AO's.

Destructive, if, in the MO obtained in this way, the contributions of the two AO's, centered on A and B, are of opposite signs at mid-point between A and B (in graphical terms, if the two "lobes" have opposite signs), leading to a higher-energy anti-bonding orbital.

A well-known application of this method to the molecule H_2^+ is shown in Fig. IV.III.15.

Let us now show the construction of water molecule MO's. The molecule belongs to the group C_{2v} with symmetry elements E, C_2, σ_v, σ'_v and irreducible representations a_1, b_1, b_2, a_2 (see Chapter II.II). The geometry and symmetry elements of water are shown in Fig. IV.III.16.

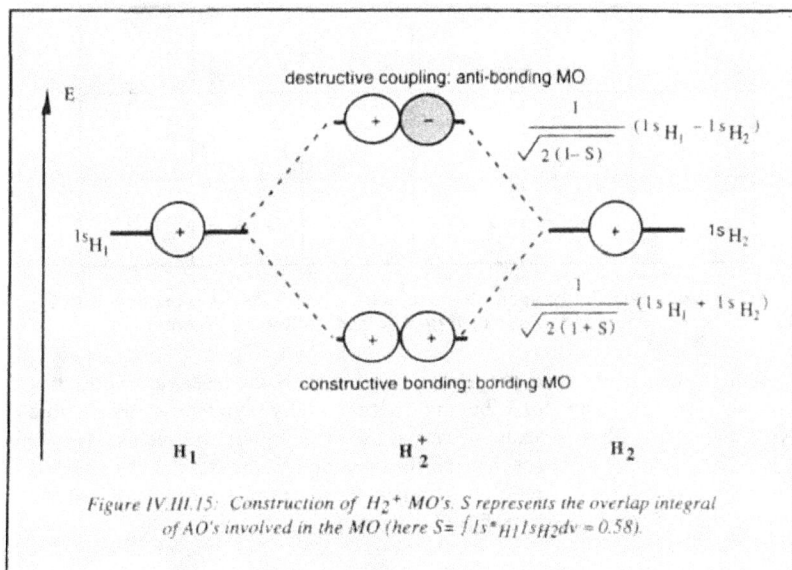

Figure IV.III.15: Construction of H_2^+ MO's. S represents the overlap integral of AO's involved in the MO (here $S = \int 1s^*_{H_1} 1s_{H_2} dv = 0.58$).

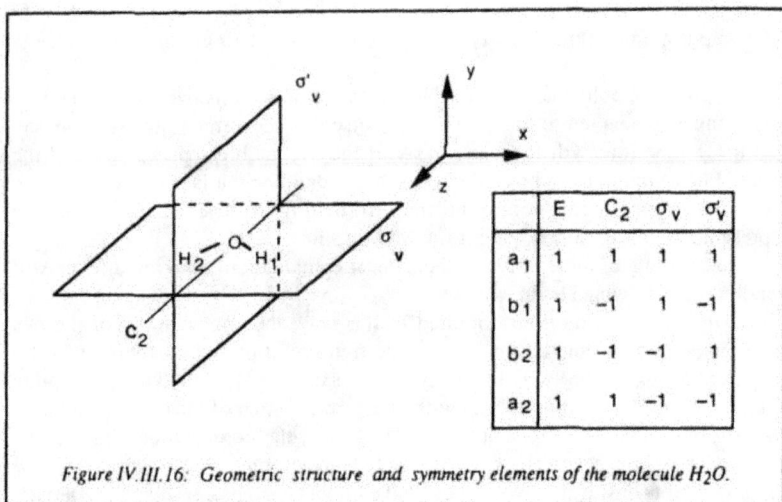

Figure IV.III.16: Geometric structure and symmetry elements of the molecule H₂O.

AO energies: the chosen AO's and their energies are given in the table below:

TABLE IV.III.3 — AO energies of atoms H_1, H_2 and O (in units of $E_H = -3.6$ eV).

orbital atom	1s	2s	$2p_x$	$2p_y$	$2p_z$
H_1	1				
H_2	1				
O	36	2.25	1.21	1.21	1.21

AO symmetries: For oxygen, the symmetry of each AO is represented in Fig. IV.III.17 (negative lobes are shaded in this and following figures).

With the help of the character table in Fig. IV.III.16, the reader can verify that orbitals $1s_O$, $2s_O$ and $2pz_O$ belong to the totally symmetric irreducible representation a_1 (these remain unchanged for all symmetry operations of group C_{2v}); $2p_{xO}$ belongs to b_1 (unchanged for operation E and σ_v, but the signs of lobes are reversed for C_2 and σ'_v); $2p_{yO}$ belongs to b_2 (unchanged for E and σ'_v, but the signs of lobes are inversed for C_2 and σ_v).

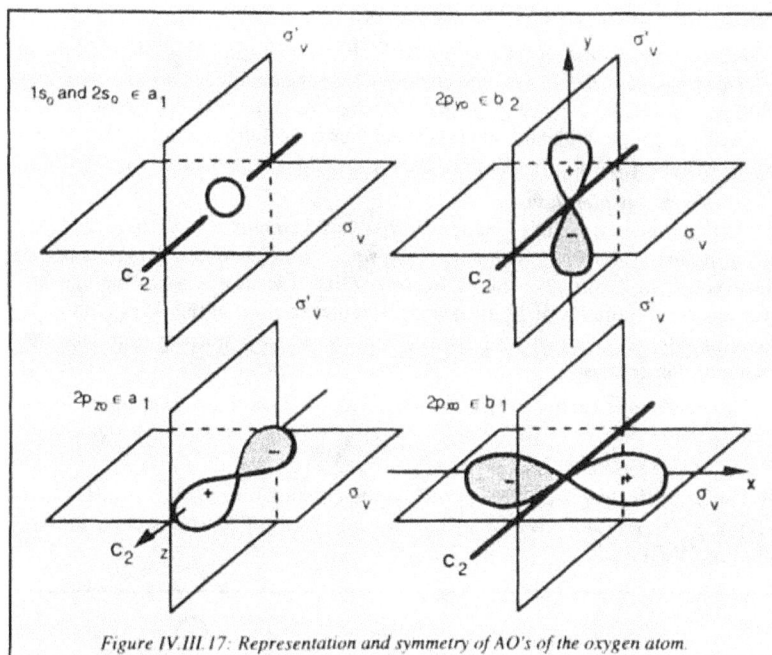

Figure IV.III.17: Representation and symmetry of AO's of the oxygen atom.

For hydrogen, $1s_{H1}$ and $1s_{H2}$, taken individually, have no group C_{2v} symmetry, but we have seen in Sec. II.II.2.2 that we may use linear combinations of these AO's which are better suited for the symmetry of the present problem. Fig. IV.III.18 specifies and illustrates this choice.

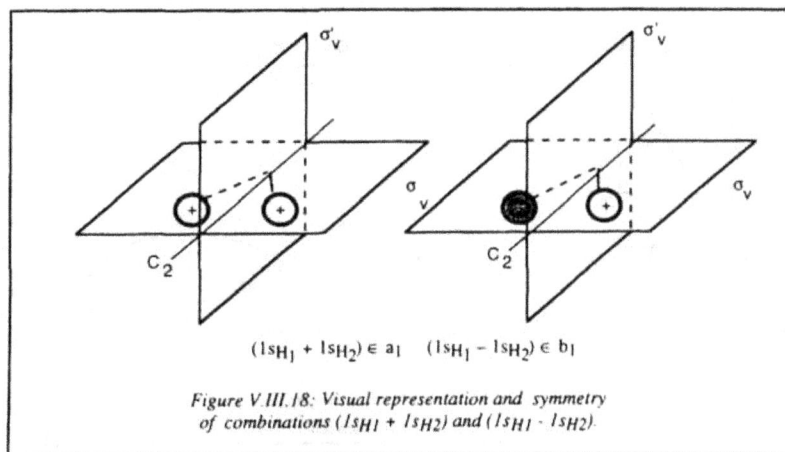

$(1s_{H_1} + 1s_{H_2}) \in a_1$ $(1s_{H_1} - 1s_{H_2}) \in b_1$

Figure V.III.18: Visual representation and symmetry
of combinations $(1s_{H1} + 1s_{H2})$ and $(1s_{H1} - 1s_{H2})$.

Further elaboration of the MO's makes use of:

— *Energy considerations*: Table IV.III.3 shows that the $1s_O$ atomic orbital has an energy of $-36 E_H$, very much lower than the energy of orbitals $1s_H$, $2s_O$ and $2p_O$, as well as the energy of the combinations $(1s_{H1} \pm 1s_{H2})$. Consequently $1s_O$ will not couple with any other AO, and thus constitutes a nonbonding "core" orbital of symmetry a_1. This will be the "lowest" MO, called $1a_1$.

— *Symmetry considerations*:

MO of symmetry a_1: The function $(1s_{H1} + 1s_{H2})$ combines with $2p_{zO}$ and $2s_O$ to form three MO's of symmetry a_1, whose energy will depend on the constructive or destructive nature of the overlap. Let us see what the various overlaps are. Figure IV.III.19 gives a visual representation of the AO's (and their combinations) of symmetry a_1, which will serve as a starting point for the "manual" construction.

— The most *constructive* overlap corresponds to maximum overlap of lobes with the same sign + : $(1s_{H1} + 1s_{H2})$ with $(2p_{zO} + 2s_O)$. The MO is designated as $2a_1$ (cf., Fig. IV.III.20).

— The most *destructive* overlap will be the combination $(1sH1 + 1sH2)$ and $-(2p_{zO} + 2s_O)$ with an overlap of opposite-sign lobes. The MO is denoted $4a_1$ (cf., Fig. IV.III.21).

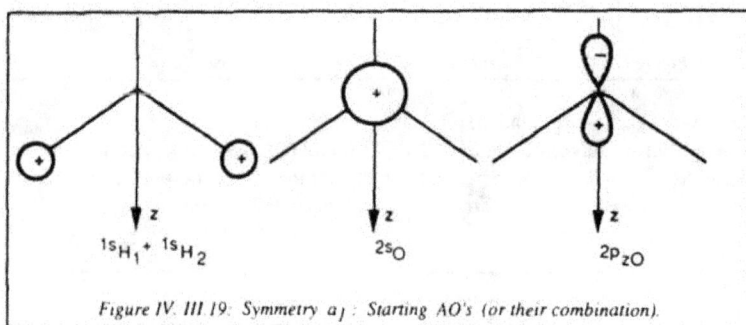

Figure IV. III 19: Symmetry a_1 : Starting AO's (or their combination).

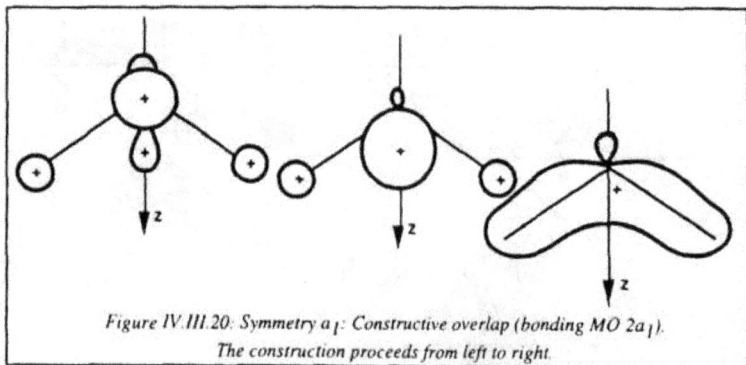

Figure IV.III.20: Symmetry a_1: Constructive overlap (bonding MO $2a_1$).
The construction proceeds from left to right.

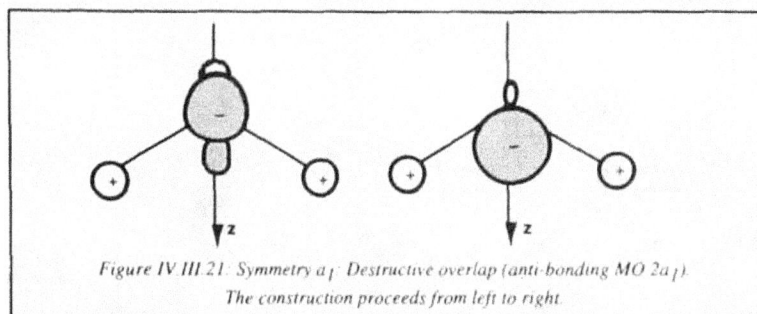

Figure IV.III.21: Symmetry a_1 Destructive overlap (anti-bonding MO $2a_1$). The construction proceeds from left to right

— A third possible overlap may be :

– Either the combination of $(1s_{H1} + 1s_{H2})$ with $+2p_zO$ and $-2s_O$, which corresponds to a very much less constructive overlap than in the case illustrated in Fig. IV.III.20, but more so than in the case illustrated in Fig. IV.III.21. It is weakly bonding.

– Or the combination $(1s_{H1} + 1s_{H2})$ with $-2p_zO$ and $2s_O$ which corresponds to a very much less destructive overlap than in the case illustrated in Fig. IV.III.21. It is weakly anti-bonding.

These two possibilities may be qualified as "nonbonding" The MO is denoted $3a_1$ (cf., Fig. IV.III.22).

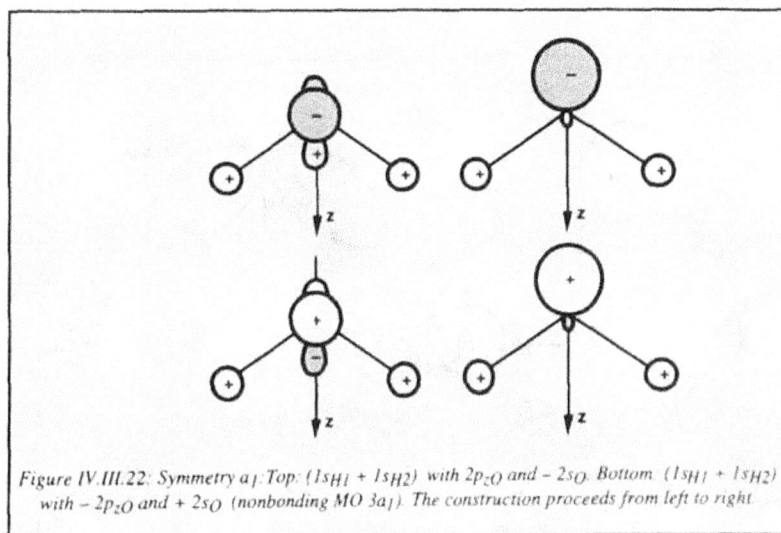

Figure IV.III.22: Symmetry a_1 Top: $(1s_{H1} + 1s_{H2})$ with $2p_zO$ and $-2s_O$. Bottom: $(1s_{H1} + 1s_{H2})$ with $-2p_zO$ and $+2s_O$ (nonbonding MO $3a_1$). The construction proceeds from left to right

The energy diagram of the three MO's of symmetry a_1 obtained in this way and of the initial AO's (or their combinations) is shown in Fig. IV.III.23.

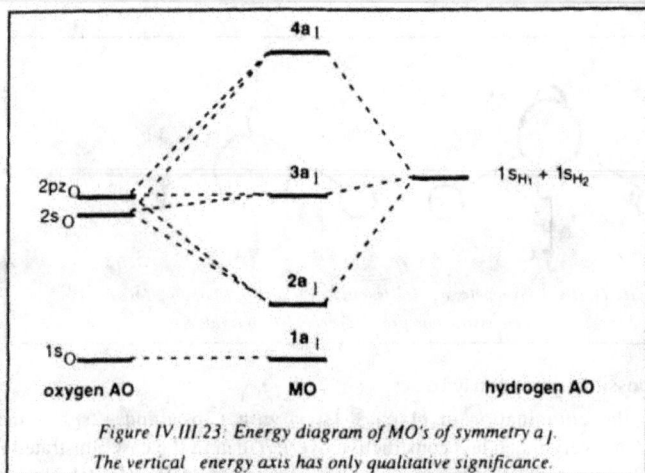

Figure IV.III.23: Energy diagram of MO's of symmetry a_1.
The vertical energy axis has only qualitative significance.

MO's of symmetry b_1: The function $(1s_{H1} \quad 1s_{H2})$ combines with $2p_{xO}$ to produce two MO's :
   – One bonding MO, corresponding to constructive overlap of same-sign lobes.
   – One anti-bonding MO, corresponding to destructive overlap of opposite-sign lobes.

The two cases are shown in Fig. IV.III.24 and the MO's are designated as $1b_1$ and $2b_1$.

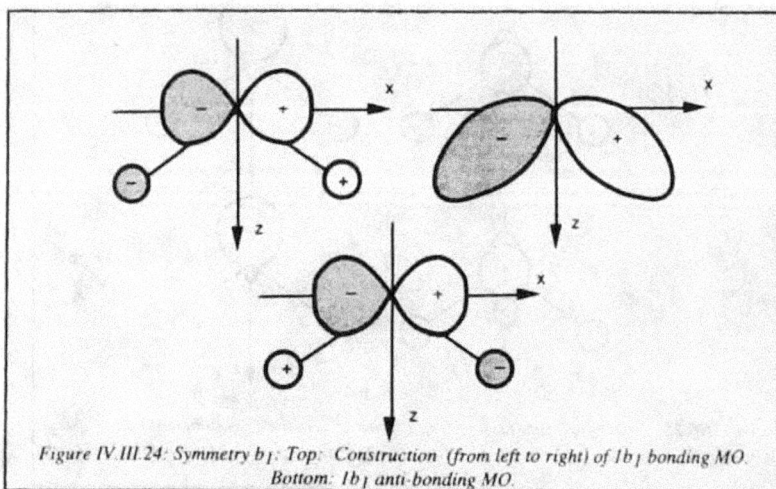

Figure IV.III.24: Symmetry b_1: Top: Construction (from left to right) of $1b_1$ bonding MO. Bottom: $1b_1$ anti-bonding MO.

MO of symmetry b_2: Only the $2p_{yO}$ AO has b_2 symmetry, so that it cannot combine with any of the hydrogen AO's and remains unchanged in the molecule.

leading to a nonbonding MO of symmetry b_2, i.e., "$1b_2$" The form of this orbital is shown in Fig. IV.III.25.

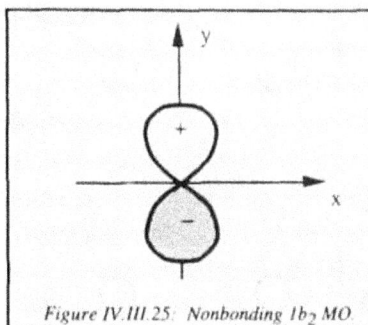

Figure IV.III.25: Nonbonding $1b_2$ MO.

The complete energy diagram and occupation of water MO's

We shall attempt to sketch an energy diagram of the MO's without any actual calculations.

First, it is necessary to find the positions of the bonding ($2a_1$ and $1b_1$) and anti-bonding ($4a_1$ and $2b_1$) MO's on the vertical energy axis. To do this, we compare the overlaps visually, at identical stages of the construction. The larger the overlap of same-sign lobes, the more stongly-bonding the MO and the lower its energy. By contrast, the larger the overlap of different-sign lobes, the more anti-bonding the MO and the higher its energy. Thus, from a simple visual examination of Fig. IV.III.26 it may be estimated that $S_{2a1} > S_{1b1}$, and hence $E_{2a1} < E_{1b1}$. Similarly we have $S_{4a1} > S_{2b1}$ and $E_{4a1} > E_{2b1}$. These results are shown in Fig. IV.III.27.

The "filling-up" of MO's proceeds in the same fashion as for atomic AO's, starting with the lowest-energy MO's. This is shown in Fig.IV.III.28.

Conclusion

From very general considerations, based on energy and symmetry, we have been able to construct "manually" the different molecular orbitals of the water molecule. We were able to identify their participation in bonding by distinguishing visually between bonding, anti-bonding and nonbonding MO's.

In addition, visual examination of the overlap makes it possible to order the corresponding energy levels. It is thus possible to describe the ground electron state of the molecule and predict the configuration of excited electron states.

Obviously, this type of method cannot be used to calculate:
— Weighting coefficients of the AO's implicated in each MO.
— First ionization energies of the MO's.

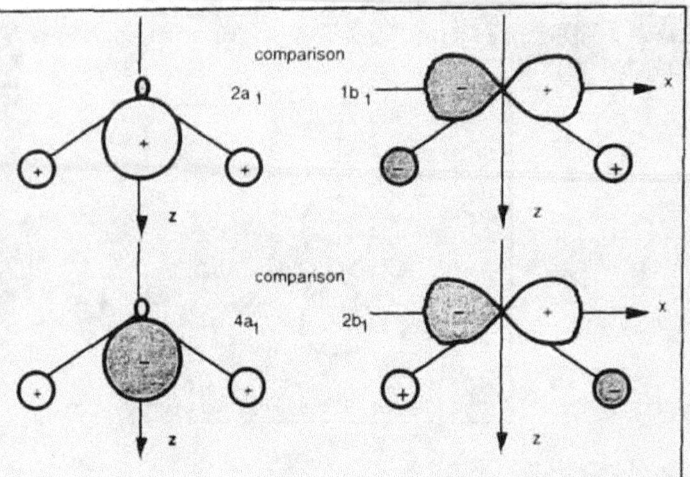

Figure IV III.26: *Visual comparison of the overlap of bonding MO's 2a₁ and 1b₁, and of anti-bonding MO's 4a₁ and 2b₁.*

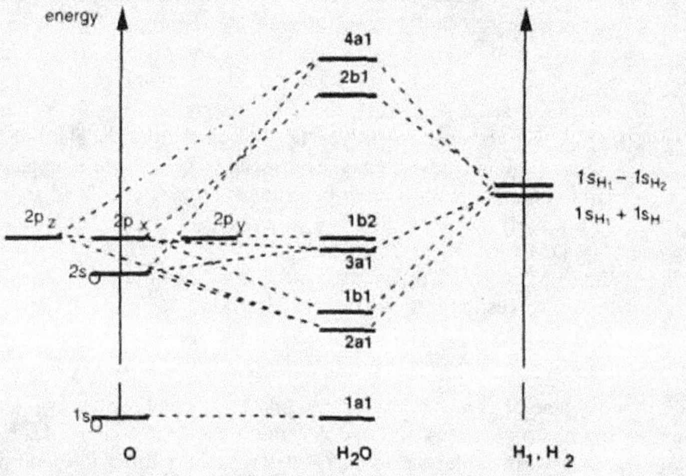

Figure IV. III.27 : *Qualitative diagram of water molecule energies, constructed "manually" and without actual calculation.*

We have seen how the H.F. LCAO method (Chapter IV.II) may be used to carry out calculations of this type as well as producing a correct quantitative representation of the corresponding MO's. It is nevertheless true that this rapid manual construction — which may be further elaborated as in the "fragment" method — is extensively used in chemistry, for example in the study of reactivity, where the geometric structure of HOMO (Highest Occupied Molecular Orbital) plays an important role. Suggested further reading : "Les orbitales moleculaires en chimie" by Yves Jean and Francois Volatron (McGraw-Hill, 1991), which offers many useful examples of manual construction of molecular orbitals.

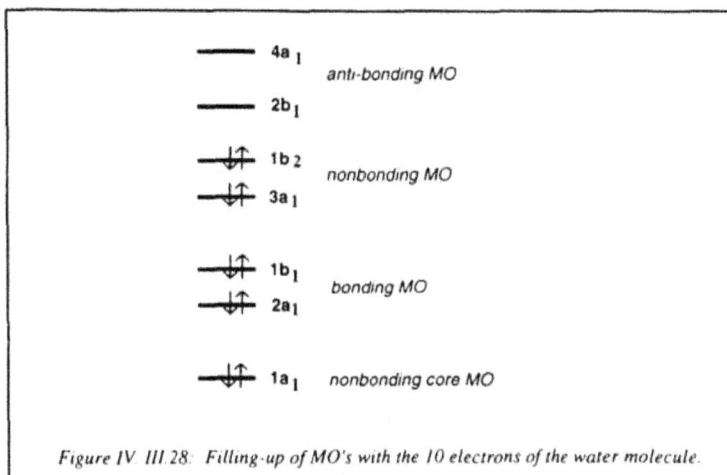

——— 4a₁	
	anti-bonding MO
——— 2b₁	
⥮ 1b₂	
	nonbonding MO
⥮ 3a₁	
⥮ 1b₁	
	bonding MO
⥮ 2a₁	
⥮ 1a₁	nonbonding core MO

Figure IV. III.28: Filling-up of MO's with the 10 electrons of the water molecule.

6.3 The Hückel method

Principles

The method proposed by Hückel in 1930 for the treatment of conjugated organic molecules represents the simplest application of the molecular orbital method. It is an interesting pedagogical tool as a preliminary to the Hartree-Fock and other more elaborate methods. In addition, despite its simplicity, it manages to describe certain essential characteristics of these molecules, and, as a result, has been widely used for many years.

The method treats π electrons of unsaturated organic molecules and is based on a number of approximations:

— Each π electron, considered as quasi-independent, is dealt with separately in the field of nuclei and other electrons, and is characterized by a one-electron effective Hamiltonian H. As we shall see, this operator need not be given explicitly.

— The ϕ molecular orbitals describing these π electrons are developed in the form of linear combinations of atomic orbitals $2p_\pi$, denoted ϕ_p, of the p different atoms of carbon, nitrogen, oxygen, etc. The $2p_\pi$ orbitals are of the 2p type ($2p_x$, $2p_y$ or $2p_z$, depending on the choice of axes), and are antisymmetric with respect to the molecular plane (method LCAO = *Linear Combination of Atomic Orbitals*).

$$\phi = c_1\,\phi_1 + c_2\,\phi_2 + \ldots = \sum_{p=1}^{n} c_p\,\phi_p \qquad \text{(IV.III.19)}$$

— The best molecular orbitals of this form (i.e., those with the best c_p coefficients) according to the variational method (cf., Secs. II.I.6 and IV.I.2) are obtained by requiring that the energy

$$E = \frac{\displaystyle\int \phi^* H \phi\, dv}{\displaystyle\int \phi^* \phi\, dv} = \frac{\displaystyle\sum_p \sum_q c_p^* c_q H_{pq}}{\displaystyle\sum_p \sum_q c_p^* c_q} \qquad \text{(IV.III.20)}$$

be minimized, i.e.,

$$\frac{\partial E}{\partial c_1^*} = \ldots = \frac{\partial E}{\partial c_p^*} = \ldots = 0 \qquad \text{(IV.III.21)}$$

The reader may verify that these coefficients satisfy the following system of linear homogeneous equations :

$$\left|\begin{array}{l} c_1(H_{11} - ES_{11}) + c_2(H_{12} - ES_{12}) + \ldots = 0 \\ c_1(H_{21} - ES_{21}) + c_2(H_{22} - ES_{22}) + \ldots = 0 \\ \ldots \end{array}\right. \qquad \text{(IV.III.22)}$$

or
$$\sum_q c_q (H_{pq} - ES_{pq}) = 0.$$

In this system E is the electron energy and the elements H_{pq} and S_{pq} have the following significance:

$$H_{pp} = \int \phi_p^* H \phi_p\, dv \qquad \text{(Coulomb integral)}$$

$$H_{pq} = \int \phi_p^* H \phi_q\, dv \qquad \text{(resonance integral)}$$

$$S_{pp} = \int \phi_p^* \phi_p\, dv \qquad \text{(normalization of } \phi_p)$$

$$S_{pq} = \int \phi_p^* \phi_q\, dv \qquad \text{(overlap integral)}$$

— This system of equations is simplified by using the Coulomb and resonance integrals as parameters, with values determined as follows:

$H_{pp} = \alpha_p$ depends only on the nature of atom p and not on its location in the molecule (α_p characterizes the electronegativity of the molecule). It is often convenient to use:

For carbon atoms $\qquad\qquad \alpha_C = \alpha$

For heteroatoms X $\qquad \alpha_X = \alpha + h_X\beta$

$H_{pq} = 0$ if atoms p and q are not bonded together.
$H_{pq} = \beta_{pq}$ if atoms p and q are bonded (the parameter β_{pq} depends only on the nature of the atoms p and q and not on the site of the p-q bond in the molecule; β_{pq} characterizes, *a priori*, the strength of the p-q bond). It is often convenient to use:

For a C-C bond $\qquad \beta_{CC} = \beta$
For a C-X bond $\qquad \beta_{CX} = k_{CX}\beta$.

Different authors have used different values for these parameters, depending on the properties studied. Reasonable values for parameters α and β are

$$\alpha = -11.26\,eV \quad \text{and} \quad \beta = -2.5\,eV.$$

The reader will note in particular the negative sign of the parameter β, which is important for the ordering of molecular orbital energies. Some of the values of h_X and k_{CX} proposed by A. Streitwieser (*Molecular Orbital Theory for Organic Chemists*, Wiley, New York 1961) are collected in Table IV.III.4.

TABLE IV.III.4 — Hückel parameters for oxygen and nitrogen atoms.

atom	h_X	bonds	k_{CX}
=O	1	C=O	1
–O–	2	C–O	0.8
=N–	0.5	C=N	1
–N<	1.5	C–N<	0.8

In addition, the overlap among basic atomic orbitals is assumed to be zero ($S_{pq} = 0$ if $p \neq q$) which amounts to assuming that base functions are orthonormal.

— Equation system IV.III.22 is solved in two stages. This system of n equations contains $n+1$ unknowns (the n coefficients c_p and energy E). Except for the obvious, trivial solution ($c_1 = c_2 = \cdots = c_p = \cdots = 0$) which has no physical interest, this system has nontrivial solutions only for certain values of E, which makes the system internally consistent. The first stage consists of finding these values of E, which we will call E_i. The condition of compatibility, $|H_{pq} - ES_{pq}| = 0$, is expressed in the form of a polynomial equation of degree n (*the secular equation*), whose roots have to be determined. In the second stage, the coefficients c_{pi}, corresponding to each previously determined value of E_i, are calculated so as to obtain the corresponding molecular orbitals ϕ_i. However it must be noted that these coefficients cannot be obtained unambiguously from Eq. IV.III.22 as the equations are not independent, as a result of the compatibility condition. In the nondegenerate case (a single root) it is only possible to express these coefficients in terms of one particular coefficient, for example c_1. This coefficient is itself determined from the normalization condition of the molecular orbital considered, taking the preceding relations into account

$$\sum_p |c_{pi}|^2 = 1 \qquad \text{(IV.III.23)}$$

(c_1 is usually taken to be positive, but this is not essential). In the degenerate case (multiple roots), the indeterminacy is even greater and the condition of orthogonality among orbitals is generally invoked, in addition to the normalization condition. In any event, it should be remembered that there is a persistent indeterminacy arising from the fact that quantum mechanical wave functions of a degenerate state can only be identified to within a linear combination.

As we have seen in Chapter II.II, it is possible, from the symmetry of a molecule, to determine *a priori* the possible symmetries of molecular orbitals, as well as the relations between the various coefficients C_p corresponding to these symmetries. Introducing these relations into Eq. IV.III.22, we may then replace this system by equivalent smaller systems (one per symmetry type), which are then easier to solve. These simplifications, used in the examples below, present the additional advantage of removing certain complications due to degeneracy through the use of molecular sub-groups.

Since the Hückel method does not initially introduce the antisymmetric condition of the many-electron wave function (see Sec. II.I.5), it is necessary to introduce the Pauli principle (cf., Sec. II.I.5.2) *a posteriori*. We shall therefore allow each molecular orbital to be occupied by *at most* two electrons of anti-parallel spins. The occupation number n_i of each molecular orbital ϕ_i may therefore take only one of three values : $n_i = 0$ (empty orbital), $n_i = 1$ and $n_i = 2$ (full orbital). These occupation numbers depend:

— On the number N of π electrons of the molecule considered, which may well be different from the number n of atomic base orbitals. If we denote by N_p the number of π electrons contributed by each atom p (one for the carbon atom, a pyridine-type nitrogen atom or ketone-type oxygen atom; two for pyrrole-type nitrogen atom or for alcohol– or furane–type oxygen atom, etc.), we should have:

$$N = \sum_p N_p = \sum_i n_i \qquad \text{(IV.III.24)}$$

— In the molecular state to be described (ground or excited state). The total energy is then given by

$$E_{tot} = \sum_i n_i E_i \qquad \text{(IV.III.25)}$$

and the wave function is represented by the product of occupied orbitals or, if required for the calculation of the molecular property, by an antisymmetrized product.

Only the lowest-energy levels are occupied in the ground state, so that the total energy is minimum. If the number N of pi electrons is even

$$E_{tot} = 2\sum_{i}^{occ} E_i \qquad\qquad (IV.III.26)$$

(where the summation only covers occupied levels).

The effect of conjugation is shown by the *delocalization energy* (or resonance energy) E_d, given by the absolute value of the difference between total energy and energy of pure double bonds, assumed to be nonconjugated (as we shall see later, the Hückel energy of an ethylene-type bond is $2\alpha + 2\beta$) as in the traditional formulae (Kekule-type, e.g., in aromatics).

Structure indices

In order to obtain results in a simpler, more concrete form, it is convenient to introduce:

— The *charge* borne by each atom p

$$Q_p = \sum_{i} n_i |c_{pi}|^2 \qquad\qquad (IV.III.27)$$

The $Q_p - N_p$ differences (net charges) indicate the displacement of charges due to conjugation.

— π *bond index* between two atoms p and q

$$P_{pq} = \left(\tfrac{1}{2}\right)\sum_{i} n_i(c_{pi}{}^*c_{qi} + c_{qi}{}^*c_{pi}) \qquad\qquad (IV.III.28)$$

The index P_{pq} measures the "force" of each π-bond after conjugation. These indices are related to total energy in the following way:

$$E_{tot} = \sum_{p} N_p\alpha_p + 2 \sum_{\text{liaisons pq}} P_{pq}\beta_{pq} \qquad\qquad (IV.III.29)$$

As β is negative, the second term shows that the contribution of the p-q bond to the lowering of energy increases with P_{pq}. Analysis of the latter at the molecular orbital level ϕ_i shows that the contributions $(c_{pi}{}^*c_{qi} + c_{qi}{}^*c_{pi})/2$ may be either positive (bonding effect), negative (anti-bonding effect), or zero (nonbonding effect). Very often, the coefficients c_{pi} are real and it is then interesting to analyze their change of sign in each molecular orbital ϕ_i, as they give the sign of the products $c_{pi} c_{qi}$ along the conjugated chain. In polyenes, for instance, the energy of molecular orbitals increases with the number of sign changes. (This property arises from the fact that same-sign coefficients c_{pi} and c_{qi} have a stabilizing effect on the molecular orbital, whereas opposite-sign coefficients have a destabilizing effect).

Examples of applications

The method will be applied below to ethylene,*cis*-butadiene and benzene molecules. The first two examples will then be used to illustrate the role symmetry may play in chemical reactivity. In order to show the simplifications due to symmetry, we shall make use of properties described in Sec. II.II.2.4. In these examples, the characters of the reducible representation Γ, obtained on the basis of atomic orbitals, are easily derived: For each group operation, they are equal to the number of orbitals that remain invariant, reduced by the number of orbitals that change sign under the effect of the symmetry operation (remember that π orbitals are antisymmetric with respect to the molecular plane). These representations are reduced using Eq. II.II.1 and the table of characters for that group, shown in Sec. II.II.5 (Appendix 4). The relations among the coefficients are given by Eqs. II.II.7 and II.II.8.

Ethylene — The base used consists of two atomic orbitals $2p\pi$ of the two carbon atoms: $\phi_1 = 2p\,\pi_1$, $\phi_2 = 2p\pi_2$ (Fig. IV.III.29).

The Hückel system of equations is

$$\begin{vmatrix} c_1\,(\alpha - E) + c_2\,\beta = 0 \\ c_1\,\beta + c_2\,(\alpha - E) = 0 \end{vmatrix} \qquad \text{(IV.III.30)}$$

or, using variable $k = (\alpha - E)/\beta$

$$\begin{vmatrix} c_1\,k + c_2 = 0 \\ c_1 + c_2\,k = 0 \end{vmatrix}$$

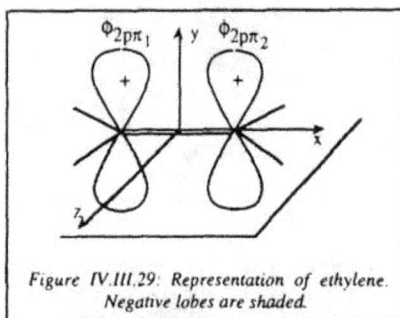

Figure IV.III.29: *Representation of ethylene. Negative lobes are shaded.*

Simplification by symmetry — The preceding base is a representation Γ of group D2h of the molecule. Its characters are given in Table IV.III.5.

TABLE IV.III.5 — **Characters of Γ representation of ethylene.**

D_{2h}	E	$C_2(z)$	$C_2(y)$	$C_2(x)$	i	$\sigma_v(xy)$	$\sigma_v(xz)$	$\sigma_v(yz)$
Γ	2	0	0	-2	0	2	-2	0

The reduction of this representation ($\Gamma = b_{1g} + b_{2u}$) shows that, from the base used, it is possible to obtain one molecular orbital of symmetry b_{1g} and another of symmetry b_{2u}.

Orbital of symmetry b_{2u}

From the relations $\sigma_v(yz)\phi_2 = \phi_1$ and $\chi_{b2u}[\sigma_v(yz)] = 1$, we deduce $c_2 = c_1$.
The preceding system reduces to $c_1(k + 1) = 0$ and the condition of compatibility leads to $k_1 = -1$.

We thus obtain $\quad E_1 = \alpha + \beta; c_2 = c_1; c_1 = [(1 + 1)]^{-1/2} = 0.707$
$$\varphi_1 = 0.707(\phi_1 + \phi_2)$$

Orbital of symmetry b_{1g}

From the relations $\sigma_v(yz)\phi_2 = \phi_1$ and $\chi_{b1g}[\sigma_v(yz)] = -1$, we deduce $c_2 = -c_1$. The preceding system reduces to $c_1(k - 1) = 0$ and the condition of compatibility leads to $k_2 = 1$.

We thus obtain $E_2 = \alpha - 1.618\,\beta; c_2 = -c_1 . c_1 = [(1 + 1)]^{-1/2} = 0.707$
$$\varphi_2 = 0.707(\phi_1 - \phi_2)$$

These results are summarized below:

sym. b_{1g}	$E_2 = \alpha - \beta$	$\varphi_2 = 0.707(\phi_1 - \phi_4)$	(anti-bonding)
sym. b_{2u}	$E_1 = \alpha + \beta$	$\varphi_1 = 0.707(\phi_1 + \phi_4)$	(bonding)

Examination of the signs of the molecular orbital coefficients shows that the coupling is constructive, hence bonding, for ϕ_1, but destructive, hence antibonding, for ϕ_2, which agrees with the order of energies.

In the ground state (Fig. IV.III.30), the two π electrons occupy the molecular orbital ϕ_1 of lowest energy (occupation numbers $n_1 = 2$ and $n_2 = 0$). The total energy is then
$$E_{tot} = 2\,E_1 = 2\alpha + 2\beta,$$
and the structure indices are, respectively:

— For the *charges*: $Q_1 = Q_2 = 2(0.707)^2 = 1$
(thus conjugation does not displace charges in this molecule)
— For the *π-bond indices* : $P_{12} = 2(0.707)^2 = 1$

This index corresponds to a "pure" ethylene-like bond.

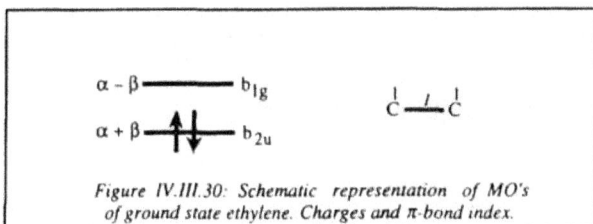

Figure IV.III.30: Schematic representation of MO's of ground state ethylene. Charges and π-bond index.

Cis-butadiene — The base used consists of four atomic orbitals $2p\pi$ of the four carbon atoms: $\phi_1 = 2p\pi_1$, $\phi_2 = 2p\pi_2$, $\phi_3 = 2p\pi_3$, $\phi_4 = 2p\pi_4$ (Fig. IV.III.31).

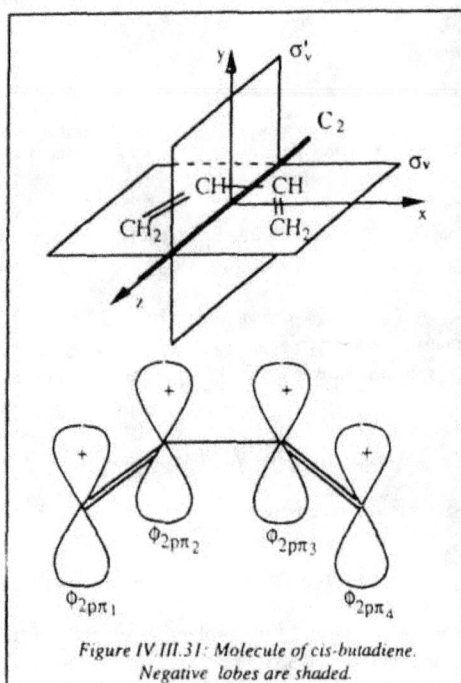

Figure IV.III.31: Molecule of cis-butadiene.
Negative lobes are shaded.

The Hückel system of equations is

$$\begin{vmatrix} c_1(\alpha - E) + c_2\beta = 0 \\ c_1\beta + c_2(\alpha - E) + c_3\beta = 0 \\ c_2\beta + c_3(\alpha - E) + c_4\beta = 0 \\ c_3\beta + c_4(\alpha - E) = 0 \end{vmatrix} \qquad \text{(IV.III.31)}$$

or, changing to variable $k = (\alpha - E)/\beta$

$$\begin{vmatrix} c_1 k + c_2 = 0 \\ c_1 + c_2 k + c_3 = 0 \\ c_2 + c_3 k + c_4 = 0 \\ c_3 + c_4 k = 0 \end{vmatrix}$$

Simplification by symmetry — The preceding base is a representation Γ of group C_{2v} of the molecule. Its characters are given in Table IV.III.6.

TABLE IV.III.6 — **Characters of Γ representation of cis-butadiene.**

C_{2v}	E	C_2	$\sigma_v(xz)$	$\sigma_v'(yz)$
Γ	4	0	-4	0

The reduction of this representation ($\Gamma = 2a_2 + 2b_2$) shows that from the base used it is possible to obtain two molecular orbitals of symmetry a_2 and two of symmetry b_2.

Orbitals of symmetry b_{2u}

From the relations: $\sigma_v'(yz)\phi_4 = \phi_1$, $\sigma_v'(yz)\phi_3 = \phi_2$ and $\chi_{b_2} [\sigma_v'(yz)] = 1$, we deduce

$$c_4 = c_1 \text{ and } c_3 = c_2$$

The preceding system reduces to

$$\begin{vmatrix} c_1 k + c_2 = 0 \\ c_1 + c_2 (k + 1) = 0 \end{vmatrix}$$

and the condition of compatibility leads to

$$\begin{vmatrix} k & 1 \\ 1 & k+1 \end{vmatrix} = k^2 + k - 1 = 0$$

We thus obtain $k_1 = -1.618$; $E_1 = \alpha + 1.618 \beta$; $c_2 = 1.618 c_1$.

$$c_1 = [2(1 + 1.618^2)]^{-1/2} = 0.3717$$
$$\varphi_1 = 0.3717 (\phi_1 + \phi_4) + 0.6015 (\phi_2 + \phi_3)$$

$k_3 = 0.618$; $E_3 = \alpha - 0.618 \beta$; $c_2 = -0.618 c_1$; $c_1 = [2(1 + 0.618^2)]^{-1/2} = 0.6015$

$$\varphi_3 = 0.6015(\phi_1 + \phi_4) - 0.3717(\phi_2 + \phi_3)$$

Orbitals of symmetry a_2

From the relations: $\sigma_v'(yz)\phi_4 = \phi_1$, $\sigma_v'(yz)\phi_3 = \phi_2$ and $\chi_{a_2} [\sigma_v'(yz)] = -1$, we deduce

$$c_4 = -c_1 \text{ and } c_3 = -c_2.$$

The system reduces to

$$\begin{vmatrix} c_1 k + c_2 = 0 \\ c_1 + c_2 (k - 1) = 0 \end{vmatrix}$$

and the condition of compatibility leads to

$$\begin{vmatrix} k & 1 \\ 1 & k-1 \end{vmatrix} = k^2 - k - 1 = 0$$

We thus obtain $\quad k_4 = 1.618$; $E_4 = \alpha - 1.618\beta$; $c_2 = -1.618c_1$

$$c_1 = [2(1 + 1.618^2)]^{-1/2} = 0.3717$$
$$\varphi_4 = 0.3717 (\phi_1 - \phi_4) - 0.6015 (\phi_2 - \phi_3)$$

$k_2 = -0.618$; $E_2 = \alpha + 0.618\beta$; $c_2 = 0.618c_1$; $c_1 = [2(1 + 0.618^2)]^{-1/2} = 0.6015$

$$\varphi_2 = 0.6015(\phi_1 - \phi_4) + 0.3717(\phi_2 - \phi_3)$$

The energies and molecular orbitals of cis-butadiene are thus

a_2	$E_4 = \alpha - 1.618\beta$	$\varphi_1 = 0.3717(\phi_1 - \phi_4) - 0.6015(\phi_2 - \phi_3)$
b_2	$E_3 = \alpha - 0.618\beta$	$\varphi_3 = 0.6015(\phi_1 + \phi_4) - 0.3717(\phi_2 + \phi_3)$
a_2	$E_2 = \alpha + 0.618\beta$	$\varphi_2 = 0.6015(\phi_1 - \phi_4) + 0.3717(\phi_2 - \phi_3)$
b_2	$E_1 = \alpha + 1.618\beta$	$\varphi_1 = 0.3717(\phi_1 + \phi_4) + 0.6015(\phi_2 + \phi_3)$

Examination of the signs of coefficients of adjacent molecular orbitals shows that the coupling is constructive for the three bonds in φ_1, but only for two in φ_2, one in φ_3 and none in φ_4. This property is related to the energy ordering of these orbitals.

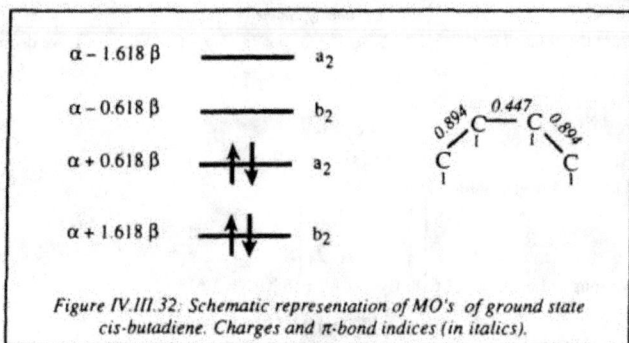

Figure IV.III.32: Schematic representation of MO's of ground state cis-butadiene. Charges and π-bond indices (in italics).

In the ground state (Fig. IV.III.32), the four π-electrons occupy the lowest-energy molecular orbitals φ_1 and φ_2 (occupation numbers $n_1 = n_2 = 2$ and $n_3 = n_4 = 0$). The total energy is thus

$$E_{tot} = 2E_1 + 2E_2 = 4\alpha + 4.472\,\beta$$

while the energy of the two ethylenic nonconjugated bonds is $4\alpha + 4\beta$. The delocalization energy is the absolute value of the difference between these two values, i.e., $E_d = |0.472\,\beta|$. It reflects the increase in stability due to delocalization (however, it should be noted that, as the Hückel method does not take into account the energy related to the elongation or shortening of σ bonds, there are some problems if this theoretical value is compared with experimental findings).

The structure indices are

— For the *charges*:

$$Q_1 = Q_4 = 2[(0.3717)^2 + (0.6015)^2] = 1$$
$$Q_2 = Q_3 = 2[(0.6015)^2 + (0.3717)^2] = 1$$

(conjugation does not displace charges in this molecule).

— For *π-bond indices:*

$$P_{12} = P_{34} = 4 \times 0.3717 \times 0.6015 = 0.894$$
$$P_{23} = 2(0.6015)^2 - 2(0.3717)^2 = 0.447$$

These bond indices show that the double-bond character of bonds 1-2 and 3-4 is stronger than that of the internal bond 2-3, which is confirmed experimentally by the fact that the external bonds (0.135 nm) are shorter than the internal bond (0.145 nm). It should also be noted that the bond index P_{23}, although fairly weak, explains why the 2-3 bond is nevertheless shorter than a simple C-C bond in a nonconjugated molecule (0.154 nm).

As an exercise, the reader may verify that trans-butadiene (more stable than the cis-isomer) produces the same energies and molecular orbitals, as the method ignores interactions between atoms other than immediate neighbors.

Benzene— The base used consists of six atomic orbitals $2p\pi$ of the six carbon atoms: $\phi_1 = 2p\pi_1$; $\phi_2 = 2p\pi_2$; $\phi_3 = 2p\pi_3$; $\phi_4 = 2p\pi_4$; $\phi_5 = 2p\pi_5$; $\phi_6 = 2p\pi_6$ (Fig. IV.III.33).

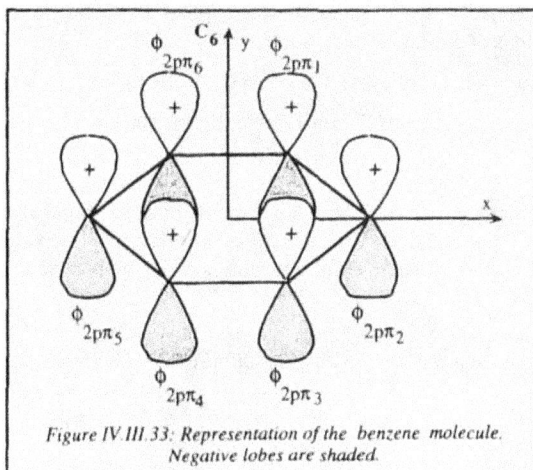

Figure IV.III.33: Representation of the benzene molecule.
Negative lobes are shaded.

The Hückel system of equations is

$$\begin{vmatrix} c_6\,\beta + c_1\,(\alpha - E) + c_2\,\beta = 0 \\ c_1\,\beta + c_2\,(\alpha - E) + c_3\,\beta = 0 \\ c_2\,\beta + c_3\,(\alpha - E) + c_4\,\beta = 0 \\ c_3\,\beta + c_4\,(\alpha - E) + c_5\,\beta = 0 \\ c_4\,\beta + c_5\,(\alpha - E) + c_6\,\beta = 0 \\ c_5\,\beta + c_6\,(\alpha - E) + c_1\,\beta = 0 \end{vmatrix} \qquad (IV.III.32)$$

or, changing to variable $k = (\alpha - E)/\beta$

$$\begin{vmatrix} c_6 + c_1\,k + c_2 = 0 \\ c_1 + c_2\,k + c_3 = 0 \\ c_2 + c_3\,k + c_4 = 0 \\ c_3 + c_4\,k + c_5 = 0 \\ c_4 + c_5\,k + c_6 = 0 \\ c_5 + c_6\,k + c_1 = 0 \end{vmatrix} \qquad (IV.III.33)$$

Simplification due to symmetry — The preceding base is a representation Γ of group D_{6h} of the molecule, but this is also the base of sub-group C_6 which, as we shall see, leads to particularly simple calculations. The characters of the representation Γ of this sub-group are given in Table IV.III.7.

TABLE IV.III.7 — **Characters of Γ representation of benzene.**

C_6	E	C_6	C_3	C_2	C_3^2	C_6^5
Γ	6	0	0	0	0	0

The reduction of this representation ($\Gamma = a + b + e_1 + e_2$) shows that from the base used it is possible to obtain one molecular orbital of symmetry a, one of symmetry b, two of symmetry e_1 and two of symmetry e_2 (in this group, the representations e_1 and e_2 actually correspond to a couple of irreducible representations of dimension one). Since

$$C_6\,\phi_2 = \phi_1 \; ; \; C_3\,\phi_3 = \phi_1 \; ; \; C_2\,\phi_4 = \phi_1 \; ; \; C_3^2\,\phi_5 = \phi_1 \; ; \; C_6^5\,\phi_6 = \phi_1,$$

it is easy to see, using Eq. II.II.7, that for each irreducible representation Γ_i

$$c_2 = \chi_i(C_6)\,c_1 \; ; \; c_3 = \chi_i(C_3)\,c_1 \; ; \; c_3 = \chi_i(C_2)\,c_1 , \quad \text{etc.}$$

The required coefficients are thus proportional to the characters of the irreducible representations of the various operations. The values of k and $E = \alpha - k\,\beta$ corresponding to each symmetry are easily obtained by writing the first equation of Eq. IV.III.33 in the form

$$k = - (c_2 + c_6)/c_1 = - (\chi_i(C_6) + \chi_i(C_6^5))$$

After normalization, we have [with $\varepsilon = \exp(2\pi i/6) = 1/2 + i\sqrt{3}/2$]

b $\quad E_6 = \alpha - 2\beta \quad\quad \varphi_6 = (1/\sqrt{6})(\phi_1 - \phi_2 + \phi_3 - \phi_4 + \phi_5 - \phi_6)$

e_2 $\quad \begin{array}{l} E_5 = \alpha - \beta \\ E_4 = \alpha - \beta \end{array}$ $\quad \begin{array}{l} \varphi_5 = (1/\sqrt{6})(\phi_1 - \varepsilon\,\phi_2 - \varepsilon^*\,\phi_3 + \phi_4 - \varepsilon\,\phi_5 - \varepsilon^*\phi_6) \\ \varphi_4 = (1/\sqrt{6})(\phi_1 - \varepsilon^*\,\phi_2 - \varepsilon\,\phi_3 + \phi_4 - \varepsilon^*\phi_5 - \varepsilon\,\phi_6) \end{array}$

e_1 $\quad \begin{array}{l} E_3 = \alpha + \beta \\ E_2 = \alpha + \beta \end{array}$ $\quad \begin{array}{l} \varphi_3 = (1/\sqrt{6})(\phi_1 + \varepsilon^*\,\phi_2 - \varepsilon\,\phi_3 - \phi_4 - \varepsilon^*\,\phi_5 + \varepsilon\,\phi_6) \\ \varphi_2 = (1/\sqrt{6})(\phi_1 + \varepsilon\,\phi_2 - \varepsilon^*\phi_3 - \phi_4 - \varepsilon\phi_5 + \varepsilon^*\,\phi_6) \end{array}$

a $\quad E_1 = \alpha + 2\beta \quad\quad \varphi_1 = (1/\sqrt{6})(\phi_1 + \phi_2 + \phi_3 + \phi_4 + \phi_5 + \phi_6)$

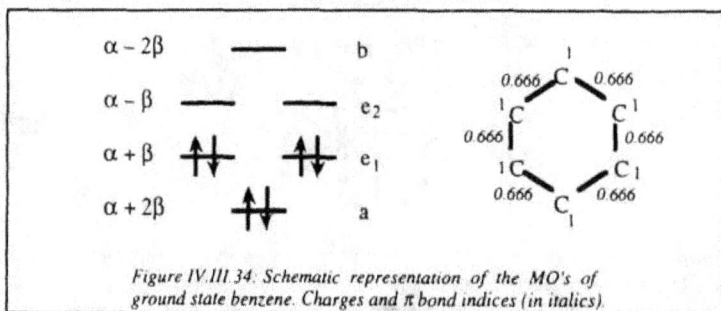

Figure IV.III.34: Schematic representation of the MO's of ground state benzene. Charges and π bond indices (in italics).

In the ground state (Fig. IV.III.34), the six π electrons occupy the three lowest energy molecular orbitals, φ_1, φ_2 and φ_3 (occupation numbers: $n_1 = n_2 = n_3 = 2$ and $n_4 = n_5 = n_6 = 0$). The total energy is thus

$$E_{tot} = 2E_1 + 2E_2 + 2E_3 = 6\alpha + 8\beta,$$

while the energy of the three ethylene-like nonconjugated bonds is $6\alpha + 6\beta$.

The delocalization energy is thus $E_d = |2\beta|$. The structure indices are

— For the charges:

$$Q_1 = Q_2 = Q_3 = Q_4 = Q_5 = Q_6 = 1$$

(conjugation does not displace charges in this molecule)

— For π-bond indices:

$$P_{12} = P_{23} = P_{34} = P_{45} = P_{56} = P_{61} = 0.666$$

These indices show that the double-bond character of benzene-like bonds is intermediate between those of the two types of bonding in butadiene. This is consistent with their bond length (0.139 nm).

Remark: Using symmetry functions in sub-group C_6 has led to the above orbitals with complex coefficients, but as orbitals φ_2 and φ_3, on the one hand, and orbitals φ_4 and φ_5 on the other, belong to a one-electron nondegenerate level, every linear combination is still an acceptable orbital. For example, the following set of orbitals with real coefficients

c_2 | $\varphi'_5 = (i/\sqrt{2})(\varphi_5 - \varphi_4) = (1/2)(\phi_2 - \phi_3 + \phi_5 - \phi_6)$
$\varphi'_4 = (1/\sqrt{2})(\varphi_4 + \varphi_5) = (1/\sqrt{12})(2\phi_1 - \phi_2 - \phi_3 + 2\phi_4 - \phi_5 - \phi_6)$

c_1 | $\varphi'_3 = (i/\sqrt{2})(\varphi_3 - \varphi_2) = (1/2)(\phi_2 + \phi_3 - \phi_5 - \phi_6)$
$\varphi'_2 = (1/\sqrt{2})(\varphi_2 + \varphi_3) = (1/\sqrt{12})(2\phi_1 + \phi_2 - \phi_3 - 2\phi_4 - \phi_5 + \phi_6)$

may replace the preceding orbitals. We would have obtained them if we had used the sub-group D_{2h} of the benzene molecule instead of the sub-group C_6. The reader may verify that these new orbitals lead to the same charges and π-bond indices.

6.4 Chemical reactivity and symmetry. Cyclization of butadiene

Some chemical reactions are guided by symmetry properties. The following is an example of so-called *concerted* reactions: The cyclization of butadiene into cyclobutene (Fig. IV.III.35).

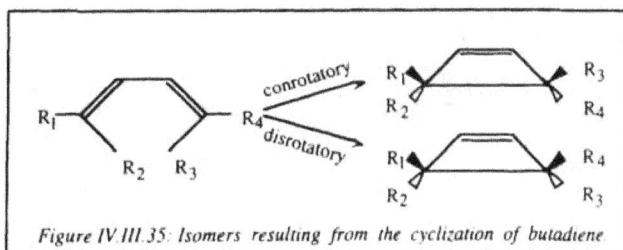

Figure IV.III.35: Isomers resulting from the cyclization of butadiene.

This reaction may be interpreted as being due mainly to the coupling of carbon orbitals $\phi_1 = 2p\pi_1$ and $\phi_4 = 2p\pi_4$ at either end of the cis-butadiene molecule (using the notations in Sec. IV.III.7.3). These two atomic orbitals are relatively far apart in cis-butadiene, so the π-type coupling is weak, but if the terminal CH_2 groups are allowed to rotate (in the same direction: *conrotatory mode*, or in opposite directions: *disrotatory mode*), the coupling of these two orbitals becomes more important and finally leads to

— A σ bond very much shorter than the original distance between the two carbon end-atoms.

— The disappearance of the two double bonds 1-2 and 3-4.

— Appearance of the double bond 2-3 (Fig. IV.III.35).

However, if the two terminal groups have different substituents, the (final) molecule formed is not the same for the two rotatory modes (Fig. IV.III.35). Experiment shows that in a thermal reaction the conrotatory mode is usually favored, while in a photochemical reaction it is the disrotatory mode. We shall interpret these results in two different ways. The first simply makes use of the bonding or antibonding character of the coupling between the orbitals $1p\pi_1$ and $2p\pi_4$, while the second applies the method of correlation diagrams.

Bonding and antibonding coupling of atomic orbitals

It is generally accepted that in a thermal reaction the principal role is played by the highest occupied molecular orbital (here orbital ϕ_2 of *cis*-butadiene (Sec. IV.III.6.3)) which guides the reaction. However the coefficients c_1 and c_4 of the two orbitals ϕ_1 and ϕ_4 are of opposite sign (as a consequence of symmetry). The result is that, initially, the positive parts of functions $c_1\phi_1$ and $c_4\phi_4$ are above and below the molecular plane, respectively. Fig.IV.III.36 shows that if the terminal groups CR_nR_n rotate in the same direction (conrotatory mode), the result is an overlap of functions with the same sign.

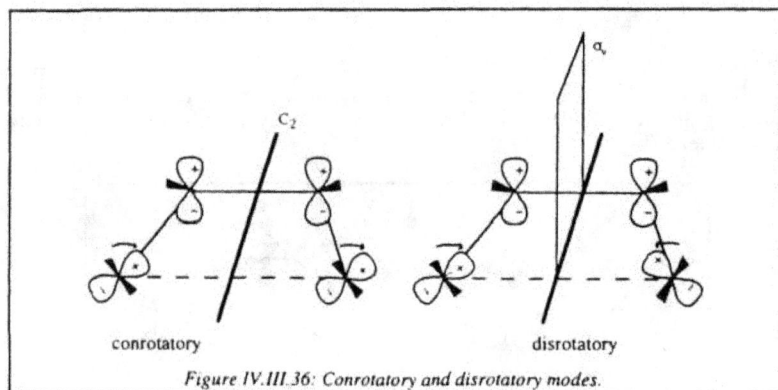

Figure IV.III.36: Conrotatory and disrotatory modes.

Electron density thus tends to increase between the two atoms: The coupling results in bonding. By contrast, if the terminal groups rotate in opposite directions (disrotatory mode), there is an overlap of functions of opposite signs: The coupling is nonbonding. We now understand why thermal reactions favor the conrotatory mode.

By contrast, when, in a photochemical reaction, an electron is promoted into the first vacant molecular orbital ϕ_3, this orbital plays a major role and guides the reaction. However the coefficients c_1 and c_4 of this orbital are equal due to symmetry, so that the conclusion is the reverse of the preceding: Photochemical reactions favor the disrotatory mode.

Correlation diagrams

In the course of the two types of transformation described above, the molecule conserves C_2 symmetry for the conrotatory mode and C_s symmetry for the disrotatory mode (if the substituents are not too bulky, they are unlikely to have much effect on the results). An important rule specifies that, in problems of this type, transformation takes place with conservation of state and molecular orbital symmetry, and that there should be no crossing of state or orbital energies belonging to the same symmetry (*no-crossing rule*).

The problem is then to determine the molecular orbital symmetry of the initial molecule (*cis*-butadiene) and of the final molecule (cyclobutene), to locate them on the energy axis in a diagram, and to study the correlation among orbitals of the same symmetry.

The molecular orbitals ϕ_i and the energies of *cis*-butadiene were previously determined in Sec. IV.III.6.3 by the Hückel method. The symmetries of these orbitals in the C_2 and C_s groups are easily obtained (Table IV.III.8). From this one may then construct the left-hand side of the correlation diagram (Fig. IV.III.37).

TABLE IV.III.8 — **Orbital symmetry and energies.**

	φ_1	φ_2	φ_3	φ_4	σ	σ^*	π	π^*
C_2	b	a	b	a	a	b	b	a
C_s	a'	a''	a'	a''	a'	a''	a'	a''
$E - \alpha$	1.618β	0.618β	-0.618β	-1.618β	$< 1.618\beta$	$> -1.618\beta$	β	$-\beta$

When *cis*-butadiene transforms into cyclobutene, π-type coupling occurs between atomic orbitals $\phi_2 = 2p\pi_2$ and $\phi_3 = 2p\pi_3$, analogous to the energies and molecular orbitals of the ethylene molecule {which we will denote here π (bonding orbital) and π^* (antibonding orbital)}, already calculated in Sec.

IV.III.6.3. The determination of their symmetry in the groups C_2 and C_s poses no problem. Butadiene atomic orbitals $\phi_1 = 2p\pi_1$ and $\phi_4 = 2p\pi_4$ produce σ-type coupling in the preceding transformation, characterized in cyclobutene by a bonding orbital, denoted σ, and an antibonding orbital denoted σ^*.

As the energy of a σ bond is clearly greater than that of a π bond, the energy of these orbitals should be below $\alpha + 1.618\,\beta$ and above $\alpha - 1.618\,\beta$, respectively. Examination of the character tables of the C_2 and C_s groups yields the symmetries in these groups (Table IV.III.8).

The correlation diagrams for the two rotatory modes are then drawn by placing the molecular orbitals of cis-butadiene and cyclobutene side by side, in order of increasing energy, taking care to correlate orbitals with the same symmetry (Fig. IV.III.37) in such a way that the corresponding lines do not cross.

It then appears that, in the conrotatory mode, the two lowest occupied levels of cis-butadiene correlate with the lowest levels of cyclobutene. This situation is favored if the initial configuration is ground-state *cis*-butadiene, i.e., in the case of a thermal reaction. In the disrotatory state, however, the ϕ_2 orbital of cis-butadiene correlates with orbital π^* of cyclobutene, which is high in energy and therefore renders the thermal reaction improbable. By contrast, orbital ϕ_3 is occupied in the case of a photochemical reaction, and correlates, in the conrotatory and disrotatory modes, respectively, with cyclobutene orbital σ^*(high-energy), and orbital π (which does not require additional energy). Thus the disrotatory mode is favored.

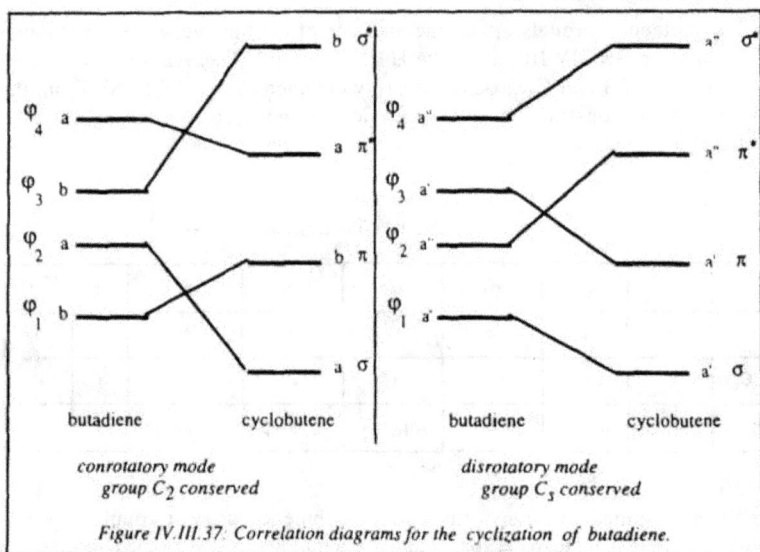

Figure IV.III.37. Correlation diagrams for the cyclization of butadiene.

7. Appendix 9: Mechanical and orbital aspects of the hydrogen molecule; uniqueness of descriptions.

We propose to develop an approximate calculation of one-electron density ρ and differential density $\Delta\rho$ of the hydrogen molecule, a problem which was dealt with formally in Sec. III.II.1.

Choice of wave functions

We shall use minimal basis wave functions formed from normalized 1s atomic orbitals of two hydrogen atoms, expressed in atomic units

$$1s_A = \exp(-r_A)/\sqrt{\pi} \qquad 1s_B = \exp(-r_B)/\sqrt{\pi} \qquad \text{(IV.III.34)}$$

using the notation

$$1s_A(1) = \exp(-r_{A1})/\sqrt{\pi}) \qquad \text{(IV.III.35)}$$

in order to indicate that orbital $1s_A$ is a function of the coordinates of electron 1.

The geometric parameters defining the molecule are shown in Fig. IV.III.38.

Although only approximate, these wave functions are of pedagogic interest as they may be obtained by simple arguments (symmetry and spin properties) and permit analytic calculations. The chosen two-function basis can only provide two orthogonal molecular orbitals, whose form is governed by symmetry considerations

$$\varphi_+ = C_+(1s_A + 1s_B) \qquad \text{and} \qquad \varphi_- = C_-(1s_A - 1s_B) \qquad \text{(IV.III.36)}$$

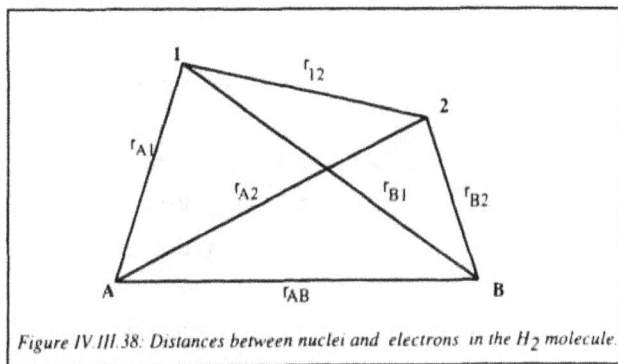

Figure IV.III.38: Distances between nuclei and electrons in the H_2 molecule.

The coefficients C_+ and C_- (assumed positive) are determined from the normalization condition. For example, for orbital ϕ_+ we obtain

$$1 = \int \varphi_+^2 \, dv = C_+^2 \int \left[1s_A{}^2 + 1s_B{}^2 + 2(1s_A \, 1s_B) \right] dv = C_+^2 (2 + 2S)$$

where S is the overlap integral between $1s_A$ and $1s_B$.

We have

$$C_+ = 1/\sqrt{2\,(1+S)}$$

and analogously

$$C_- = 1/\sqrt{2\,(1+S)}$$

The S integral may be obtained analytically in elliptic coordinates

$$S = \int 1s_A\,1s_B\,dv = \exp(-r_{AB})\,(1 + r_{AB} + r_{AB}^2/3) \qquad \text{(IV.III.37)}$$

Taking $r_{AB} = 1.4$ Bohr (1 Bohr = 52.9 nm), we find $S = 0.753$. (All numbers and functions are real in the preceding and following expressions, and the conjugation sign has been omitted.) From these two molecular orbitals, it is possible to construct the Slater determinant by respecting the principle of anti-symmetrization (Eq. IV.III.38).

These wave functions must be eigenfunctions of the spin operator S^2. As this is not the case for Δ_3 and Δ_4, we have to replace them by appropriate linear combinations. The basis used thus provides six quantum states. A more detailed analysis shows that three of these states are nondegenerate of zero total spin (singlet states), which we will denote in order of increasing energy as $\psi_0{}^S$, $\psi_1{}^S$ and $\psi_2{}^S$, while the three remaining states are triply degenerate and of spin $S = 1$ (triplet states).

$$\Delta_1 = \frac{1}{\sqrt{2}} \begin{vmatrix} \varphi_+(1)\alpha(\sigma_1) & \varphi_+(1)\beta(\sigma_1) \\ \varphi_+(2)\alpha(\sigma_2) & \varphi_+(2)\beta(\sigma_2) \end{vmatrix}$$

$$\Delta_2 = \frac{1}{\sqrt{2}} \begin{vmatrix} \varphi_+(1)\alpha(\sigma_1) & \varphi_-(1)\alpha(\sigma_1) \\ \varphi_+(2)\alpha(\sigma_2) & \varphi_-(2)\alpha(\sigma_2) \end{vmatrix}$$

$$\Delta_3 = \frac{1}{\sqrt{2}} \begin{vmatrix} \varphi_+(1)\alpha(\sigma_1) & \varphi_-(1)\beta(\sigma_1) \\ \varphi_+(2)\alpha(\sigma_2) & \varphi_-(2)\beta(\sigma_2) \end{vmatrix}$$

$$\Delta_4 = \frac{1}{\sqrt{2}} \begin{vmatrix} \varphi_+(1)\beta(\sigma_1) & \varphi_-(1)\alpha(\sigma_1) \\ \varphi_+(2)\beta(\sigma_2) & \varphi_-(2)\alpha(\sigma_2) \end{vmatrix} \qquad \text{(IV.III.38)}$$

$$\Delta_5 = \frac{1}{\sqrt{2}} \begin{vmatrix} \varphi_+(1)\beta(\sigma_1) & \varphi_-(1)\beta(\sigma_1) \\ \varphi_+(2)\beta(\sigma_2) & \varphi_-(2)\beta(\sigma_2) \end{vmatrix}$$

$$\Delta_6 = \frac{1}{\sqrt{2}} \begin{vmatrix} \varphi_-(1)\alpha(\sigma_1) & \varphi_-(1)\beta(\sigma_1) \\ \varphi_-(2)\alpha(\sigma_2) & \varphi_-(2)\beta(\sigma_2) \end{vmatrix}$$

Taking account of the projection of total spin on the Oz axis, we will denote these as $\psi_1{}^T$, $\psi_0{}^T$ and $\psi_{-1}{}^T$.

$$\psi_0{}^S(1,2) = \Delta_1 = \varphi_+(1)\varphi_+(2)\,(1/\sqrt{2})\,[\alpha(\sigma_1)\beta(\sigma_2) - \beta(\sigma_1)\alpha(\sigma_2)]$$

$$\psi_1{}^S(1,2) = (1/\sqrt{2})(\Delta_3 - \Delta_4) = (1/\sqrt{2})\,[\varphi_+(1)\varphi_-(2) + \varphi_-(1)\varphi_+(2)]$$

$$\times (1/\sqrt{2})\,[\alpha(\sigma_1)\beta(\sigma_2) - \beta(\sigma_1)\alpha(\sigma_2)]$$

$$\psi_2{}^S(1,2) = \Delta_6 = \varphi_-(1)\varphi_-(2)\,(1/\sqrt{2})\,[\alpha(\sigma_1)\beta(\sigma_2) - \beta(\sigma_1)\alpha(\sigma_2)]$$

$$\psi_1{}^T(1,2) = \Delta_2 = (1/\sqrt{2})\,[\varphi_+(1)\varphi_-(2) - \varphi_-(1)\varphi_+(2)]\alpha(\sigma_1)\alpha(\sigma_2)$$

$$\psi_0{}^T(1,2) = (1/\sqrt{2})\,(\Delta_3 + \Delta_4)$$

$$= (1/\sqrt{2})\,[\varphi_+(1)\varphi_-(2) - \varphi_-(1)\varphi_+(2)](1/\sqrt{2})[\alpha(\sigma_1)\beta(\sigma_2)+\beta(\sigma_1)\,\alpha(\sigma_2)]$$

$$\psi_{-1}{}^T(1,2) = \Delta_5 = (1/\sqrt{2})\,[\varphi_+(1)\varphi_-(2) - \varphi_-(1)\varphi_+(2)]\,\beta(\sigma_1)\beta(\sigma_2)$$

The energies of these states are shown schematically in Fig. IV.III.39. The "filling-up" of the two orbitals is suggested by arrows drawn inside the parentheses.

Remark: This is an adequate basis for an approximate representation of $\psi_0{}^S$, but totally insufficient for the other states. The corresponding functions are certainly not anywhere near the actual excited state functions.

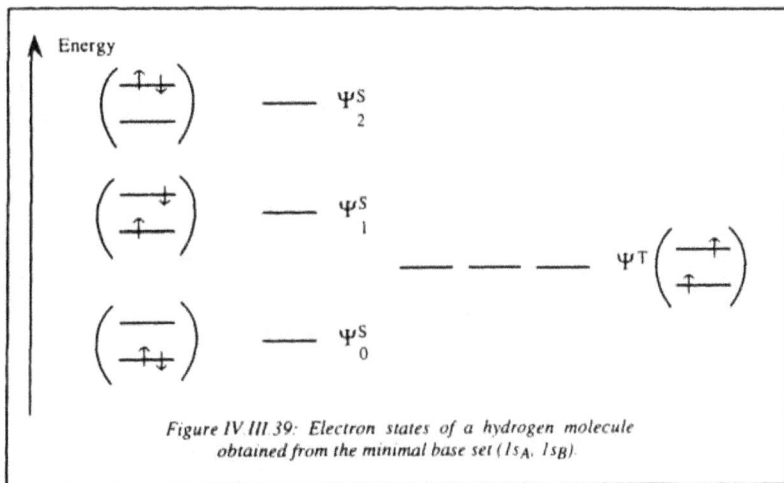

Figure IV.III.39: Electron states of a hydrogen molecule obtained from the minimal base set ($1s_A$, $1s_B$).

One-electron volume density of various states

We shall now discuss the electron density calculation of ground state $\psi_0{}^S$ starting with the determination of volume density $\rho_1(1)$ of electron 1 when electron 2 may be anywhere in space (cf., Sec. III.I.2).

$$\rho_1(1) = \sum_{\sigma_1} \sum_{\sigma_2} \int |\psi_0^s|^2 \, dv_2$$

(IV.III.39)

$$= |\varphi_+(1)|^2 \int |\varphi_+(2)|^2 dv_2 \sum_{\sigma_1} \sum_{\sigma_2} (1/2)\left[\alpha(\sigma_1)\beta(\sigma_2) - \beta(\sigma_1)\alpha(\sigma_2) \right]^2$$

The orthonormal character of spin functions $\alpha(\sigma)$ and $\beta(\sigma)$ means that the double sum over σ_1 and σ_2 is equal to 1. The integral is likewise equal to 1, due to normalization of φ_+.

We have

$$\rho_1(1) = |\varphi_+(1)|^2 = \frac{1}{2(1+S)}\left[1s_A(1)^2 + 1s_B(1)^2 + 2\left(1s_A(1)1s_B(1)\right) \right]$$

and, replacing the coordinates of electron 1 by a general distance variable

$$\rho_1 = \varphi_+^2 = \frac{1}{2(1+S)}\left[1s_A^2 + 1s_B^2 + 2\left(1s_A 1s_B\right) \right] \qquad \text{(IV.III.40)}$$

The reader may verify that calculation of ρ_2 leads to the same result (this is a consequence of wave function antisymmetry, ensuring the indistinguishability of electrons). The total one-electron volume density in state ψ_0^s is at any point

$$\rho_0^S = \rho_1 + \rho_2 = \frac{1}{1+S}\left[1s_A^2 + 1s_B^2 + 2\left(1s_A 1s_B\right) \right] \qquad \text{(IV.III.41)}$$

Similar calculations, taking into account the orthonormal character of molecular orbitals φ_+ and φ_- and spin functions α and β, lead to the following densities in the other states

$$\rho_1^S = \frac{1}{2(1+S)}\left(1s_A^2 + 1s_B^2 + 2\left(1s_A 1s_B\right) \right)$$

$$+ \frac{1}{2(1-S)}\left(1s_A^2 + 1s_B^2 - 2\left(1s_A 1s_B\right) \right)$$

(IV.III.42)

$$\rho_2^S = \frac{1}{1-S}\left(1s_A^2 + 1s_B^2 - 2\left(1s_A 1s_B\right) \right)$$

In view of the preceding remarks, these densities are likely to be quite far from the actual densities.

One-electron differential density contours

The one-electron differential density $\Delta\rho$ is the difference between the one-electron density of the molecule and the sum of electron densities of hypothetical atoms, assumed to be nonbonded and in the same positions as in the molecule.

$$\Delta\rho = \rho - (\rho_A + \rho_B) \qquad\qquad (IV.III.43)$$

with
$$\rho_A = 1s_A{}^2 \text{ and } \rho_B = 1s_B{}^2$$

For example, for the ground state $\psi_0{}^S$, we obtain

$$\Delta\rho_0{}^S = \frac{1}{\pi}\left[\frac{1}{1+S}\left(\exp(-2r_A) + \exp(-2r_B) + 2\exp(-r_A-r_B)\right) - \exp(-2r_A) - \exp(-2r_B)\right]$$

and for the doubly excited state $\psi_2{}^S$

$$\Delta\rho_2{}^S = \frac{1}{\pi}\left[\frac{1}{1-S}\left(\exp(-2r_A) + \exp(-2r_B) - 2\exp(-r_A-r_B)\right) - \exp(-2r_A) - \exp(-2r_B)\right]$$

It is then possible to trace countour curves for the one-electron differential density of these states using an appropriate computer program (Figs. IV.III.40 and IV.III.41).

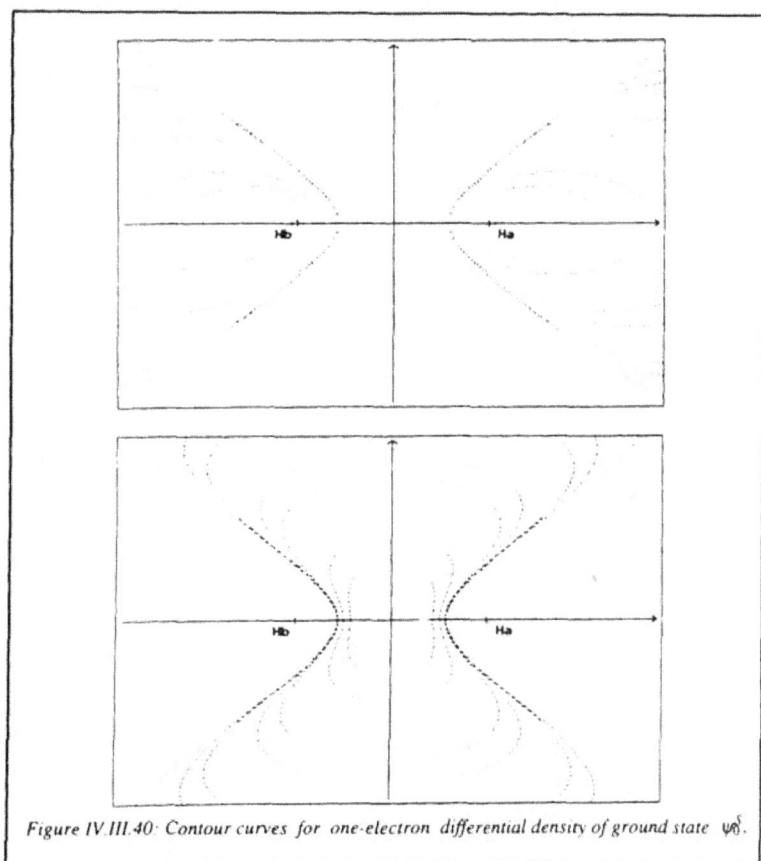

Figure IV.III.40· Contour curves for one-electron differential density of ground state $\psi_0{}^S$.

These curves show that:

— In the ψ_0^S ground state, $\Delta\rho$ is positive in the region of space that corresponds approximately to the Berlin region of bound states (Sec. III.I.2). There is thus an increase in one-electron density in this region; $\Delta\rho$ is negative in a region of space that corresponds approximately to the anti-bonding Berlin zone. It is interesting to compare this result with the form of the curve representing electron energy as a function of distance r_{AB} (Fig. IV.III.42). In fact, we have seen (cf., Sec. III.I) that this electron energy may be considered to represent the potential energy $U(r_{AB})$ of the nuclei in the presence of an electron cloud and that, in this case, it reveals the existence of a stable bond.

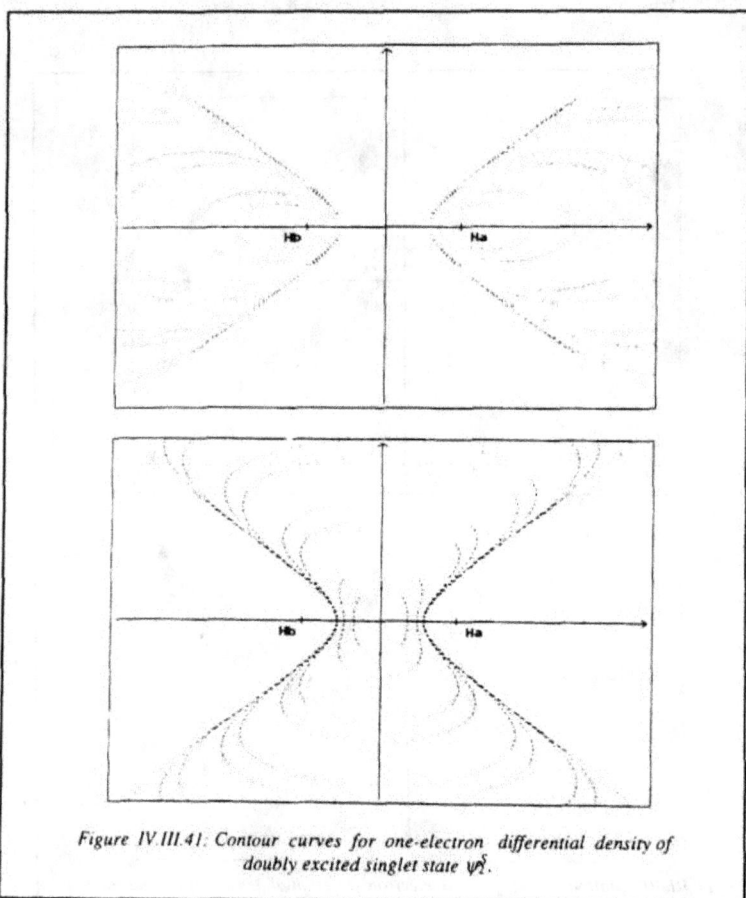

Figure IV.III.41: Contour curves for one-electron differential density of doubly excited singlet state ψ_1^S.

The one-electron differential volume density, calculated simply from well-known expressions for bonding and anti-bonding molecular orbitals of the hydrogen molecule, provides a correct illustration of the connection between the mechanical and orbital aspects of a covalent bond, as developed in Chapters III and IV.

— By contrast, in the doubly excited ψ_2^S state, $\Delta\rho$ is, to a first approximation, positive in the anti-bonding Berlin zone, and negative in the bonding zone, corresponding only to an anti-bonding state, whatever the value of r_{AB}.

From a qualitative point of view, these results may be extended to more complex molecules. Consider for example an occupied molecular orbital, where the part ... $+ c_p\phi_p + c_q\phi_q + ...$ stands for the coupling between orbital $c_p\phi_p$ of atom P and orbital $c_q\phi_q$ of a neighboring atom Q. If in the region of space between the nuclei the functions $c_p\phi_p$ and $c_q\phi_q$ are of the same sign (positive or negative), the electron density in this region will tend to be high, and the molecular orbital will therefore make a bonding contribution to the P-Q bond.

On the other hand, if the functions $c_p\phi_p$ and $c_q\phi_q$ are of opposite signs, electron density will be lower and the contribution will be anti-bonding. This qualitative rule also applies to the preceding discussion of the hydrogen molecule and facilitates comprehension of certain properties, such as the change of bond length in ionization, or the energy ordering of molecular orbitals.

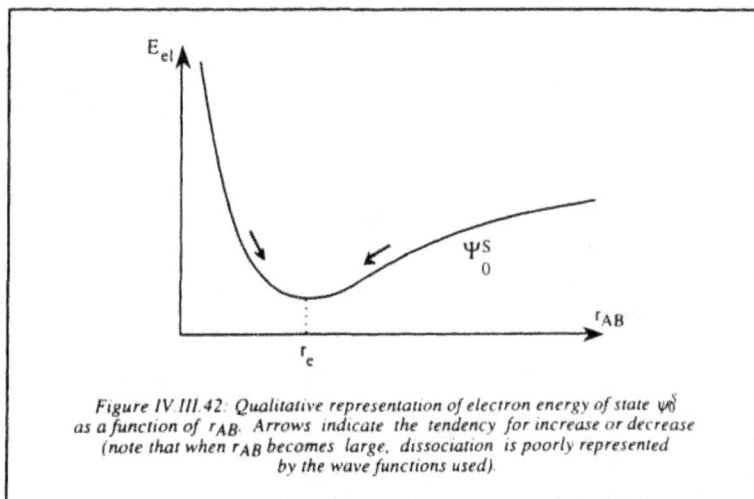

Figure IV.III.42: Qualitative representation of electron energy of state ψ_0^S as a function of r_{AB}. Arrows indicate the tendency for increase or decrease (note that when r_{AB} becomes large, dissociation is poorly represented by the wave functions used).

8 - Appendix 10: *Different representations of the basic molecules in organic and biological chemistries, deduced from Hartree-Fock wave functions expanded in a minimal basis (localized and delocalized orbitals).*

We study methane, ethylene, acetylene and benzene in turn and give the Hartree-Fock results obtained from minimal, nonorthogonal bases. Then, we illustrate the concepts developed in Chapter IV.III.1. For each molecule, we give the occupied molecular orbitals (MOs given by normal data using diagonal Fock matrices). These MOs have a spatial extension covering the whole molecule and are, for this reason, called "delocalized". As indicated in Chapter II.II, they form a basis for irreducible representations of the molecule symmetry group. Moreover, their associated energies (eigenvalues of the Fock matrix) may be successfully linked to the photoelectronic spectrum of the molecule. However, as already mentioned in Chapter IV.II.3, we obtain the same wave function by replacing the set of delocalized MOs by any new set obtained from unitary transforms. It is then possible to obtain sets of localized MOs, picturing chemical bonding in a way close to the representation usually assumed by chemists, and including mesomerism and hybridization. It is, however, very important to point out that these localized MOs do not have the same properties as delocalized ones (symmetry and energy).

At the Hartree-Fock level of description, all the sets of MOs are equivalent: A more correct description would be obtained by a closer approach to the exact wave function, for instance, by using the configuration interaction method. In this case, however, we would lose the MO concept.

Methane

This molecule has 10 electrons and its geometry is given in Fig. IV.III.43. The minimal base set is $(1s_C; 2s_C; 2px_C; 2py_C; 2pz_C; 1s_{H_1}; 1s_{H_2}; 1s_{H_3}; 1s_{H_4})$.

Delocalized MOs

$$\varphi_1 = 1sC$$
$$\varphi_2 = 0.32\,(1\,s_{H1} + 1\,s_{H2} + 1\,s_{H3} + 1\,s_{H4}) + 0.55\,(2s_C)$$
$$\varphi_3 = 0.32\,(1\,s_{H1} + 1\,s_{H2} - 1\,s_{H3} - 1\,s_{H4}) + 0.55\,(2px_C)$$
$$\varphi_4 = 0.32\,(1\,s_{H1} - 1\,s_{H2} + 1\,s_{H3} - 1\,s_{H4}) + 0.55\,(2py_C)$$
$$\varphi_5 = 0.32\,(1\,s_{H1} - 1\,s_{H2} - 1\,s_{H3} + 1\,s_{H4}) + 0.55\,(2pz_C)$$

As expected, these functions do not show the usual representation of the C-H bonds in the molecule.

Localized MOs

$$\varphi'_1 = 1s_C$$
$$\varphi'_2 = 0.29\,(2s_C) + 0.29\,(2px_C + 2py_C + 2pz_C) + 0.57\,(2s_{H1})$$

$$\varphi'_2 = 0.29 \ (2s_C) + 0.29 \ (2px_C - 2py_C - 2pz_C) + 0.57 \ (2s_{H2})$$

$$\varphi'_3 = 0.29 \ (2s_C) + 0.29 \ (- 2px_C + 2py_C - 2pz_C) + 0.57 \ (2s_{H3})$$

$$\varphi'_4 = 0.29 \ (2s_C) + 0.29 \ (- 2px_C - 2py_C - 2pz_C) + 0.57 \ (2s_{H4})$$

$$\varphi'_5 = 0.29 \ (2s_C) + 0.29 \ (- 2px_C - 2py_C + 2pz_C) + 0.57 \ (2s_{H5})$$

We immediately notice that, in both sets of MOs (φ_1 and φ'_1), the atomic orbital $1s_C$ appears to remain uncoupled. It corresponds to the low-energy core orbital of the carbon atom. The energy gap between this orbital and the other AOs is very large (similar to the property of the MO $1s_O$ in the water molecule H_2O, see Sec. IV.II.2).

The four MOs φ'_2, φ'_3, φ'_4 and φ'_5 are very similar:

— They all include a part originating from the carbon atom (the first four terms of each MO). AO $2s_C$ has the same weight (0.29) in each MO. The different contributions of the three AOs $2p_C$ lead to the well-known tetrahedral structure illustrated in Fig. IV.III.44.

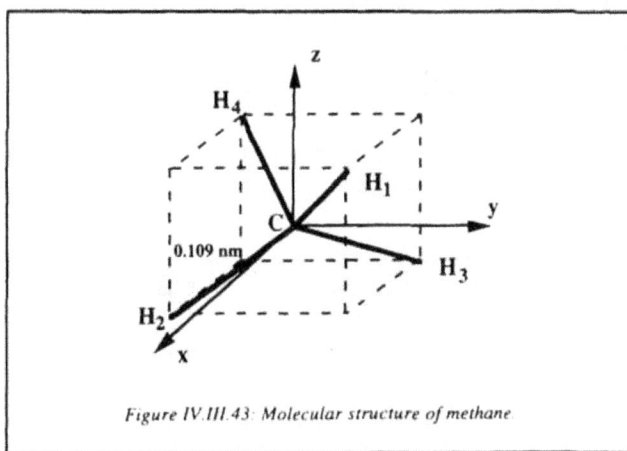

Figure IV.III.43: Molecular structure of methane

— Each of these includes the AO $1s_H$ of a hydrogen atom (weight 0.57).

Once again, we find the well-known image of the four tetrahedral bonds.

These four MOs have exactly the same mathematical forms as those obtained by coupling the OA $1s_H$ of each hydrogen atom to a *"hybrid"* AO sp^3 of the carbon atom. The localized MOs, derived from the Hartree-Fock treatment, include the hybridization concept introduced by Pauling.

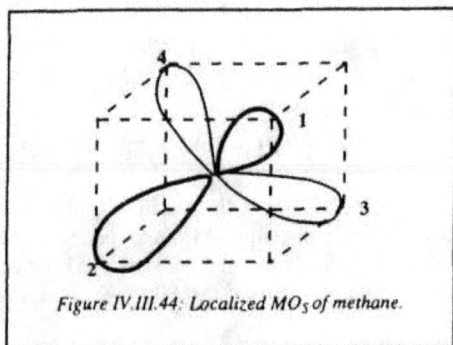

Figure IV.III.44: Localized MOs of methane.

Ethylene

This molecule has 16 electrons. Its geometry is shown in Fig.IV.III.45. The minimal base set used is $(1s_{C_1}; 2s_{C_1}; 2px_{C_1}; 2py_{C_1}; 2pz_{C_1}; 1s_{C_2}\ 2s_{C_2}; 2px_{C_2}; 2py_{C_2}; 2pz_{C_2}; 1s_{H_1}; 1s_{H_2}; 1s_{H_3}; 1s_{H_4})$.

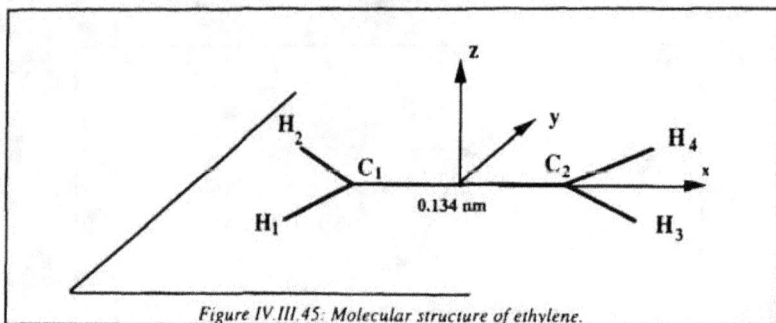

Figure IV.III.45: Molecular structure of ethylene.

Delocalized MOs

$$\varphi_1 = 0.70\,(1s_{C1} + 1s_{C2})$$

$$\varphi_2 = 0.70\,(1s_{C1} - 1s_{C2})$$

$$\varphi_3 = -0.16\,(1s_{C1} + 1s_{C2}) + 0.47\,(2s_{C1} + 2s_{C2}) + 0.109\,(2px_{C1} - 2px_{C2})$$
$$+ 0.10\,(1s_{H1} + 1s_{H2} + 1s_{H3} + 1s_{H4})$$

$$\varphi_4 = -0.13\,(1s_{C1} - 1s_{C2}) + 0.44\,(2s_{C1} - 2s_{C2}) + 0..20\,(2px_{C1} + 2px_{C2})$$
$$+ 0.21\,(1s_{H1} + 1s_{H2} - 1s_{H3} - 1s_{H4})$$

$$\varphi_5 = -0.41\,(2py_{C1} + 2py_{C2}) - 0.24\,(1s_{H1} - 1s_{H2} + 1s_{H3} - 1s_{H4})$$

$$\varphi_6 = -0.01\,(1s_{C1} + 1s_{C2}) - 0.52\,(2px_{C1} - 2px_{C2})$$
$$+ 0.20\,(1s_{H1} + 1s_{H2} + 1s_{H3} + 1s_{H4})$$

$$\varphi_7 = -0.42\,(2py_{C1} - 2py_{C2}) + 0.33\,(1s_{H1} - 1s_{H2} - 1s_{H3} + 1s_{H4})$$

$$\varphi_8 = -0.63\,(2pz_{C1} + 2pz_{C2})$$

We identify the two MOs (φ_1 and φ_2). The difference between the weights of $1s_{C_2}$ (0.7 for φ_1 and -0.7 for φ_2, respectively) is directly linked to the molecular symmetry.

The six other orbitals are called valency orbitals. Note that the nonzero values for the weights of $1s_{C_1}$ and $1s_{C_2}$ do not correspond to any physical effect but are only a mathematical correction for the uncertainty of the weights of orbitals $2s_{C_1}$ and $2s_{C_2}$. The first seven MOs are symmetrical with respect to the molecular plane and are called σ-*orbitals*, while the eighth is antisymmetrical and is called a π-*orbital*.

Localized MO

As previously indicated (see Chapter IV.II.3), there are an infinite number of MOs corresponding to the same wave function. Consequently, the localized MOs obtained depend on the choice of a localization criterion. We have only presented two examples here:

σ-π *localization*

We find, respectively
— Two "core" orbitals $\varphi'_1 = 1s_{C_1}$ and $\varphi'_2 = 2s_{C_2}$
— Four σ valency carbon-hydrogen MOs

$\varphi'_3(C_1\text{-}H_1) = 0.37\,(2s_{C_1}) - 0.26\,(2px_{C_1}) - 0.41\,(2py_{C_1}) + 0.49\,(1s_{H_1})$

$\varphi'_4(C_1\text{-}H_2) = 0.37\,(2s_{C_1}) - 0.26\,(2px_{C_1}) + 0.41\,(2py_{C_1}) + 0.49\,(1s_{H_2})$

$\varphi'_5(C_2\text{-}H_3) = 0.37\,(2s_{C_2}) + 0.26\,(2px_{C_2}) - 0.41\,(2py_{C_2}) + 0.49\,(1s_{H_3})$

$\varphi'_6(C_2\text{-}H_4) = 0.37\,(2s_{C_2}) - 0.26\,(2px_{C_2}) + 0.41\,(2py_{C_2}) + 0.49\,(1s_{H_4})$

— One σ valency carbon-carbon MO
$\varphi'_7{}^\sigma\,(C_1\text{-}C_2) = 0.35\,(2s_{C_1} + 2s_{C_2}) + 0.42\,(2px_{C_1} - 2px_{C_2})$
— One π valency carbon-carbon MO
$\varphi'_8{}^\pi\,(C_1\text{-}C_2) = 0.63\,(2pz_{C_1} + 2pz_{C_2})$

The last six MOs are shown in Fig. IV.III.46.

This localization leads to the chemists' usual representation: Four σ (C – H) bonds, one σ (C – C) bond and one π (C – C) bond.

Localized "banana" MOs

If we leave the σ-π partition already used, φ'_7 and φ'_8 may be coupled to give:

$$\begin{cases} \varphi''_7 = \dfrac{1}{\sqrt{2}}\,(\varphi'_7{}^{(\sigma)}(C_1-C_2) + \varphi'_8{}^{(\pi)}(C_1-C_2)) \\ \varphi''_8 = \dfrac{1}{\sqrt{2}}\,(\varphi'_7{}^{(\sigma)}(C_1-C_2) - \varphi'_8{}^{(\pi)}(C_1-C_2)) \end{cases}$$

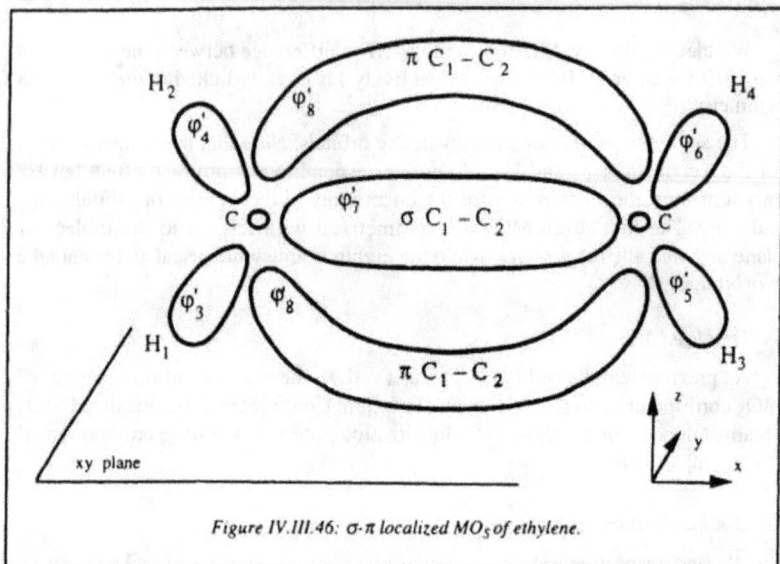

Figure IV.III.46: σ-π localized MOs of ethylene.

Then, we built two new MOs, symmetrical with respect to the molecular plane. These "banana" MOs are shown in Fig. IV.III.47.

Within the Hartree-Fock treatment, the two sets (φ'_7 and φ'_8) and (φ''_7 and φ''_8) are strictly equivalent, as they correspond to the same wave function.

However, in this treatment, only the delocalized MOs, *implying a σ-π partition*, may be fruitfully associated with the first ionization energy. It should be remembered that the higher energy delocalized MOs (frontier orbitals) play an essential role in chemical reactivity. Nevertheless, a good image of chemical bonding implies the use of localized MOs. As they are infinite in number, none may be privileged and the greatest care must be taken in the *description of the double bond.*

Acetylene

This molecule has 14 electrons. Its geometry is given in Fig. IV.III.48. The minimal base set is ($1s_{C_1}$; $1s_{C_2}$; $2s_{C_1}$; $2s_{C_2}$; $2p_{xC_1}$; $2p_{xC_2}$; $2p_{yC_1}$; $2p_{yC_2}$; $2p_{zC_1}$; $2p_{zC_2}$; $1s_{H_1}$; $1s_{H_2}$)

Delocalized MOs

We find seven delocalized MOs:
— Two σ "core" MOs

$$\varphi_1 = -0.70\,(1s_{C1} + 1s_{C2})$$
$$\varphi_2 = -0.70\,(1s_{C1} - 1s_{C2})$$

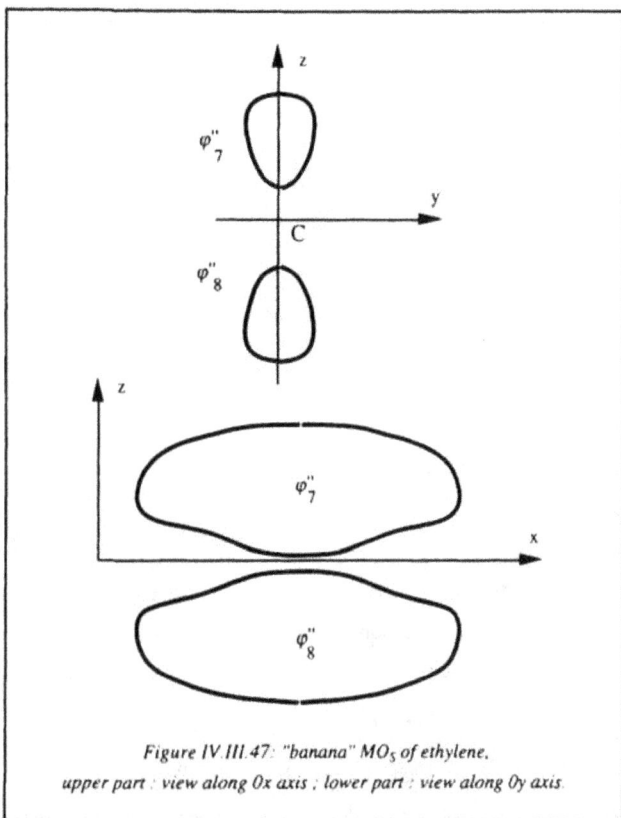

Figure IV.III.47: "banana" MOs of ethylene,
upper part : view along Ox axis ; lower part : view along Oy axis.

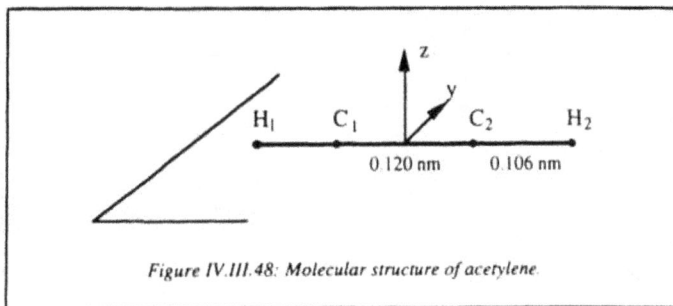

Figure IV.III.48: Molecular structure of acetylene.

— Three σ valency MOs

$\varphi_3 = -0.17 \, (1s_{C1} + 1s_{C2}) + 0.48 \, (2s_{C1} + 2s_{C2}) + 0.18 \, (2px_{C1} - 2px_{C2})$
$\qquad + 0.11 \, (1s_{H1} + 1s_{H2})$

$\varphi_4 = -0.11 \, (1s_{C1} - 1s_{C2}) + 0.34 \, (2s_{C1} + 2s_{C2}) - 0.29 \, (2px_{C1} - 2px_{C2})$
$\qquad + 0.31 \, (1s_{H1} - 1s_{H2})$

$\varphi_5 = -0.09 \, (2s_{C1} + 2s_{C2}) + 0.46 \, (2px_{C1} - 2px_{C2}) - 0.32 \, (1s_{H1} + 1s_{H2})$

— Two π MOs

$$\varphi_6 = 0.61 \, (2py_{C1} + 2py_{C2})$$
$$\varphi_7 = 0.61 \, (2pz_{C1} + 2pz_{C2})$$

As previously noted, no simple picture of the bonds inside the molecule emerges.

Localized MOs

We also choose to limit ourselves to two sets : "σ-π" and "banana".

"σ-π" localized MOs

The five σ MOs are:

— Two "core" MOs given by the carbon atoms $\varphi'_1 = 1s_{C1}$ and $\varphi'_2 = 1s_{C2}$

— Two valency MOs corresponding to $C_1 - H_1$ and $C_2 - H_2$ bonds

$$\varphi'_3 \, (C_1\text{-}H_1) = 0.44 \, (2s_{C1}) - 0.46 \, (2px_{C1}) + 0.45 \, (1s_{H1})$$

$$\varphi'_4 \, (C_2\text{-}H_2) = 0.44 \, (2s_{C2}) + 0.46 \, (2px_{C2}) + 0.45 \, (1s_{H2})$$

— One valency MOs corresponding to $C_1 - C_2$ bond

$$\varphi'_5{}^{\sigma} \, (C_1\text{-}C_2) = 0.45 \, (2s_{C1}) + 0.35 \, (2px_{C1}) + 0.45 \, (2s_{C2}) - 0.35 \, (2px_{C2})$$

The last three MOs (φ'_3, φ'_4 and φ'_5) are represented in the Fig. IV.III.49 .

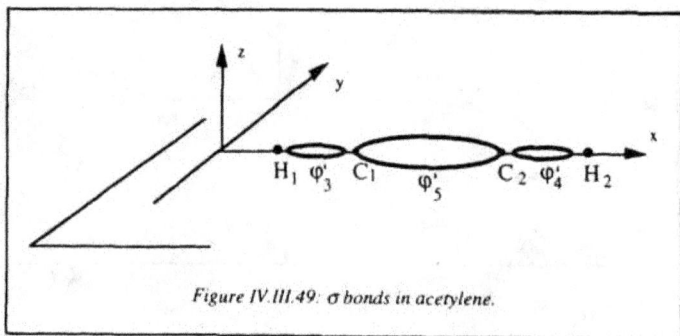

Figure IV.III.49: σ bonds in acetylene.

The two π MOs $\varphi'_6{}^{(\pi)}$ and $\varphi'_7{}^{(\pi)}$ are shown in Fig. IV.III.50. $\varphi'_7{}^{(\pi)}$ is practically identical to $\varphi'_8{}^{(\pi)}$ of the ethylene molecule. $\varphi'_6{}^{(\pi)}$ may be easily deduced from $\varphi'_7{}^{(\pi)}$ by a 90° rotation about the $0x$ axis.

We obtain the usual representation (one σ and two π MO) of a triple bond.

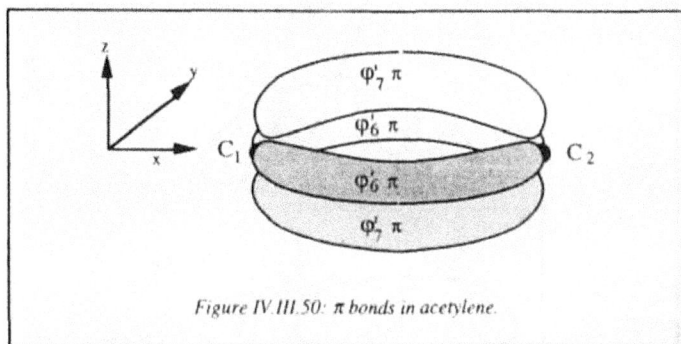

Figure IV.III.50: π bonds in acetylene.

Delocalized "banana" MOs

Here we find again the two "core" orbitals of the carbon atoms, φ_1 and φ_2, and the two orbitals φ_3 and φ_4 describing the C_1 -H_1 and C_2 - H_2 bonds. The three "σ-π" MOs are replaced by three identical "banana" MOs, located in three planes at 120 degrees from each other, also describing the triple bond (see Fig. IV.III.51).

The reader will note that, because of the degeneracy due to the C_∞ axis, no privileged direction exists for the "banana" MOs. If φ''_5 is assumed to point in the $0z$ direction, we obtain

$$\varphi''_5 = 0.23 \ (2s_{C_1}) + 0.20 \ (2p_{xC_1}) + 0.5 \ (2p_{zC_1}) + 0.23 \ (2s_{C_2})$$
$$- 0.20 \ (2p_{xC_2}) + 0.5 \ (2p_{zC_2})$$

$$\varphi''_6 = 0.23 \ (2s_{C_1}) + 0.20 \ (2p_{xC_1}) + 0.25 \ (2p_{zC_1}) + 0.43 \ (2p_{yC_1})$$
$$+ 0.23 \ (2s_{C_2}) - 0.20 \ (2p_{xC_2}) - 0.25 \ (2p_{zC_2}) + 0.43 \ (2p_{yC_2})$$

$$\varphi''_6 = 0.23 \ (2s_{C_1}) + 0.20 \ (2p_{xC_1}) - 0.25 \ (2p_{zC_1}) - 0.43 \ (2p_{yC_1})$$
$$+ 0.23 \ (2s_{C_2}) - 0.20 \ (2p_{xC_2}) - 0.25 \ (2p_{zC_2}) + 0.43 \ (2p_{yC_2})$$

Then, as in the case of the $C \equiv C$ triple bond in ethylene, we see that there are many representations of the triple bond. Besides the indetermination due to the use of the Slater determinant, there is also that of molecular symmetry. Once more, the greatest caution is required in this subject.

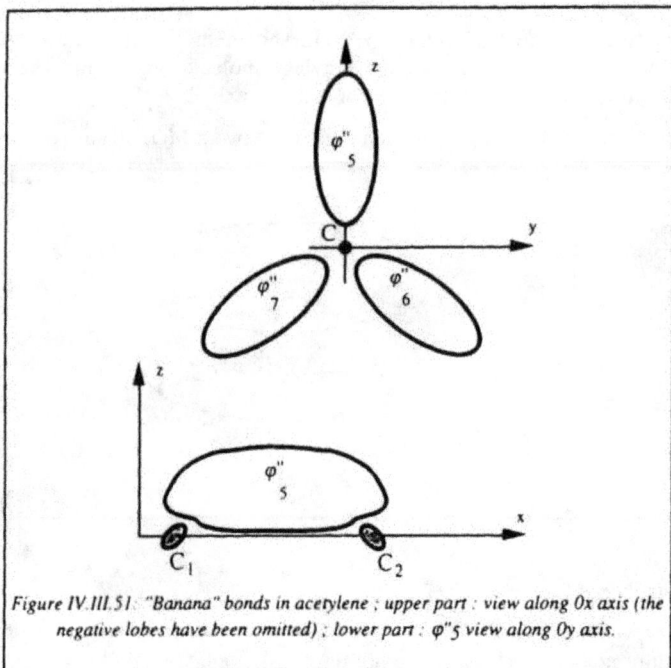

Figure IV.III.51: "Banana" bonds in acetylene ; upper part : view along Ox axis (the negative lobes have been omitted) ; lower part : φ''_5 view along Oy axis.

Benzene

This molecule has 42 electrons. Its geometry is indicated in Fig. IV.III.52.

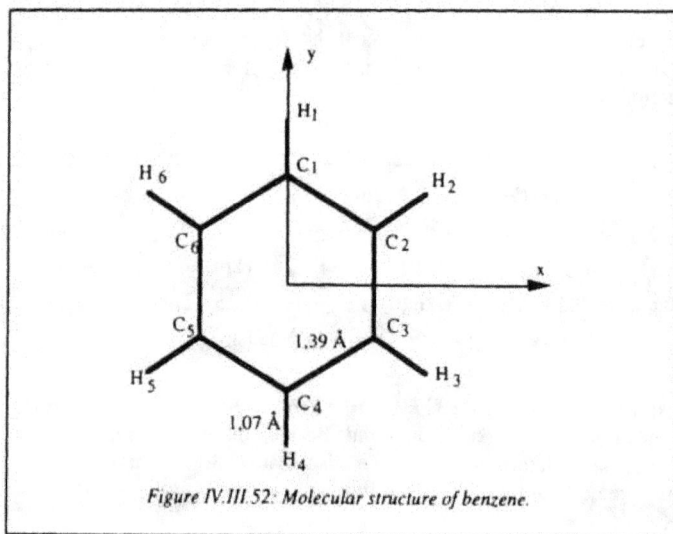

Figure IV.III.52: Molecular structure of benzene.

Delocalized MO_S

Calculations give:

— 18 σ MO_S built with the $1s_H$ AO of the six hydrogen atoms and with the AO $1s_C$, $2s_C$, $2px_C$ and $2py_C$ of the six carbon atoms.

— 3 π (φ_{19}, φ_{20}, φ_{21}) built with the AO $2pz_C$ of the six carbon atoms. They have already been obtained by neglecting the overlap, using the Hückel method (see 6.3) and may be expressed:

$$
\begin{cases}
\varphi_{19} = \dfrac{1}{\sqrt{6}} \, (2pz_{C_1} + 2pz_{C_2} + 2pz_{C_3} + 2pz_{C_4} + 2pz_{C_5} + 2pz_{C_6}) \\[2mm]
\varphi_{20} = \dfrac{1}{\sqrt{3}} \, (2pz_{C_1} - 2pz_{C_4}) + \dfrac{1}{\sqrt{12}} \, (2pz_{C_2} - 2pz_{C_3} - 2pz_{C_5} + 2pz_{C_6}) \\[2mm]
\varphi_{21} = \dfrac{1}{2} \, (2pz_{C_2} + 2pz_{C_3} - 2pz_{C_5} - 2pz_{C_6})
\end{cases}
$$

We observe that the two φ_{20} and φ_{21} MO_S are not consistent with the hexagonal symmetry of the molecule. In fact, and as already noted in Sec. IV.III.6.3, these two MO_S correspond to the same, doubly degenerate eigenvalue of the Fock matrix and form a basis for the irreducible representation e_1 (dimension 2) of the C_{6v} reduced symmetry group of the molecule. Consequently, all the linear combinations of these two MO_S still produce suitable MO_S.

Localized MO_S

$\sigma - \pi$ localized MO_S

We may build 18 s MO_S: The six "core" MO_S given by the six carbon atoms, 6 valency MO_S given by the six C - H bonds, and six valency MO_S describing the six C – C bonds.

However, in the case of the π MO_S, we only reach a very poor localization whatever the criterion used. It would appear to be impossible to reduce the three MO_S φ_{19}, φ_{20} and φ_{21} to something representing a bond between atoms. For instance, we find from (φ_{18}, φ_{19}, φ_{20}) by using the unitary* transformation represented by matrix (T) :

$$
(T) = \begin{pmatrix}
1/\sqrt{3} & 1/\sqrt{6} & 0 \\
1/\sqrt{3} & -1/\sqrt{6} & 1/\sqrt{2} \\
1/\sqrt{3} & -1/\sqrt{6} & -1/\sqrt{2}
\end{pmatrix}
$$

---

* A unitary matrix is defined by $(M)^{+} = (M)^{-1}$ (+ : transposition plus conjugation). We may easily check that (T) is a unitary matrix. A unitary operator keeps the norm of the transformed function Indeed, if $\Psi' = A\Psi$ is the transformed function of Ψ

$$
N' = \int \Psi'^{*}\Psi' d\tau = \int \Psi^{*}A^{*}A\Psi d\tau = \int \Psi^{*}A^{-1}A\Psi d\tau = \int \Psi^{*}\Psi d\tau = N
$$

$$\varphi_{2,3} = \frac{1}{3\sqrt{2}} [(1 + \sqrt{3})(2pz_{C_2} + 2pz_{C_3}) + (2pz_{C_1} + 2pz_{C_4})$$
$$+ (1 - \sqrt{3})(2pz_{C_5} + 2pz_{C_6})]$$

$$\varphi_{4,5} = \frac{1}{3\sqrt{2}} [(1 + \sqrt{3})(2pz_{C_4} + 2pz_{C_5}) + (2pz_{C_3} + 2pz_{C_6})$$
$$+ (1 - \sqrt{3})(2pz_{C_1} + 2pz_{C_2})]$$

$$\varphi_{6,1} = \frac{1}{3\sqrt{2}} [(1 + \sqrt{3})(2pz_{C_4} + 2pz_{C_5}) + (2pz_{C_3} + 2pz_{C_6})$$
$$+ (1 - \sqrt{3})(2pz_{C_5} + 2pz_{C_4})]$$

▶

These MOs are not perfectly localized, but we notice that the weights of the pairs $(2pz_{C_2}, 2pz_{C_3})$, $(2pz_{C_4}, 2pz_{C_5})$ and $(2pz_{C_6}, 2pz_{C_1})$ are largest in MOs $\varphi_{2,3}$, $\varphi_{4,5}$ and $\varphi_{6,1}$, respectively. These MOs may be considered as approximate representations of the three π bonds in Kekule's formula (see Fig. IV.III.53).

In the same way, by using the following matrix :

$$(T) = \begin{pmatrix} 1/\sqrt{3} & 0 & 2/\sqrt{6} \\ 1/\sqrt{3} & 1/\sqrt{6} & -1/\sqrt{6} \\ 1/\sqrt{3} & -1/\sqrt{2} & -1/\sqrt{6} \end{pmatrix}$$

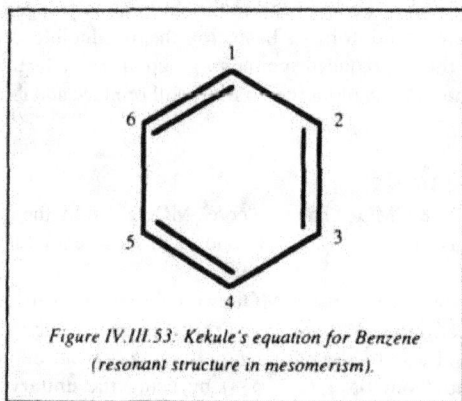

Figure IV.III.53: Kekule's equation for Benzene
(resonant structure in mesomerism).

We obtain, respectively,

$$\varphi_1 = \frac{1}{3\sqrt{2}} [3(2pz_{C_1}) + 2(2pz_{C_2}) - (2pz_{C_4}) + 2(2pz_{C_6})]$$

$$\varphi_3 = \frac{1}{3\sqrt{2}} [3(2pz_{C_3}) + 2(2pz_{C_2}) - (2pz_{C_6}) + 2(2pz_{C_4})]$$

$$\varphi_5 = \frac{1}{3\sqrt{2}} [3(2pz_{C_5}) + 2(2pz_{C_4}) - (2pz_{C_2}) + 2(2pz_{C_6})]$$

This new MO_S may be pictured (see Fig. IV.III.54) by reporting the weight (defined by the squared value of the corresponding coefficient) of each AO of the six carbon atoms.

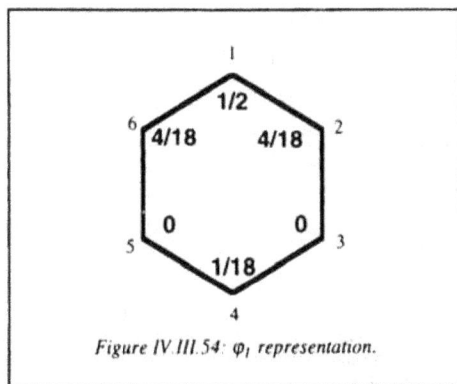

Figure IV.III.54: φ_1 representation.

"banana" MO_S

As in previous cases, by coupling σ and π MO_S, one may obtain "banana" MO_S, located on either side of the nuclei plane and always poorly localized.

Conclusion

Although the Hartree-Fock method, as described here, always leads to delocalized MO_S, it often appears possible to obtain the same complete wave function by using localized MO_S, thus making it possible to describe chemical bonding. It should always be remembered that there is, in fact, an infinity of sets, so it is not correct to privilege one of them on the basis of scientific arguments in the context of the Hartree-Fock treatment.

In the case of benzene, we must also note that, whatever the transformation matrix used, it is impossible to obtain a correct localization for π MO_S. This property may be considered as a characteristic of "delocalized" bonds in conjugated molecules.

9. Appendix 11: Bibliography

ALLAN M.— Electron Spectroscopy Methods in Teaching, in : *J. Chem. Educ.* **64**, 419 (1987).

ANH N.T.— *Les Règles de Woodward-Hoffmann*, Ediscience, Paris (1970).

ANH N.T. THANH B.T. — Parlez vous Chimie Théorique ? II. La Chimie Quantique. Pourquoi faire ? dans : *l'Actualité Chimique*, p.1 (janv. - fév. 1987).

BARTELL L.S. BROCKWAY L.O. — The investigation of Electron Distribution in Atoms by Electron Diffraction, in : *Phys. Rev.* **90**, 833 (1953).

BARTLETT R.J.— Many Body Perturbation Theory and Coupled Cluster Theory for Electron Correlation in Molecules, in : *Ann. Rev. Phys. Chem.* **32**, 359 (1981).

BIGOT B. VOLATRON F. — Parlez vous Chimie Théorique ? I. Méthodes de calcul, dans :*l'Actualité Chimique* , p. 43 (novenbre 1984).

BLINDER S.M.— Basis concepts of Self-Consistent-Field Theory, in : *Am. J. Phys.* **33**, 431 (1965).

BOBROWICZ F.W. GODDARD III W.A. — The Self-Consistent Field Equations for Generalized Valence Bond and Open-Shell Hartree-Fock Wave Functions, in : *H. F. Schaefer III - Methods of Electron Theory*, Plenum Press, New York, p. 79 (1977).

CHAQUIN P.— Faut-il "tordre son cou" à l'hybridation ?, dans : *Bulletin de l'Union des Physiciens* 664, 995 (1984).

CHAVY C.— Théorie de l'optimisation des orbitales dans les calculs SCF and MCSCF, dans :*Thèse de l'Université Paris XI* (1991).

COFFEY P. JUG K.— A Pedagogic Approach to ConFiguration Interaction, in : *J. Chem. Educ.* **51**, 252 (1974).

COOPER,D.L. GERRATT J. RAIMONDI M.— Modern Valence Bond Theory, in *Adv. Chem. Phys.* **69**, 319 (1987).

COOPER D.L. GERRATT J. RAIMONDI M.— Applications of Spin-Coupled Valence Bond Theory, in : *Chem. Rev.* **91**, 929 (1991).

DAUDEL R.— Théorie Quantique de la Liaison Chimique, P.U.F., Paris (1971).

DUCASSE A. LALANNE J.R. LALANNE P. RAYEZ J.C.— Concepts modernes sur la structure électronique des molécules and la liaison chimique, dans : *Bulletin de l'Union des Physiciens* **678**, 129 (1985).

DULIEU O.— Etude des systèmes atomiques à deux électrons externes par la méthode de la fonction d'onde corrélée de Pluvinage. Application à l'étude des ions alcalins négatifs, dans : *Thèse de l'Université Paris VI* (1987).

DULIEU O. LESECH C.— Study of Doubly Excited States of Alkali Negative Ions with the Pluvinage Method, in : *Europhysics Lett.* **3**, 975 (1987).

DUNNING JR. T.H.— Gaussian base functions for use in molecular calculations. I. Contraction of (9s5p) atomic base sets for the first row atoms, in : *J. Chem. Phys.* **53** , 2823 (1970).

DUNNING T.H. PITZER R.M. AUNG S.— Near Hartree-Fock Calculations on the Ground State of the Water Molecule : Energies, Ionization Potentials, Geometry, Force Constants, and One-Electron Properties, in : *J. Chem. Phys.* **57**, 5044 (1972).

DUNNING T.H. HAY P.J. — Gaussian Basis Sets for Molecular Calculations, in : H. F.Schaefer III - *Methods of Electron Theory*, Plenum Press, New York , p. I (1977).

FOCK V.— Näherungsmethode zur Lösung des quantenmechanischen Mehrkörperproblems, in : *Z.Physik* **61**, 126 (1930).

GILLESPIE R.J.— The Valence-Shell Electron-Pair Repulsion (VSEPR) Theory of Directed Valency, in : *J. Chem. Educ.* **40**, 295 (1963).

HARTREE D.R.— The Wave Mechanics of an Atom with a NonCoulomb Central Field, in *Proc. Cambridge Phil. Soc.* **24** (1928)."*Part I* - Theory and Methods, p. 89,"*Part II* - Some Results and Discussion , p. 111, "*Part III* - Term Values and Intensities in Series" in : Optical Spectra, p. 426.

HEHRE W.J. RADOM J.L. SCHLEYER P.V.R. POPLE J.A.— *Ab initio Molecular Orbital Theory*, J. Wiley, New York (1986).

HENRIET A. AUBERT - FRECON M. LESECH C. MASNOU - SEEUWS F.— The Pluvinage Method for Alkalide Dimers. I. One and three conFigurations calculations for the ground states of Li_2, Na_2 and K_2, in : *J. Phys. B : At. Mol. Phys.* **17**, 3417 (1984).

HENRIET A. MASNOU-SEEUWS F. — The Pluvinage Method for Alkalide Dimers. III. Potential Energy Curves for the Excited States of Na_2 up to the (3p + 3p) Dissociation Limit, in : *J. Phys. B : At. Mol. Phys.* **20**, 671 (1987).

HERZBERG G.— The Dissociation Energy of the Hydrogen Molecule, in : *J. Mol. Spectry.* **33**, 47 (1970).

SHAIK S. HIBERTY P.— Curve Crossing Diagrams as General Models for Chemical Reactivity and Structure, in : *"Theoretical Models of Chemical Bonding"* Ed. by Z. B. Maksic, Springer Verlag, Heidelberg (1991).

HURON B. MALRIEU J.P. RANCUREL P.— Iterative Perturbation Calculations of Ground and Excited State Energies from MulticonFigurational Zeroth-Order Wavefunctions, in : *J. Chem. Phys.* **58**, 5745 (1973).

HUZINAGA S.— Gaussian type functions for polyatomic systems, in : *J. Chem. Phys.* **42**, 1293 (1965).

HYLLERAAS E.A.— Neue Berechnung der Energie des Heliums im Grundzustande, sowie des tiefsten Terms von ortho-Helium, in : *Z. Physik* **54**, 347 (1928) ; ROOTHAAN C.C.J. WEISS A.W.— Correlated Orbitals for the Ground State of Heliumlike Systems, in : *Rev. Mod. Phys.* **32**, 194 (1960).

JAMESH.M. COOLIDGE A.S.— The Ground State of the Hydrogen Molecule, in : *J. Chem. Phys.* **1**, 825 (1933).

KOLOS W. ROOTHAAN C.C.J.— Accurate Electron Wave Functions for the H$_2$ Molecule, in : *Rev. Mod. Phys.* **32**, 219 (1960).

KOLOS W. WOLNIEWICZ L.— Accurate Adiabatic Treatment of the Ground State of the Hydrogen Molecule, in : *J. Chem. Phys.* **41**, 3663 (1964).

KOLOS W. WOLNIEWICZ L.— Improved Theoretical Ground-State Energy of the Hydrogen Molecule, in : *J. Chem. Phys.* **49**, 404 (1968).

KOOPMANS T.— Uber die Zuordnung von Wellenfunktionen und Eigenwerten zu den Einzelnen Elektronen eines Atoms, in : *Physica* **1**, 104 (1934).

LEVASSEUR N. MILLIÉ P. ARCHIREL P. LEVY B.— Bond formation between positively charged species. Nonadiabatic analysis and valence-bond model in the CO $^{++}$ case, in : *Chem. Phys.* **153**, 387 (1991).

LEVINE I.N.— *Molecular Spectroscopy*, Wiley-Interscience, New-York (1975).

LEVINE I.— *Quantum Chemistry*, 4th ed., Prentice Hall, New Jersey (1991).

LEVY B.— Best choice for the coupling operators in the open-shell and multiconFiguration SCF methods, in : *J. Chem. Phys.* **48** , 1994 (1968).

LEVY B. BERTHIER G.— Generalized Brillouin theorem for multiconFiguration SCF theories, in : *Int. J. Quantum Chem.* **2**, 307 (1968) and *ibid.* **3**, 247 (1969).

LEVY B.— Etude de méthodes variationnelles à plusieurs conFigurations. Applications à quelques molécules organiques, *Thèse d'Etat*, Paris (1971).

LIANG J.H.— Improved Methods of Calculation of Energy Localized Molecular Orbitals : Methods of Steepest and Principal Ascents and Applications to the H$_2$O, NH$_3$, CH$_4$, HCHO and CH$_2$OH Molecules, in : *Ph. D. Thesis Ohio State University* (1970), voir aussi F. Franks ed. ·*Water* Vol 1, Plenum Press, New York , p. 42 (1972).

M^C DONALD J.K.L.— Successive Approximations by the Rayleigh-Ritz Variation Method, in : *Phys. Rev.* **43**, 830 (1933) ; voir aussi : Young R.H.- New Proof of the Minimum Principle for Excited States, in : *Int. J. Quantum Chem.* **6**, 596 (1972).

M^C LEAN A.D. WEISS A. YOSHIMINE M.— ConFiguration Interaction in the Hydrogen Molecule - The Ground State, in : *Rev. Mod. Phys.* **32**, 211 (1960).

MALLI G.L.— Accurate Analytical self-consistent field (SCF) Hartree-Fock (H-F) Wave Functions for Second-Row Atoms, in : *May. J. Phys.* **44**, 3121 (1966).

MALRIEU J.P. MAYNAU D.— Un parcours initiatique au problème à n-corps, à l'intention des physico-chimistes. Proposition pédagogique, dans : *J. Chim. Phys.* **75**, 31 (1978).

MØLLER C. PLESSET M.S.— *Note on an Approximation Treatment for Many-Electron Systems*, in : *Phys. Rev.* **46**, 618 (1934).

MOUMENI A. DULIEU O. LESECH C.— Correlated wavefunctions for two electron systems using new screened hydrogen-like orbitals, in : *J. Phys. B : At. Mol. Phys.* **23**, L739 (1990).

OHANESSIAN G. MAITRE P. HIBERTY P.C. LEFOUR J.M.— Parlez vous Chimie Théorique ? III. La méthode VB, dans : *l'Actualité Chimique* , p. 33 (mars - avril 1989).

PAULING L. WILSON E.B.— *Introduction to Quantum Mechanics*. M^C Graw-Hill, New York (1935).

PAUNCZ R.— *Spin Eigenfunctions*, Plenum Press, New York (1979).

PEKERIS C.L.— 1 ^{1}S and 2 ^{3}S States of Helium, in : *Phys Rev.* **115**, 1216 (1959).

PILAR F.— *Elementary Quantum Chemistry*, 2nd ed. Mc Graw-Hill, New York (1990).

PITZER R.M. MERRIFIELD D.P.— Minimum Basis Wavefunctions for Water, in : *J. Chem. Phys.* **52**, 4782 (1970).

RIVAIL J.L.— *Eléments de Chimie Quantique à l'usage des chimistes*, Inter-Editions / Editions du CNRS, Paris (1989).

ROETTI C. CLEMENTI E.— Simple base sets for molecular wavefunctions containing atoms from $Z = 2$ to $Z = 54$, in *J. Chem. Phys.* **60**, 4725 (1974) and Roothaan-Hartree-Fock Atomic Wavefunctions, in : *Atomic Data and Nuclear Data Tables*, **14**, 177 (1974).

ROOS B.O.— The Complete Active Space Self-Consistent Field Method and its Applications in Electron Structure Calculations, in : *Adv. Chem. Phys.* **69**, 399 (1987).

ROOTHAAN C.C.J.— New developments in molecular orbital theory, in : *J. Chem. Phys.* **23** , 69 (1951).

ROTHENBERG S.— Localized Orbitals for Polyatomic Molecules. I. The Transferability of the C-H Bond in Saturated Molecules, in : *J. Chem. Phys.* **51**, 3389 (1969).

ROTHENBERG S.— Localized Orbitals for Polyatomic Molecules. II. The C-H Bond Transferability in Insaturated Molecules, in : *J. Am. Chem. Soc.* **93** 68 (1971).

SAEBØ S. PULAY P.— Local ConFiguration Interaction : An Efficient Approach for Larger Molecules, in : *Chem. Phys. Lett.* **113**, 13 (1985).

SHAVITT I.— "The method of ConFiguration Interaction" in : *Methods of Electron Structure Theory*, H. F. Schaefer III ed. Plenum Press, New York, p. 189 (1977).

SHEPARD R.— The MulticonFiguration Self-Consistent Field Method, in : *Adv. Chem. Phys.* **69**, 63 (1987).

SINANOGLU O.— Many-Electron Theory of Atoms and Molecules, in : *Proc. Nat. Acad. Sci.* **47**, 1217 (1961).

SLATER J.C.— Note on Hartree's Method, in : *Phys. Rev.* **35**, 210 (1930).

SLATER J.C.— Atomic Shielding Constants, in : *Phys. Rev.* **36**, 57 (1930).

SZABO A. OSLUND N.S.— *Modern Quantum Chemistry* , in : Rev. ed. McMillan Publishing Co. Inc., New York (1989).

TURNER, D.W. BAKER C. BAKER A.D. BRUNELLE C.R.— *Molecular Photoelectron Spectroscopy*, in : Wiley - Interscience, New-York (1970).

VAN DUIJNEVELDT F.B.— Gaussian Basis Sets for the atoms H-Ne for use in molecular calculations, in : *IBM Research Report* RJ 945, San José (1971).

VON NIESSEN W.— Density Localization of Atomic and Molecular Orbitals, in : *Theor. Chim. Acta* **29**, 29 (1973).

WAHL A.C. DAS G.— The MulticonFiguration Self-Consistent Field Method, in F. Schaefer III - *Methods of Electron Theory*, Plenum Press, New York, p. 51 (1977).

ZENER C.— Analytic Atomic Wave Functions, in : *Phys. Rev.* **36**, 51 (1930).

Index